RESEARCH METHODOLOGY AND DATA ANALYSIS

STATISTICAL MODELLING OF COMPLEX CORRELATED AND CLUSTERED DATA USING SURVEYS IN AFRICA

RESEARCH METHODOLOGY AND DATA ANALYSIS

Additional books and e-books in this series can be found on Nova's website under the Series tab.

RESEARCH METHODOLOGY AND DATA ANALYSIS

STATISTICAL MODELLING OF COMPLEX CORRELATED AND CLUSTERED DATA USING SURVEYS IN AFRICA

NGIANGA-BAKWIN KANDALA
AND
LAWRENCE N. KAZEMBE
EDITORS

nova
science publishers
New York

Library of Congress Cataloging-in-Publication Data

ISBN: 978-1-53615-981-3
Library of Congress Control Number:2019946602

Published by Nova Science Publishers, Inc. † New York

CONTENTS

PREFACE

This book was borne out of the need to document and show-case outputs generated from postgraduate training in Biostatistics at a number of African countries. The postgraduate training at these Universities is a flagship programme of the Sub-Saharan Africa Consortium for Advanced Training in Biostatistics (SSACAB).

The two editors for this volume acted as supervisors and examiners in the course of such training. When meeting at one annual general meeting of SSACAB in September 2018, we realized that in the past three years since the consortium came to life, a remarkable research that have been produced which ought to have been published in the form of either scholarly papers or an edited book bringing together previously uncollected essays in biostatistics in Africa. The decision was then made to go along with the latter option.

As a consequence of this departure we launched a call for chapter abstracts for which various contributions were received and the book title selected. We then approached various publishers enlisting their cooperation to have the book published. Nova Science Publishers was the first to show interest; and thanks to their positive feedback, the book finally came out from publishing house following a fairly lengthy and intense process of peer-reviewing and corrections, especially between September and December 2018.

What follows is, therefore, a series of essays showcasing statistical techniques for handling complex clustered and correlated secondary data. The uniqueness of the book is that it goes beyond the standard methods of logistic regression and survival analysis, to exploit the nested structure or variable dependence in the data. The book addresses four main areas of topical research in health sciences: (i) structural equation modeling; (ii) spatial and spatio-temporal modeling; (iii) correlated or clustered copula modeling; and (iv) survival analysis.

N-B. Kandala, Newcastle, UK
L. N. Kazembe, Windhoek, Namibia

ACKNOWLEDGMENTS

The editors would like to express sincere thanks to all reviewers who dedicated their valuable time to provide critical comments that helped shape the presentation of various chapters.

AN OVERVIEW

In order to assist hospitals in managing its resources and patients, modelling the length of stay is highly important. Recent health scholarship and practice has largely remained empirical, dwelling on primary data. This is critically important, first, because health planners generally rely on data to establish trends and patterns of disease burden at national or regional level. Secondly, epidemiologists depend on data to investigate possible risk factors of the disease. The use of routine or secondary data has in recent years proved increasingly significant in such endeavours. Various units within the health systems collect such data primarily as part of the process for surveillance, monitoring and evaluation. Such data is sometimes periodically supplemented by population-based sample survey datasets. Thirdly, coupled with statistical tools, public health professionals are able to analyze health data and breathe life into what may turn out to be meaningless data.

The main focus of this book is to present and showcase advanced modelling of routine or secondary survey data. Studies demonstrate that statistical literacy and knowledge are needed to understand health research outputs. The advent of user-friendly statistical packages combined with computing power and widespread availability of public health data resulted in more reported epidemiological studies in literature. However, analysis of secondary data has some unique challenges. These widely reported health literature, so far has resulted in inappropriate analysis, and erroneous conclusions.

This book presents the application of advanced statistical techniques to real examples emanating from routine or secondary survey data. These are essentially datasets in which the two editors have been involved, demonstrating how to tackle these challenges. Some of these challenges are: the complex sampling design of the surveys, the hierarchical nature of the data, the dependence of data at the sampled cluster, missing data among many more challenges. Using data from the Health Management Information System (HMIS), and Demographic and Health Survey (DHS), we provide various approaches and techniques of dealing with data complexity, how to handle correlated or clustered data.

The book considers four main topics in the field of health sciences research: (i) structural equation modeling; (ii) spatial and spatio-temporal modeling; (iii) correlated or clustered copula modeling; and (iv) survival analysis.

The book has potential to impact on methodologists, including students undertaking a Master's or Doctoral level programme as well as on other researchers seeking some related reference on quantitative analysis in public health or health sciences or other areas where data of similar nature would be applicable. Further the book can be a resource to public health professionals interested in quantitative approaches to answer questions of epidemiological nature. Each chapter starts with a motivating background, review of statistical methods, analysis and results, ending discussion and possible recommendations.

In: Statistical Modelling …
Editors: N-B. Kandala and L. N. Kazembe

ISBN: 978-1-53615-981-3

Chapter 1

ANALYSIS AND MODELLING OF COMPLEX SECONDARY DATA: AN OVERVIEW OF METHODOLOGICAL ISSUES AND CHALLENGES

***Lawrence N. Kazembe*[1] *and Ngianga-Bakwin Kandala*[*2]**
[1]Department of Statistics and Population Studies,
University of Namibia, Windhoek, Namibia
[2]Department of Mathematics,
Physics and Electrical Engineering and Environment,
Northumbria University, Newcastle, UK

ABSTRACT

The emerging landsacpe of evidence-based decision making has seen an increased demand for both statistical methods, and statistical data. The availability of different data through health surveys or routine hospital data and other administrative data has made it possible for scholars and reseachers to incorporate statistical practices into the analysis and interpretation of public health. The data, while presenting an opportunity to advance public health, however, has it's challenges which have to be acknowledged. This chapter provides an overview of complexities and how to handle such by introducing a general framework for analyzing correlated and clustered data.

Keywords: routine data, secondary data, health information system, electronic health records, generalized linear mixed models

[*] Corresponding Author's Email: lkazembe@yahoo.com.

INTRODUCTION

Health scholarship remains largely empirical. First, health planners generally rely on data to establish trends and patterns of disease burden at national or regional level. Epidemiologists depend on data to investigate possible risk factors of the disease. The use of routine or secondary data has proved increasingly important in such endeavours. Such data are collected on a daily basis at health facilities, by various units of the health system, primarily, as part of the integrated disease surveillance and response, or for monitoring and evaluation. Periodically these are supplemented by population-based sample survey data. Second, coupled with statistical tools, public health professionals are able to analyze health data and breathe life into meaningless data.

A couple of years back this move could have been impossible. However, the advent of high computing power and technological progress, together with accessible statistical methodology, combined with regulatory demand for transparency, and need for precision medicine have created the need for statistical insights in public health. Moreover evidence-based decision making in public health has been propelled by availability of secondary data which have made it easy to foster the importance of statistical methods in public health. This chapter aims to provide an overview of some of the methodological challenges one can face when dealing with secondary data. Most reported health literature, so far, has failed to recognize such challenges. As a result of this failure, inappropriate analysis has been undertaken and erroneous conclusions reached. These need to be considered when analyzing secondary data. Among others, these include:

1. Incompleteness, measurement error over-reporting and under-reporting in data;
2. Clustered or hierarchical or multilevel data;
3. Correlated dependent variables;
4. Covariates which are time-dependent or spatially correlated;
5. Covariates that are subject or group-specific which can be modeled as fixed or random effects.

In the next section we provide an overview of some of most common data available in developing countries, particularly in Africa, followed by a section that outlines challenges and approaches of handling such issues.

SOURCES OF HEALTH DATA

Various sources of secondary data are available for use in epidemiology and public health. These include the household health and demographic surveys, routine hospital data, or administrative data. Table 1.1 presents a summary of data available in most African countries.

Health Surveys

In a number of countries, as a result of USAID supported funding, repeated population surveys have been conducted to measure, monitor and evaluate progress made on a number of health indicators, mainly in three diseases: AIDS, Tuberculosis and Malaria, through global fund initiatives. This has resulted in open data available which would enable to establish trends and compare patterns across countries and permit measuring of within-country variability. The widely used data sets are listed in Table 1.1.

The health surveys provide a range of information on health, treatment seeking behaviour, behavioural determinants of health including feeding patterns for children under five years of age, alcohol and smoking behaviour, as well as maternal and reproductive health. Added, the survey included information on socio-economic factors such as education, household composition and size, household wealth ranking and occupation. For some surveys, they are targetted for some demographic groups and certain diseases. For example, AIDS Informations Surveys (AIS) and malaria information surveys (MIS) are targetted for HIV/AIDS and malaria respectively. The survey design employed is multi-stage sampling, and for some time, geographical data including geo-coordinates of enumeration areas or households have been captured.

Table 1.1. Common health data in Africa and other developing countries

Secondary data	Examples
Health surveys	Demographic and Health Surveys (DHS); Malaria Indicator Surveys (MIS); AIDS Indicator Surveys (AIS); Multi-Cluster Indicator Surveys (MICS); Global Schools Health survey (GSHS); Living Standards Measurement surveys/Integrated Household surveys (IHS)/Household Income and expenditure surveys (HIES)
Health care data	Health Management Information System (HMIS)/Health information system (HIS); Health Facility Census (HCF); Service Provision Assessment; Service Availability and Readiness Assessment (SARA); Population-based HIV Impact Assessment (PHIA)
Morbidity data	Cancer registers; ART registers; TB registers
Mortality data	Population Registers; Civil Registration and Vital Statistics (CRVS) Registers
Other administrative health data	Population and Housing Census; Demographic Surveillance System (DSS) data

Health Care Data

Health facility data provide a wide range of information on use of health care including length of hospital stay, availability of services and readiness of health facilities to offer care. The data is captured at health facility level through a tally of patient visits, and services demanded and reported routinely to health district level, and aggregated up to national level. The HMIS/HIS is atypical routine data aggregated through this process. In most countries, there is an online national system, now DHIS2, to which data clerks are able to feed the data at health district level, for reporting to the national level. In most countries, HMIS has collected data since 2000 (Chilagua et al. 2001), and now includes data from private and public health facilities (Chikumba 2017; Moahi et al. 2017).

Morbidity Data

These are databases that are specific to certain diseases. These are basically registers for cancer, TB, ART, among others. Each contains a range of information about an individual patient's disease progression and hospital admission including:

- Clinical information about diagnoses and operations;
- Patient demographic data such as age group, and sex;
- Geographical information on where the patient was treated and the area in which they live;
- Administrative information such as time waited, and dates and methods of admission and discharge.

Other databases exist that are specific to notifiable diseases including cholera, meningistis, Ebola, measles among others. The notifiable diseases list may vary from country to country, for example, diseases like malaria are notifiable in African countries where it is geared for elimination (WHO 2017).

Mortality Data

In countries where the civil registration system is functional, mortality data are available as part of the population register within the Ministry of Home Affairs or coroner. Data is captured on death certificates, and include information on age at death, sex, place of death, and place of domicile, and causes of death. The National Statistical

Agency use such data to derive mortality statistics. Causes of death have raised some challenges, and are subject to misclassification.

The DHS and population census also have mortality data collected through sisterhood and orphanhood approach. These are indirect methods for estimation of maternal and adult mortality.

Other Administrative Data

Health data is also available in many different administrative data. For example routine Extended Programme of Immunization (EPI) contains data on childhood mordidity like pneumonia, malaria, diarrhoea and coughing, as well as on growth monitoring, nutrition, micro-nutrients deficiencies, deworming and on vaccine uptake and timing. Other data include antenatal care (ANC) registers that contain data on both maternal and child health.

Methodological Issues and Challenges

Methodological issues are critical in evidence-based decision making, or research synthesis to allow for the best evidence to emerge. A good understanding of the issues that may arise is important. These may include sampling errors, non-sampling errors, selection bias, systematic errors, measurement errors, clustering and dependence. Table 1.2 summarizes some of the challenges faced when dealing with secondary data.

Table 1.2. Methodological challenges when handling secondary data

Secondary data types	Challenges
Routine data	Completeness
	Accuracy (e.g., recording and reporting errors; under-reporting
	Timeliness
	Participation bias due to treatment seeking behaviour
	Non-sampling error introduce important sources of uncertainity
Survey data	Coverage
	Limited scope of survey (and variables)
	Missingness
	Non-response
	Non-participation
	Complex survey design

Evidently, in applying statistical methods to any data, the complexities that arise should be known and possible solutions applied. In practice, routine data such as those from the HMIS have a number of limitations and quality problems, such as missing data values, bias, and computational errors. Furthermore, when HMIS data are used to estimate population coverage rates, assumptions have to be made about the relevant demoninators or target populations. These assumptions are often prone to errors.

General Framework for Modelling Clustered and Correlated Data

The above issues not withstanding, an interesting aspect in analyzing secondary data arises when these are clustered or correlated. We present a general modelling framework for analysis of data from hierarchical structure - including small areas analysis, spatio-temporal analysis, and measurement errors.

Generalized linear mixed models (GLMM) has some useful structures that can account for correlations due to relatedness induced by spatial or temporal processes, or bias introduced by measurement errors. Assume $Y_i, i = 1, \ldots, n$ be a response variable, then GLMM, with a random effect, takes the form (Zhou et al. 2006; Congdon 2010):

$$Y = X\beta + b + \varepsilon \tag{1}$$

where β are fixed effects corresponding to X. The random effect b and the error term variance ε are assigned Gaussian distributions:

$$b \sim N(0, \boldsymbol{\eta}\sigma^2\phi),\ \varepsilon \sim N(0, (1-\boldsymbol{\eta})\sigma^2\boldsymbol{I})$$

such that ϕ is a known positive semi-definite and symetric covariance, $\boldsymbol{I}$ is the identity matrix and parameters $\boldsymbol{\sigma^2}$ and $\boldsymbol{\eta}$ determine how the variance is split between b and ε. The joint density of Y is a multivariate normal:

$$Y|(\beta, \boldsymbol{\eta}, \sigma^2) \sim N(X\beta, \boldsymbol{\eta}\sigma^2\phi + (1-\boldsymbol{\eta})\sigma^2\boldsymbol{I}), \tag{2}$$

The random effect, in equation (2), allows for different covariance structures used in longitudinal data modelling, smoothing and spatial statistics. For example, in a spatial modelling approach, one may assume district-specific structured random effects, while in spatio-temporal analysis one may assume district-specific and temporal-specific random effects to capture differences across districts and years. In case of under-reporting and incompleteness encountred in routine data, a bias term is introduced to account for expected change or smooth out any temporal variability in reported cases.

REFERENCES

Chaulagai, C. N., Moyo, C. M., Koot, J., Moyo, H. B., Sambakunsi, T. C., Khunga, F. M., Naphini, P. D. *Design and implementation of a health management information system in Malawi: issues, innovations and results, Health Policy and Planning,* Volume 20, Issue 6, 1 November 2005, Pages 375 - 384, https://doi.org/10.1093/heapol/czi044.

Chikumba, P. A. (2017). Management of Health Information in Malawi: Role of Technology. *Advances in Science, Technology and Engineering Systems Journal,* Vol. 2, No. 1, (2007): 157 - 166.

Congdon, P. B. *Applied Bayesian hierarchical methods*. CRC Press: Boca Raton, 2010.

Moahi, K. H., Bwalya, K. J., Sebina, P. M. (2017). *Health information systems and the advancement of medical practice in developing countries*. IGI global.

World Health Organization. *A framework for malaria elimination*. Geneva: World Health Organization; 2017. Licence: CC BY-NC-SA 3.0 IGO. (Accessed 24 July, 2017). Available at http://apps.who.int/iris/bitstream/10665/2547/61/ 1/9789241511988-eng.pdf.

Zhou, Y., Staudenmayer, J., Coull, B. A., Wand, M. P. General design Bayesian generalized linear mixed models. *Statistical science,* 2006, Vol. 21(1): 35 - 51.

In: Statistical Modelling …
Editors: N-B. Kandala and L. N. Kazembe

ISBN: 978-1-53615-981-3

Chapter 2

A MIXED DISCRETE-TIME SURVIVAL ANALYSIS OF LENGTH OF HOSPITALIZATION: APPLICATIONS TO MALARIA ADMISSIONS AMONG PEADIATRIC CHILDREN IN MALAWI

Lawrence N. Kazembe*
Department of Statistics and Population Studies, University of Namibia, Windhoek, Namibia

ABSTRACT

Length of hospital stay (LOS) is an important indicator in health services research probably because it is a main source of hospital costs. Modelling LOS can assist hospital managers and clinicians in decision-making towards resource allocation and workforce planning. We applied a mixed discrete-time survival model to analyze the length of hospitalization of children admitted for malaria in peadiatric ward at a referral hospital in Malawi. Our approach invoked a multinomial logistic regression model that accounted for multiple exits, in consistency with competing risk models. To account for unobserved heterogeneity or frailty and possible clustering within health centers, random intercepts were incorporated in the model. Results showed that age was an important predictor on LOS, while hospital-specific effects (referral patterns) were not evident. This implies that health centers that feed into the referral hospital do not differ with respect to delaying referring cases to the district hospital.

Keywords: discrete-time survival model, length of hospital stay, malaria, competing-risk model, Malawi

* Corresponding Author's Email: lkazembe@yahoo.com.

Introduction

Malaria is a disease of public importance in much of the sub-Saharan Africa. It is estimated that 3 million children are sick of the disease per year in the continent (Snow et al., 1999; Snow et al., 2003). In Malawi alone, malaria accounts for 33.2% of all outpatient visits, with 40% of all under-five hospitalizations attributed to malaria (Government of Malawi, 2008). This puts enormous pressure on the limited resources individuals, households and governments have in trying to control the disease. Arguably, individuals, households and government are poorer because resources are channeled to fight the disease towards good health. Indeed direct illness costs borne by the government attributed to malaria are projected to be US$1.10 per consultation in outpatient care, while for hospital admission at US$35 per admission (Government of Malawi, 2008).

Analysis of length of hospitalization due to malaria can probably give an insight into the cost of malaria control. Studies have shown that the costs of hospitalization are directly related to the length of hospital stay (LOS) (e.g., Polverejan et al., 2003; Berry et al., 2013; Baek et al., 2018). For example, with so many children at greater risk of malaria during the rainy season more resources would be required if most of these are hospitalized. When LOS is for few days then the hospital would cope with an increased number with fewer resources demanded than when LOS is much longer. Practically, if a health facility has limited beds as it is in many African settings, it would mean sick children would be overcrowded, becoming yet another health risk to the children in the hospital or even discharged early when they have not fully recovered. Assessing LOS would provide important information for decision-making on resource allocation and workforce planning, more critically in developing countries where expenditures on health are slim and dwindling and the work force is in shortfall (Bradbury et al., 2000; Lawrence and Delaney, 2013).

Several studies on tropical pediatrics have focused on pediatric health care or the prognosis of the disease with malaria considered as a risk factor or cause of child mortality (Jima et al., 2012; Okiro et al., 2013; Aregawi et al., 2014). Very few studies have considered analysis of hospital admission due to malaria (Kum et al., 2013; Keene et al., 2018) and possible covariates (Ariyo et al., 2012; Cairns et al., 2015). None to our knowledge have focused on duration of hospitalization as a discrete outcome.

This chapter develops a discrete-time survival models to study LOS due to malaria, with duration time considered as days spent in the hospital ward before released home or death. As in many applications, duration of hospital stay is not observed continuous but is only known to lie between a pair of consecutive follow-ups (Sá et al., 2007; Chaou et al., 2017; Zaccardi et al., 2017). For example, a child is only observed at a particular point in a day during regular check-ups. This naturally would mean time is treated as discrete and not continuous (Fahrmeir and Wagenpfeil, 1996; Fahrmeir and Tutz, 2001; Tutz, 2016). Further, in our studies, the outcome of each LOS is known with three state space $\{0, 1, 2\}$.

That is, a child hospitalized of malaria would either stay in hospital i.e., continued admission (state 0); or released home after improved prognosis (state 1); or leave the hospital through death, probably after developing severe malaria (state 2). To account for these two terminating states (state 1 and 2) we considered survival models with competing risks, since one event happening precludes the other from occurring.

We proposed to jointly model the length of stay and terminating outcomes using discrete-time competing risk models by adjusting for known explanatory variables. Models with discrete-time competing risks have proposed using multinomial logit/probit (MNL/MNP) models, with exit states treated as categories in the response variable (Fahrmeir and Wagenpfeil, 1996). To account for unobserved heterogeneity or frailty and possible hospital/health centre clustering effects, random intercepts are introduced in the model. Inference of such models can be performed either as fully Bayesian or empirical Bayesian approach. Based on Fahrmeir, Kneib and Lang (2004), we implemented an empirical Bayesian (EB) methodology along the generalized linear mixed models (GLMM) representations developed in Lin and Zhang (1999). For EB inference, variance or smoothing parameters are considered as unknown constants. They are estimated by using (approximate) restricted maximum likelihood (REML). For given or estimated smoothing parameters, unknown functions and covariate effects are obtained as posterior mode estimators by maximizing the posterior density.

The objectives of the present study were as follows. The first objective was to analyze the length of stay using survival modeling techniques. Some studies have assumed LOS as normally distributed and used analysis of variance and yet simple plots showed LOS depicting an exponential decay (Stevens et al. 2001; Sá et al. 2007; Chaou et al. 2017). The second goal was to jointly analyze LOS and discharge outcomes while adjusting for some possible factors. Modelling discharge outcomes and LOS separately would result in loss of information. Thirdly, the aim was to investigate the effects of referral health system on the LOS and its outcomes. We conducted our studies at a central hospital. This hospital is part of a network of referral hospitals aimed at providing primary care at a district level.

Hospital specific effects are entered into the model as random effects and treated within the paradigm of random intercepts. The inclusion of hospital/cluster specific effects as random effects is of substantive statistical interest (Berry et al., 2013). From modeling viewpoint, clustering induces dependence that affects the standard errors and significance of the results and these have to be adjusted. On the other hand, these effects are said to be frailties. They account for extra heterogeneity which was not observed or unmeasured.

The rest of this chapter is structured as follows. Section 2 describes the sources of the data and methods used for collection. Section 3 proposes the model used to analyze the data, and outlined the generalized mixed model procedure used for inference. Section 4 gives the results of our analysis and we conclude in section 5.

METHODS

Data

The study was carried out at the peadiatric ward of Zomba central hospital in Zomba district, Malawi. Zomba is located in the south-eastern region of Malawi and it is one of the 27 administrative districts in the country. Health care services in the district are mainly provided at government and mission owned health centres. The district has a network of clinics and health centres that provided primary care level, with complicated cases referred to the district hospital.

We obtained discharge records for all pediatric hospital admissions (patients of up to 14 years of age) at the hospital for the period January 1, 2002, to December 31, 2003. Patients referred to the paediatric ward from all clinics in the district were included in this study. Data sets included hospital admissions linked by patient. For the purposes of this study, we included only admissions due to malaria. Hospital discharge records were entered following procedures and diagnoses coded with Health Management Information System classification of diseases nomenclature. The unique CODE 8 identified patients admitted for malaria, either as first diagnosis or second diagnosis. However, we limited our analysis on cases of first diagnosis.

As part of the primary health care system, Malawi health system developed a network of referral hospitals. Clinics, dispensaries, health centres and rural hospitals were developed to provide care for uncomplicated cases, referring most serious cases to a central hospital. District hospitals served in this capacity in all districts in the country. More complicated cases would be referred further to a regional/central hospital. In Malawi's hospital referral system, district hospitals were to provide general inpatient services, accepting referrals from urban and rural health centres, dispensaries and clinics. Regional hospitals, which were at third level, were to receive patients from the district hospital and provide specialist services (Government of Malawi, 1999a, 1999b).

For each case in the register we obtained the following: age, sex, LOS, the outcome at discharge (either healed and released home, or released for home care or absconded or died), month of admission, year of admission, and location (village and referring health center). For the study period, 3937 children were admitted at the hospital with malaria. The mean and median LOS were 3.04 and 3.00 days respectively. Figure 2.1 shows the survival function with the cumulative probability of hospital stay (cumulative survival function). The figure clearly showed a reduced probability of stay by day 5.

For each health center we obtained details on: owner (either government or mission operated), type (either clinic, rural or dispensary) and distance to the central hospital. Summaries of these variables are given in Table 2.1. We considered these covariates in the model discussed in the next Section.

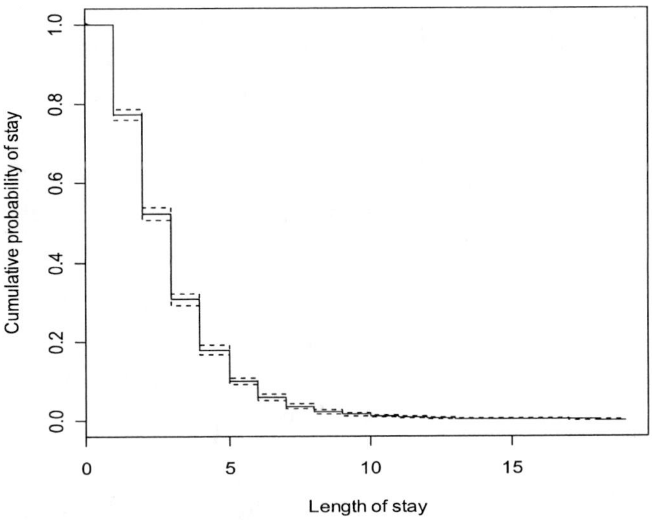

Figure 2.1. A cumulative survival plot (solid line) with 95% confidence band (dashed lines).

Table 2.1. Summary of explanatory variables considered in equation 4

Variable	N(%)	Mean (Standard deviation)	Coefficient
Age		30.2 (30.3)	f_{ti}
<6 months	1183 (30.0)	-	β_0
6-11 months	2123 (53.9)	-	β_1
1-4 years	516 (13.1)	-	β_2
5-14 years	115 (2.9)	-	0.0
Sex			
Male	2272 (57.7)	-	0.0
Female	1665 (42.3)	-	β_3
Owner of facility			
Government	3340 (84.8)	-	β_4
Mission	597 (15.2)	-	0.0
Type of facility			
Dispensary	1869 (47.5)	-	β_5
Rural hospital	93 (2.4)	-	β_6
Clinic	1975 (50.2)	-	0.0
Season			
Dry	1121 (28.5)	-	0.0
Wet	2236 (56.8)	-	β_7
Distance		8.4 (7.4)	β_8

Reference category where applicable is denoted by 0.0. f_{ti} denotes a smooth function.

The Model

In this section we outline the discrete-time survival model for analyzing LOS. Following Fahrmeir and Wagenpfeil (1996) and Brezger et al. (2004), we define the following model. Let days since admission be the time scale. Then for each child admitted, there were two discharge destinations, either the child died or was released home. These alternative destinations are mutually exclusive and are assumed to exhaust the possible destinations. Thus a natural approach to study the length of a hospital stay that takes into account the likely correlation of LOS with discharge destinations, is to use a duration model with, m, multiple exits (competing risks model). Often the duration or survival times are not observed continuously, instead, discrete time $T \in \{1, \ldots k\}$, where $T = t$ denotes end of duration within the interval $[b_t, b_{t+1})$. Distinct terminating causes, competing risks or types of failure are denoted by $R \in \{1, \ldots m\}$. Classical formulations of competing risks models introduce latent durations $T_1 \ldots T_m$, one for each terminating cause. The observed duration T and terminating cause R are then viewed as $T = min(T_1 \ldots T_m)$ and $R = r$ if $T = T_r$.

Accordingly, the duration and competing risks models can be formulated in terms of cause-specific hazard functions as the basic characteristics for the observables T and R. The cause-specific discrete hazard function resulting from cause or risk r is given by the conditional probability

$$\lambda_r(t|x_t) = Pr(T = t, R = r|T \geq t, x_t) \quad (1)$$

where $r = 1, \ldots, m$ and $t = 1, \ldots, k$ and is a vector of covariates x_t. The overall hazard function regardless of cause is given by

$$\lambda(t|x_t) = \sum_{r=1}^{m} \lambda_r(t|x_t) = \sum_{r=1}^{m} Pr(T = t, R = r|T \geq t, x_t) \quad (2)$$

For a sample of individuals $i = 1, \ldots, n$, with T_i duration times and C_i right censoring indicator and a set of risks $R \in \{1, \ldots, m\}$, we assume that the censoring is non-informative and occurs at the end of the interval so that each individual is exposed to the risk set R_t in that interval t. Then we define a failure indicator (Farhrmier and Tutz, 2001; Brezger, Kneib and Lang, 2004):

$$y_{itr} = \begin{cases} r & \text{failure of type } r \text{ occurred in the interval } [b_t, b_{t+1}). \\ 0 & \text{no failure of type } r \text{ occurred in the interval } [b_t, b_{t+1}). \end{cases} \quad (3)$$

where $r = 0, 1, 2$; $t = 1, \ldots, t_i$ and $i = 1, \ldots, n$. Then the duration process of individual i can be considered as a sequence of binary decisions between remaining in the transient state $y_{itr} = 0$ or leaving for the absorbing state $y_{itr} = 1$ or $y_{itr} = 2$ for the states of death ($r = 1$) or that of being recovered and released home ($r = 2$) at the end of duration at t. For $i \in R$, the cause-specific hazard function for individual i can be modelled based on multicategorical ($r = 0, 1, 2$) response models. Further assuming i.i.d. error variables following the extreme value distribution, $F(z) = exp(-exp(-z))$ yields the multinomial logit model since events are unordered

$$\lambda_r(t|x_t) = Pr(T = t, R = r|T \geq t, x_t, u_t) = \frac{\exp(\eta_r)}{1+\sum_{r=0}^{2}\exp(\eta_r)} \tag{4}$$

for $r = 0, 1, 2$ and where η_r is the predictor for the the terminating state r, u_t is the risk set and x_t is the vector of explanatory variables. The predictor in additive structure is given by

$$\eta_r = f_{0r}(t) + f_{1r}(age) + w'\gamma_r + h_r \tag{5}$$

where f_{0r} is the nonparametric function of the baseline effect for the terminating effect r. Similarly f_{1r} is the nonparametric effect of age. The vector w' represents a vector of categorical covariates in the model (see Table 1). Frailty effects are captured through h_r by assuming that these are spatially unstructured. These effects are specified as random intercepts with $h_r \sim N(0, \tau^2)$.

Rewriting equation (5) in matrix form we have,

$$\boldsymbol{\eta} = \boldsymbol{Z}_1\boldsymbol{\beta}_1 + \boldsymbol{Z}_2\boldsymbol{\beta}_2 + \boldsymbol{Z}_{h_r}\boldsymbol{\beta}_{hr} + \boldsymbol{W}\boldsymbol{\gamma} \tag{6}$$

where the nonparametric effects f is expressed as a product of design matrix $\boldsymbol{Z}_j$ and a vector of regression coefficients $\boldsymbol{\beta}_j$ and the fixed terms associated with $\boldsymbol{\gamma}$ expressed using the usual design matrix $\boldsymbol{W}$. Within the framework of mixed models methodology, the parameter vector $\boldsymbol{\beta}_j$ is partitioned into a penalized ($\boldsymbol{\beta}_j^{pen}$) and unpenalized ($\boldsymbol{\beta}_j^{unpen}$) parts yielding a variance component model (Fahrmeir et al., 2003)

$$\boldsymbol{\beta}_j = \Psi_j^{pen}\boldsymbol{\beta}_j^{pen} + \Psi_j^{unpen}\boldsymbol{\beta}_j^{unpen} \tag{7}$$

Appropriate priors assumed are as follows. For the penalized part, an i.i.d Gaussian prior is suitable, while for the unpenalized part we assume a flat prior, that is

$$p(\boldsymbol{\beta}_j^{pen}) \sim N(0, \Lambda) \text{ and } p(\boldsymbol{\beta}_j^{unpen}) \propto const. \tag{8}$$

where $\Lambda = diag(\tau_1^2, \ldots, \tau_k^2)$, where τ_k^2 is a smoothing parameter which controls the trade-off between smoothness and flexibility.

Applying decomposition (Eqn. 7) to all the components of predictor (Eqn.6) yields

$$\boldsymbol{\eta} = \boldsymbol{X}_j^{pen}\boldsymbol{\beta}_j^{pen} + \boldsymbol{X}_j^{unpen}\boldsymbol{\beta}_j^{unpen}. \tag{9}$$

We have obtained in (9) a GLMM with fixed effects $\boldsymbol{\beta}^{unpen}$ and random effects $\boldsymbol{\beta}^{pen}$. The posterior, in terms of the GLMM representation, is given by

$$p(\boldsymbol{\beta}^{unpen}, \boldsymbol{\beta}^{pen} \mid y) \propto L(y, \boldsymbol{\beta}^{unpen}, \boldsymbol{\beta}^{pen}) \prod_{j=1}^{L} \left(p\left(\boldsymbol{\beta}_j^{pen} \middle| \tau_h^2 \right) \right) \quad (10)$$

where $L(\bullet)$ denotes the likelihood which is the product of individual likelihood contributions and $p\left(\boldsymbol{\beta}_j^{pen} \middle| \tau_h^2\right)$ is as defined above.

Estimation of regression coefficients and variance parameters is carried out using iteratively weighted least squares and approximate restricted maximum likelihood. At the first iteration the default (starting) values are assumed for the penalized, unpenalized and variance parameters. Then updates for $\bar{\beta}_j^{unpen}$ and $\bar{\beta}_j^{pen}$ are obtained in the first step by solving a system of linear equations given estimates for the variance parameters. In the second step updates of the variance parameters are obtained by maximizing the approximate restricted log-likelihood. The restricted log-likelihood is maximized through a Fisher scoring technique. The two steps above are iterated until convergence. Details are given in Lin and Zhang (1999); Fahrmeir and Tutz (2001) and Fahrmeir, Kneib and Lang (2003). Fahrmeir et al. (2003) derived numerically efficient formulae that allow for handling large data sets.

Implementation

We fitted a multinomial logistic model for the response states of 0, 1, 2 with reference category 0, while 1=death and 2= released home. The baseline hazard function was fitted nonparametrically using equidistant knotted P-spline of second order random walk. Age was also fitted nonlinearly, and estimated nonparametrically as P-spline of second-order random walk. Two other specifications on effect of age were tried. In model M1, age is assumed a fixed continuous variable, while model M2 assumed a fixed categorical variable. Age effects are nonlinearly fitted in model M3. Implementation of these models was carried out using a public domain software, BayesX (Brezger et al. 2004).

RESULTS

The three models fitted for pediatric malaria were:

$$M1: \eta = \beta_0 + \beta_1 Age + \cdots + f_{hospital}(random)$$

$$M2: \eta = \beta_0 + \beta_1 Age1 + \beta_2 Age2 + \beta_3 Age3 + \cdots + f_{hospital}(random)$$
$$M3: \eta = \beta_0 + \cdots + f_{age}A(ge) + f_{hospital}(random)$$

The best model is selected based on Akaike Information Criterion (AIC). The best model is M3 with AIC= 14554.4 (see Table 2.2).

Table 2.2. AIC for multinomial logistic models fitted

Model	-2*log-likelihood	Df	AIC
M1	14480.5	40.91	14562.3
M2	14460.0	49.91	14559.8
M3	14460.7	46.89	14554.4

Table 2.3. Posterior estimates (in log odds ratios) of the fixed-effect parameters in the semi-parametric competing risk hazard models. 95% credible intervals are given in the parantheses

Variable	Discharge Destination	
	Death	**Home**
Intercept	-2.881 (-3.629,-2.132)	-0.911 (-1.234, -0.584)
Male child	0.073 (-0.055, 0.202)	0.011 (-0.034, 0.055)
Government hospital	-0.099 (-0.283, 0.085)	0.002 (-0.072, 0.077)
Dispensary	-0.142 (-0.424, 0.141)	-0.051 (-0.169, 0.066)
Rural hospital	0.137 (-0.360, 0.635)	-0.008 (-0.217, 0.202)
Distance to Zomba central hospital	0.022 (-0.005, 0.049)	-0.001 (-0.012, 0.010)
Time (P-spline prior)		
Variance	0.578	0.090
Smoothing parameter (smoothing parameter=1/variance)	1.731	11.056
Age (P-spline prior)		
Variance	0.017	1.135e-05
Smoothing parameter (smoothing parameter=1/variance)	58.364	88071.2
Health effects (i.i.d. Gaussian random effects)		
Variance	0.018	0.004
Smoothing parameter	56.344	236.053

Table 2.3 shows results of the posterior estimates of the fixed effect parameters in the discharge category-exited with death in model M3. There were no significant effects of fixed parameters in this category except the intercept, based on the 95% credible intervals. The intercept can be interpreted as that on average the odds of staying long were reduced for those who died (coeff = -2.881), relative to those discharged home (coeff = -0.911), confirming worsening prognosis for those who would eventually die.

Interesting to note also is the sign of coefficients for the variable distance. The positive sign for this coefficient implied that duration of stay was positively related to distance from the referral hospital.

The estimated baseline hazard function for category-death is presented in Figure 2.2(a). Figure 2.2(a) shows a variation over duration time, with a decreasing trend from day 0 up to day 3 and then rises for the rest of the survival time. There is a smaller uncertainty attached at up to day 3 or 4, but then the increasing trend has a larger uncertainty attached to it.

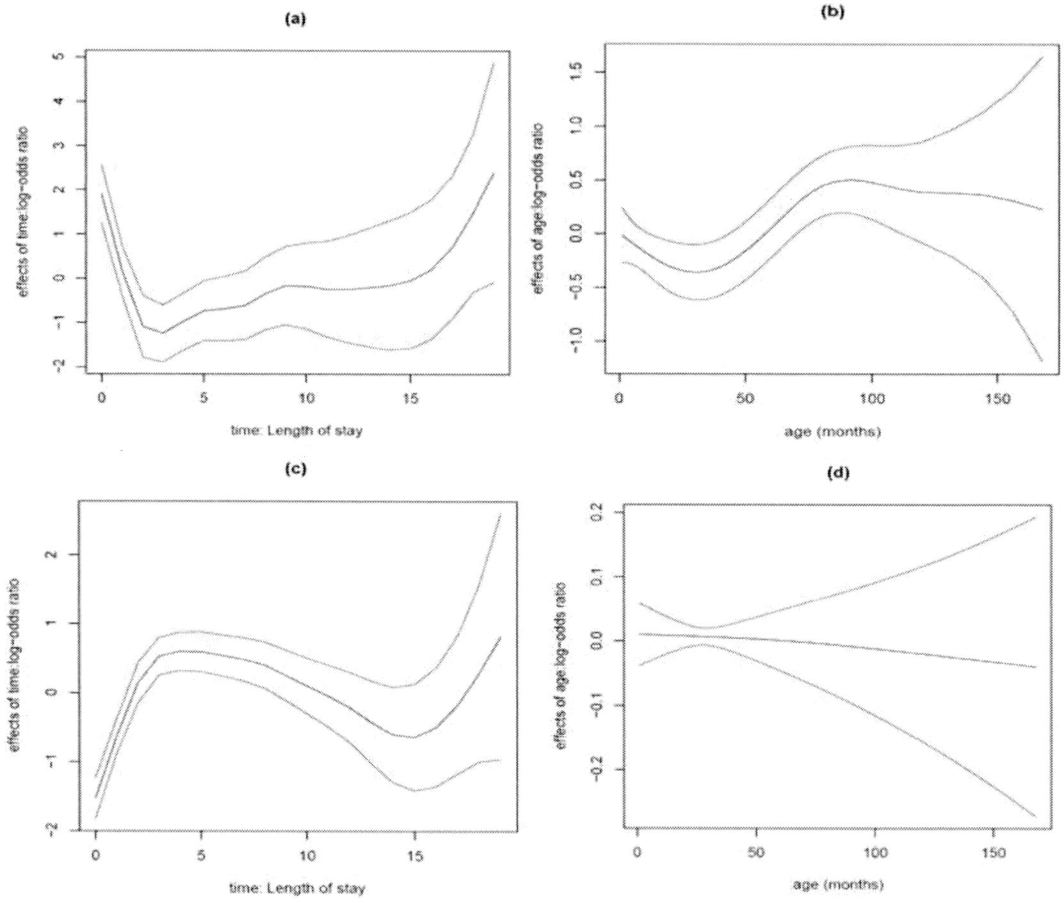

Figure 2.2. Nonlinear effects. Shown are the posterior modes (log odds-ratios) together with 95% pointwise credible intervals (in red). (a) of time in the category-death; (b) of age-death category; (c)- time for discharge category; (d) age for discharge category.

The age effect is shown in Figure 2.2(b). The effects decreased up to about 25th/30th month, and then increased to approximately the 90th month above the 0 log-odds ratio mark. From here (at about 100th month) the trend remained fairly constant and dropped slightly. Levels of uncertainty were largest at the later months of life, indicating no significant effects of age with LOS that ended in death.

For the category that was discharged home, results for the fixed parameters are also given in Table 2.3. Again, none of the fixed effects are significant except the intercept. The baseline hazard estimates for this category is shown in Figure 2.2 (panel c). The estimated effect has an increasing trend from day 0 up to day 3, and then falls steadily up to day 15, and then rises again. The uncertainty envelopes are smaller at the beginning and wider at the end. Effects of age for this category are also shown in Figure 2.2(d). There is a decreasing trend in the effect of age on LOS among those who were released home. The increasing uncertaininty as age increases, may be indicative of fewer children in the older age group.

The plots (Figures 2.2(a) and 2.2(c)) have interesting features. The two plots indicated that we are able to explain LOS discharge destination better in the first 3 days than in the later days (as verified by narrower credible intervals). Among those who died, the odds of dying decreased up to day 3 with clear certainty and not on the days thereafter. Similarly among those who were discharged home, the odds increased with greater certainty up to again day 3 or 4. The same picture is depicted in the age plots (Figures 2.2(b) and 2.2(d)).

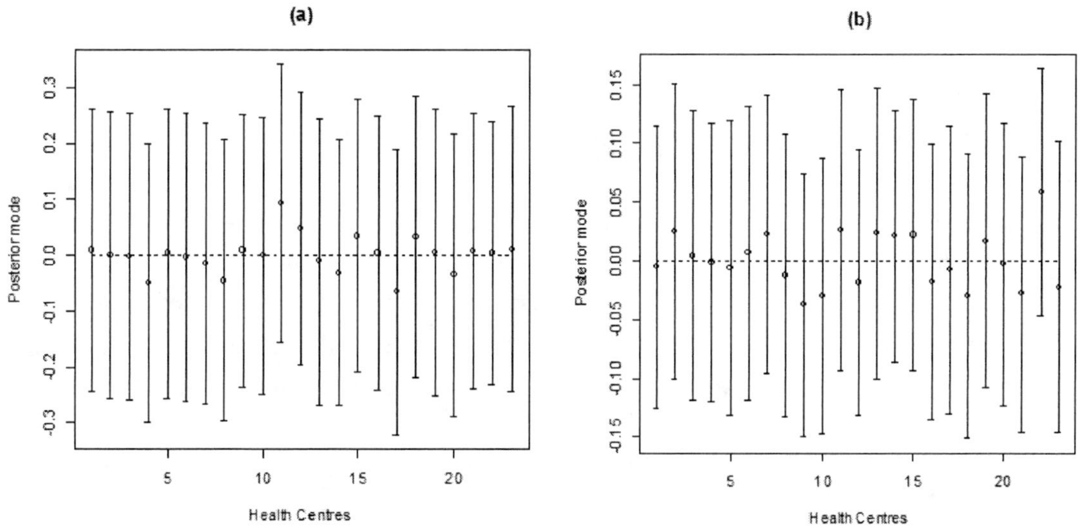

Figure 2.3. Random intercepts for the health centres effects. The solid vertical lines indicate a 95% credible interval: (a) category –death; (b) category- discharged home.

Hospital effects on LOS are shown in Figures 2.3(a) and 2.3(b). For both categories, hospital effects do not contribute to the variation in the duration of hospitalization. All random effects span the zero line (log OR = 0), which implies that all hospital effects are independent of the variation in LOS. Methodologically, this means that the assumption of clustering due to hospital of origin is not supported by the data. The spatial distribution of these effects is shown in Figures 2.4(a) and 2.4(b). However the pattern is difficult to interpret. Large and small effects seem to exhibit a spatial randomness.

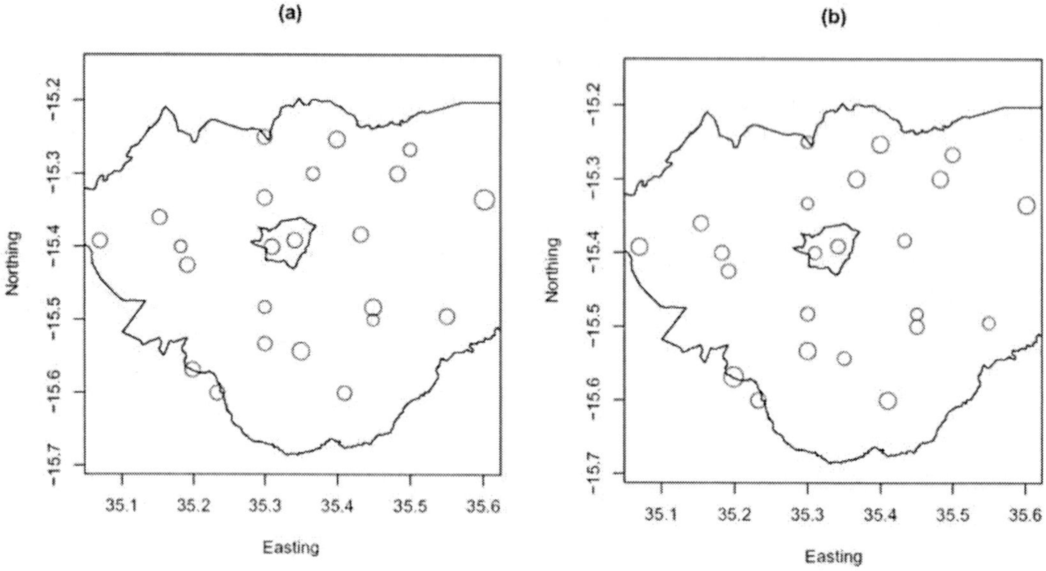

Figure 2.4. Hospital-specific effects. Shown are the random intercepts for each health center proportional to size: (a) for the category-death; (b) for category-discharged home.

Discussion and Conclusion

We have undertaken to analyze the length of hospitalization of children with malaria in at a referral hospital. Analysis of LOS is important for hospital managers and clinicians, probably because of its clear meaning as one of the main sources of hospital costs and thus LOS has been used as an indicator for predicting bed occupancy and resource allocation (Brownell et al. 1995; Bradbury et al., 2000; Jemenez et al. 2004; Lawrence and Delaney, 2013). In some instances, LOS has been used to explain the quality of hospital care and efficiency in inpatient care (Lorch et al. 2004; Lawrence and Delaney, 2013).

Use of models in analyzing LOS can help with decision-making such as bed management and work force planning. This is important in developing countries, where in recent years there has been an increased demand for health care with reduced or same expenditure and understaffing. While the use of models in modeling bed occupancy is not new in general (Sorensen, 1996; Pendergast and Vogel, 1988; Wang et al. 2002; Picone et al., 2003; Lawrence and Delaney, 2013), and survival models in particular (Picone et al., 2003; Lorch et al., 2004), here we proposed use of discrete-time survival models. The advantage of this approach is that, firstly, duration time is treated as discrete and not continuous because terminating events are observed at the end of the interval only. Secondly, within the context of Cox regression the baseline hazard is unspecified and this was fitted non- parametrically either using random walk or P-splines. Thirdly, the survival model (Fahrmeir and Wagenpfeil, 1996; Tutz and Schmid, 2016) considered here knows more than two states. The patient will either stay in the hospital (state 0) or will leave either through death (state 1) or released home (state 2). The commonest approach is to consider two states only (stay or leave). In a study like ours, where how they leave is important, it was necessary to apply the competing risk models (through multinomial logistic regression). Mortality attributable to malaria (state 1) is of primary importance in paediatric health care in developing countries, and has to be explained in the model as we have done.

The duration of hospitalization due to malaria was on average for 3.04 days (see Figure 2.1). This resulted in either death or release home. Of the 3,937 children hospitalized for malaria in the period 2002-2003, 7.7% resulted in death. This proportional is relatively higher than that found in other parts of Africa (Greenberg et al., 1989; Greenwood et al., 1987). It is also worth noting that those who died stayed at the hospital for an average of 2.08 days. Probably this short latency could be because these children developed severe malaria due to delayed treatment. In Africa, where a greater majority, seek malaria treatment first at home, children would only be reported for hospital care if the disease is severe enough for hospital admission than when the condition is not (de Savigny, 2004).

Our analysis revealed that age was able to explain the variation in LOS. The age effect is of epidemiological importance in malaria. Because of intense transmission, people tend to develop immunity against the disease as they grow. It is expected that those at greater risk of the disease are young children. It has been argued that children within 6 months of age or up to 1 year are under the shadow of maternal immunity and are of lower risk than of subsequent years until the age of 9 or 10 years when they develop visible immunity (Bruce-Chwatt 1985). The proportion hospitalized by age were 30% (<1 year), 54% (1-4 years), 13% (5-9 years) and 3% (10-14 years) with case fatality rate of 2.6%, 3.4%, 1.4% and 0.3% in the respective age categories. This is supported in our analysis (Figures 2.2(b) and 2.2(d)) that the probability of staying longer in the hospital reduced with age.

Although the effect of distance to the referral hospital was found not to be significant, in medical geography distance is an important aspect on access to care. Studies have shown that geographical accessibility of care is directly related to distance to the health facility (Jones et al., 1998; Haynes, et al., 1999, Muller et al., 2000). The type and owner of facility variables provided an indicator of quality expected at the hospital (Berry et al., 2013). Rural hospitals are expected to give quality care that dispensaries and clinics in terms of diagnosis and appropriate treatment. Often is the case than not, that facility lack antimalarials and patients are given antipyretics for antimalarials. People would then resort to travel long distances for a facility that would provide adequate care. Who operates or owns the hospital is another variable whose effect is not well documented. There is a tendency somewhat that mission hospitals are often preferred to government hospitals especially in urban and peri-urban centers. Understaffing and lack of resources in recent years has contributed to this trend.

Adjusting for hospital level clustering, although not supported here, was important for two reasons. Firstly, because of similarities in practice styles and organization of the hospital, children originating from the same hospital are likely to receive the same type of care than those coming from another hospital (Berry et al., 2013; van Diepen et al., 2015). By adjusting for each hospital clustering, the standard errors are also adjusted, thus correcting for the correlation in the data. This implies that statistical significance in the results is also adjusted for (Fahrmeir and Wagenpfeil, 1996; Lorch et al., 2004). Secondly, by specifying random intercepts we directly modeled any unobserved or unmeasured heterogeneity.

This study was limited by its retrospective nature and absence of a comparing control disease which is also of significant public health in children such as acute respiratory illnesses (ARI). The impact of malaria has to be measured and compared against other diseases. Despite these limitations, an important objective was to understand the trend in malaria admissions and deaths and predict the length of bed occupancy. To address these issues, we applied a regression model that adjusted for risk factors.

A longitudinal surveillance design could have been an ideal framework for this study. However we took advantage of hospital discharge records as previously demonstrated by Lorch et al. (2004), Okiro et al. (20113) and Aregawi et al. (2014). Hospital discharge records are an important source of data, especially in developing countries, because they are readily available in the health facilities and are useful not only for planning and evaluating hospital services, but also for epidemiological surveillance. Of course hospital data have their own limitations because hospital records are not designed for research but rather for patient care, and they may be incomplete, illegible, or missing (Aregawi et al. 2014; Lorch et al. 2004).

REFERENCES

Aregawi, M., Lynch, M., Bekele, W., Kebede, H., Jima, D., Taffese, H. S., Yenehun, M. A., Lilay, A., Williams, R., et al. (2014). Time Series Analysis of Trends in Malaria Cases and Deaths at Hospitals and the Effect of Antimalarial Interventions, 2001-2011, Ethiopia. *PLoS ONE* 9(11):e106359; DOI: 10.1371/journal.pone.0106359.

Ariyo, J. I., Abiodun, A. A., Oyejola, B. A. (2012). Application of survival analysis on malaria among children: A case study of University of Ilorin Teaching Hospital. *Scientia Africana,* 11(2).

Baek, H., Cho, M., Kim, S., Hwang, H., Song, M., Yoo, S. (2018). Analysis of length of hospital stay using electronic health records: A statistical and data mining approach. *PLoS ONE* 13(4): e0195901.

Berry, J. G., Toomey, S. L., Zaslavsky, A. M., et al. (2013). Pediatric readmission prevalence and variability across hospitals. *JAMA,* 309:372–80.

Bradbury, R. C., Golec, J. H., Steen, P. M. (2000). Linking health outcomes and resource efficiency for hospitalized patients: do physicians with low mortality and morbidity rates also have low resource expenditures? *Health Serv Manage Res,* 13(1): 57-68.

Brezger, A., Kneib, T., Lang, S. (2004). *BayesX-Software for Bayesian Inference based on Markov Chain Monte Carlo simulation techniques.* Available under http://www.stat.uni-muechen.de/lang/BayesX/.

Brock, G. N., Barnes, C., Ramirez, J. A., Myers, J. (2011). How to handle mortality when investigating length of hospital stay and time to clinical stability. *BMC Med Res Methodol.,* 11: 144.

Brownell, M. D., Roos, N. P. (1995). Variation in length of stay as a measure of efficiency in Manitoba hospitals. *CMAJ,* 152(5): 675-682.

Cairns, M., Cheung, Y. B., Xu, Y., et al. (2015). Analysis of Preventive Interventions for Malaria: Exploring Partial and Complete Protection and Total and Primary Intervention Effects. *Am J Epidemiol.,* 181(12):1008-17.

Chaou, C-H, Chen, H-H, Chang, S-H, Tang, P, Pan, S-L, Yen, AM-F, et al. (2017). Predicting Length of Stay among Patients Discharged from the Emergency Department—Using an Accelerated Failure Time Model. *PLoS ONE* 12(1): e0165756.

de Savigny, D., Mayombana, C., Mwageni, E., Masanja, H., Minhaj, A., Mkilindi, Y., Mbuya, C., Kasale, H., Reid, G. (2004). Care-seeking patterns for fatal malaria in Tanzania. *Malaria Journal 3:27..*

Fahrmeir L, Kneib T, Lang S. (2004). *Penalized additive regression for space-time data: A Bayesian perspective, Statistica Sinica, 14:731-761 .*

Fahrmeir L, Tutz G. (2001). *Multivariate Statistical Modelling based on General- ized Linear Models,* SpringerVerlag, New York.

Fahrmeir L, Wagenpfeil S. (1996). Smoothing Hazard Functions and Time-Varying Effects in Discrete Duration and Competing Risk Models. *Journal of the American Statistical Association,* 91 (436), 1584-1594.

Gallup JL, Sachs JD. (2001). The economic burden of malaria. *Am. J Trop Med Hyg,* 64S:85-96.

Government of Malawi. (1999*). To the year 2020: A Vision for the Health Sector in Malawi.* Ministry of Health and Population. pp: 51-72.

Government of Malawi. (2008). *Malaria Policy. National Malaria Control programme. Community Health Science Unit. Division of Preventive Health Services.* Ministry of Health and Population.

Government of Malawi. (2017*). Health Sector Strategic Plan II (2017-2022),* 2017-2022. Ministry of Health and Population. p: 13-14, 16.

Greenberg AE, Ntumbanzondo M, Ntula N. (1989). Hospital-based surveillance of malaria related paediatric morbidity and mortality in Kinshasa, Zaire. *Bulletin of WHO,* 67:189-196.

Greenwood BM. (1987). Mortality and morbidity from malaria among children in a rural area of the Gambia, West Africa. *Trans. Royal Soc of Trop Medicine and Hygiene,* 81: 478-486.

Haynes, R, Bentham, C. G., Lovett, A., Gale, S. (1999). Effects of distances to hospital and GP surgery on hospital inpatient episodes, controlling for needs and provision. *Soc Sci Med*, 49:425-433.

Jima, D., Wondabeku, M., Alemu, A., Teferra, A., Awel, N., Deressa, W., et al. (2012). Analysis of malaria surveillance data in Ethiopia: what can be learned from the Integrated Disease Surveillance and Response System? *Malar J.,* 11:330. doi: 10.1186/1475-2875-11-330.

Jimnez, R. E., Lam, R. M., Marot, M., Delgado, A. (2004). Observed-predicted length of stay for an acute psychiatric department, as an indicator of inpatient care inefficiencies. Retrospective case-series study. *BMC Health Services Research,* 4:4.

Jones, A. P., Bentham, G., Harrison, B. D. W., Jarvis, D., Badminton, R. M., Wareham, N. J. (1998). Accessibility and health service utilization for asthma in Norfolk, England. *J Pub Health Med.*, 20: 312-317.

Keene, C. M., Dondorp, A., Crawley, J., Ohuma, E. O., Mukaka, M. A. (2018). Competing-Risk Approach for Modeling Length of Stay in Severe Malaria Patients in South-East Asia and the Implications for Planning of Hospital Services. *Clin Infect Dis.,* 67(7):1053-1062.

Kum, C. K., Thorburn, D., Ghilagaber, G., Gil, P., Bjorkman, A. (2013). A Nonparametric Bayesian Approach to Estimating Malaria Prophylactic Effect After Two Treatments. *International Journal of Statistics in Medical Research, 2:76-87.*

Lawrence, J., Delaney, C. P. (2013). Integrating hospital administrative data to improve health care efficiency and outcomes: "the Socrates story". *Clinics in colon and rectal surgery*, 26(1):56-62.

Lin, X., Zhang, D. (1999). Inference in generalized additive mixed models by using smoothing splines, *Journal of the Royal Statistical Society Series B,* 61: 381-400.

Lorch, S. A., Zhang, X., Rosenbaum, P. R. (2004). Non-rural hospitals equivalent lengths of stay of pediatric patients hospitalized in rural and non-rural hospitals. *Peadiatrics,* 114;400-408.

Okiro, E. A., Kazembe, L. N., Kabaria, C. W., Ligomeka, J., Noor A. M., Ali D., Snow R. W. (2013). Childhood malaria admission rates to four hospitals in Malawi between 2000 and 2010. *PLoS One,* 8(4):e62214. doi: 10.1371/journal.pone.0062214.

Pendergast, J. F., Vogel, W. B. (1988). A multistage model of hospital bed requirements. *Health Services Research,* 23, 381-399.

Picone, G., Wilson, R. M., Chou, S. (2003). Analysis of hospital length of stay and discharge destination using hazard functions with unmeasured heterogeneity, *Health Economics,* 12: 1021-1034.

Polverejan, E., Gardiner, J. C., Bradley, C. J., Holmes-Rovner, M., Rovner D. (2003). Estimating mean hospital cost as a function of length of stay and patient characteristics. *Health Econ.,* 12: 935-947.

Sá, C., Dismuke, C. E., Guimarães, P. (2007). Survival analysis and competing risk models of hospital length of stay and discharge destination: the effect of distributional assumptions. *Health Services and Outcomes Research Methodology,* 7(3–4):109–124.

Snow, RW, Craig M, Deichmann U, Marsh K. (1999). Estimating mortality, morbidity and disability due to malaria among Africa's non-pregnant population. *Bulletin of the World Health Organization,* 77: 624-640.

Snow, RW, Craig, M. H., Newton C. R. J. C., Sketee, R. W. (2003). *The public health burden of Plasmodium falciparum malaria in Africa: Deriving the numbers.* Working Paper No. 11, Disease Control Priorities Project. Bethesda, Maryland: Fogarty International Center, National Institutes of Health.

Sorensen, J. (1996). Multi-phased bed modelling. *Health Services Management Research* 9, 61-67.

Stevens, A, Hammer, K, Buchkremer, G. (2001). A statistical model for length of psychiatric in-patient treatment and an analysis of contributing factors. *Acta Psychiatrica Scand.,* 103:203-211.

Tutz, G., Schmid, M. (2016). *Modeling Discrete Time-to-Event Data.* Springer: Berlin.

van Diepen, S., Bakal, J. A., Lin, M., Kaul, P., McAlister, F. A., Ezekowitz, J. A. (2015). Variation in critical care unit admission rates and outcomes for patients with acute coronary syndromes or heart failure among high- and low-volume cardiac hospitals. *Journal of the American Heart Association,* 4(3): e001708.

Wang, K., Yau, K. K. W., Lee, A. H. (2002). A hierarchical Poisson mixture regression model to analyse maternity length of hospital stay. *Statistics in Medicine,* 21:3639-3654.

Zaccardi, F., Webb, D. R., Davies, MJ., et al. (2017). Predicting hospital stay, mortality and readmission in people admitted for hypoglycaemia: prognostic models derivation and validation. *Diabetologia,* 60(6):1007-1015.

In: Statistical Modelling …
Editors: N-B. Kandala and L. N. Kazembe

ISBN: 978-1-53615-981-3

Chapter 3

BIVARIATE MODEL OF HEALTH SEEKING BEHAVIOUR AMONG WOMEN FOR THEIR UNDER-FIVE CHILDREN WITH FEVER

Bamidele O. Oguntoyinbo** and Lawrence N. Kazembe
Mathematical Sciences Department, Chancellor College,
University of Malawi, Zomba, Malawi
Department of Statistics and Population Studies,
University of Namibia, Windhoek, Namibia

ABSTRACT

Malaria is a major public health problem in Malawi, contributing to large numbers of morbidity and mortality among children under five. Not seeking timely treatment and not seeking appropriate care in health facilities, often degenerates into fatal emergencies. Using bivariate binary logistic regression model, this study jointly analysed timing/delay in seeking treatment and choice of health care provider of mothers for their under-five children reported with malarial fever. The results confirmed that timely treatment seeking was significantly enhanced in wealthiest household (OR = 1.535; 95% CI = 1.102, 2.138), for mothers currently working (OR = 0.662; 95% CI = 0.559, 0.784), in male- headed households (OR = 1.286; 95% CI = 1.102, 1.615), and for children aged between 12 and23 months (OR = 0.703; 95% CI = 0.513, 0.962). There was also a negative association between delay in seeking treatment and age of the child, with children aged 38-47 months less likely to be delayed (OR = 0.655; 95% CI = 0.465, 0.921). However, the choice of healthcare provider was significantly associated with mother's level of education for those with primary education (OR = 1.382; 95% CI = 1.033, 1.848), place of residence (OR = 1.756; 95% CI = 1.278, 2.413), poorer household (OR = 0.768; 95% CI = 0.494, 0.874), wealthier household (OR = 0.631; 95%

* Corresponding Author's Email: deleoguns0406@gmail.com.

CI = 0.463, 0.859), wealthiest household (OR = 0.372; 95% CI = 0.258, 0.535), mothers currently working (OR = 0.712; 95% CI = 0.582, 0.870), and southern region (OR = 1.296; 95% CI = 1.016,1.654). It was concluded that a greater effort is needed to improve the quality of fever treatment or expand health facility utilization, at all levels of administration if the goal of reducing malaria/fever burden is to be realised in Malawi.

Keywords: bivariate regression analysis, childhood fever, treatment seeking, health-care utilization, Malawi

INTRODUCTION

The Sustainable Development Goals (SDGs), adopted by the United Nations in 2015, aim to ensure healthy lives and promote well-being for all children. Particularly, SDGs goal 3 target 3.2 is set to end preventable deaths of new-borns and under-5 children by 2030 (UN, 2017). Malaria remains a major killer of children, mostly in sub-Saharan Africa (SSA), taking the life of a child every 2 minutes (WHO, 2016). According to the latest estimates, 212 million new cases of malaria occurred globally in 2015 and the disease led to 429,000 deaths (WHO, 2016). Children under five years are particularly susceptible to malarial illness, infection and death. In 2015, malaria killed an estimated 303 000 under-fives globally, including 292,000 in the African Region including Malawi (WHO, 2016). In Malawi, malaria was the leading cause of death for children under the age of five years in 2015, accounting for 21% of all deaths of the under-five children (WHO, 2016). Approximately one-third of all outpatient visits and 40% of hospitalizations of children under-five years of age are reported to be due to malaria in the country (HMIS, 2010).

The World Health Organization recognizes that early diagnosis and prompt treatment, within 24 hours of onset of symptoms, is an essential element of malaria control (WHO, 2016). Evidence shows that timely and appropriate treatment has resulted into reduced severe morbidity and mortality among children under the age of 5 years (Sharma, 2008). The ability of caregivers to recognize and seek appropriate care for the common childhood illnesses like malaria is instrumental in reducing child deaths in low- and middle income countries. Early treatment of childhood malaria depends upon mothers' perception about malaria and prompt recognition of signs and symptoms of the disease and treatment-seeking behaviour (Oyekale, 2015). Mothers/caretakers play a very important and significant role in managing childhood illness, since they spend more time with the children and child health care practices have been recognized as a significant factor behind reduced mortality rates among under-five age group children (Shaikh and Hatcher, 2005). More, importantly a large proportion of malaria/fever cases are treated outside the formal health care system, thus care-givers' knowledge and practice in health care utilization remain critical (McCombie, 1996).

Delay in seeking timely health care or not seeking care at all, and ineffective treatment choices can compromise the outcome of a child's health status. Poor health care-seeking behaviour for fever among caretakers of under-five children is still a major concern (Chukwuocha et al., 2009) and this has a huge bearing on the morbidity and mortality in this population group. Ordinarily, children are dependent by nature on their mothers/caregivers in seeking health care. The decisions mothers make affect their children when seeking treatment for any child illness, and more childhood deaths would be averted if mothers/caregivers would: (i) appropriately seeking care immediately; (ii) reaching an appropriate facility; (iii) receiving and applying treatment correctly. This is called the "Three Delays" framework for examining factors associated with morbidity and mortality (Upadhyay et al., 2013).

The analysis in this study is considered within different choices of health-seeking behavioural events by mothers/caregivers, from detection of child illness, to the decision to seek treatment, either same day or not, and to the facility where they sought advice or treatment, either public or not public facility. Thus our analysis considers the first two delays. Some studies have been done in Malawi on health care seeking behaviour, for example (Kazembe et al., 2007; Oyekale, 2015), but to our knowledge none of these studies have jointly assessed delay in seeking treatment and choice of first healthcare provider for fever in Malawi, in the context of bivariate binary regression model. Applications of these models in the analysis of healthcare seeking behaviour for under-five children with fever in Malawi, to our knowledge, are lacking in literature in Malawi.

Bivariate binary analysis of the data may provide insights into mother's/caregiver health seeking behavioural patterns and suggest ways to improve the medical choices made by mothers/caregivers of under-five children with fever. This motivated the present study, and the use of these models may help identify factors influencing the decisions and choices made by mothers /caregivers for their under-five children with fever which is important for policy development (Upadhyay et al., 2013). Therefore, this study aimed to jointly analyse those factors responsible for delay in seeking treatment and choice of first healthcare provider by mothers/caregivers of under-five children with fever in Malawi using bivariate binary regression model.

Bivariate Regression Models

Logistic Regression

Logistic regression allows one to predict a discrete outcome, especially when the outcome variables are measured on a binary scale. The relationship between the predictor variables and expectation of response variables is not a linear function in logistic regression; instead the logistic regression function is used, which is the logit transformation of the probability of occurrence θ,

$$\theta = \frac{exp(\beta_0+\beta_1 x_1+\cdots+\beta_k x_k)}{1+exp(\beta_0+\beta_1 x_1+\cdots+\beta_k x_k)} \quad (1)$$

where β_j is the coefficient corresponding to the predictor variables $X_j, j = 1, \ldots, k$. An alternative form of the ordinary logistic regression equation is given by,

$$log\left[\frac{\theta}{1-\theta}\right] = \alpha + \beta_1 x_1 + \beta_2 x_2 + \cdots + \beta_k x_k. \quad (2)$$

The goal of logistic regression is to correctly predict the category of outcome for individual cases using the most parsimonious model. The binary regression model (Collet, 1994), is used to explain the probability of a binary response variable as function of some covariates. According to McCullagh and Nelder (1989), binary logistic regression model which has two interrelated dependent variables can be modelled into a model called bivariate binary logistic regression.

Bivariate binary logistic regression is a development of binary logistic regression where we have two binary dependent variables jointly modelled together. Bivariate logistic regression (BLR) differs from ordinary logistic regression in the sense that related response variables are not assumed to be independent of one another. In this regard, bivariate logistic regression is a useful procedure with advantages that include (i) individual modelling of the marginal probability distribution of the bivariate binary responses, and (ii) modelling the odds ratio describing the pairwise association between the two binary responses in relation to several covariates (Darda, 2014).

Assume two binary dependent variables, Y_1 *and* Y_2, corresponding to delay and choice of first healthcare provider, each of which takes the value of either '0' or '1'. Then joint outcome (Y_1, Y_2), for a set of n-paired observation with their occurrence probabilities is given in Table 3.1. It is evident from Table 3.1 that $P_{rs} = P(Y_1 = r, Y_2 = s), r, s, = 0,1$, are the joint probabilities and $p_j = P(Y_j = 1), j = 1,2$ are the marginal probabilities for each of the response variables. It is assumed that the observations within pairs are correlated but observations from different pairs are independent.

Table 3.1. Joint association of probabilities on a cross-classification of the outcomes of delay and choice of first healthcare provider

	Healthcare provider	$Y_2 = 0$	$Y_2 = 1$	$Total$
Delay	$Y_1 = 0$	P_{00}	P_{01}	$1 - P_1$
	$Y_1 = 1$	P_{10}	P_{11}	P_1
	$Total$	$1 - P_2$	P_2	1

When there are k independent variables $x_1, x_2, \ldots, x_k$ then the value P_1 *and* P_2 are

$$P_1(x) = \frac{exp(\beta_{01}+\beta_{11}x_1+\cdots+\beta_{k1}x_k)}{1+exp(\beta_{01}+\beta_{11}x_1+\cdots+\beta_{k1}x_k)} \tag{3}$$

$$P_2(x) = \frac{exp(\beta_{02}+\beta_{12}x_1+\cdots+\beta_{k2}x_k)}{1+exp(\beta_{02}+\beta_{12}x_1+\cdots+\beta_{k2}x_k)} \tag{4}$$

The bivariate logistic model or bivariate logistic odds-ratio model described by McCullagh and Nelder(1989), Palmgren(1989) and later by Cessie & Houwellingen (1994) is specified by modelling the marginal distribution of each of Y_j, and also the odds ratio. The odds ratio $\varphi = \frac{P_{11}/P_{10}}{P_{01}/P_{00}}$ is used to describe the association between the two responses i.e., φ can be seen as the ratio of the odds of $Y_1 = 1$ given that $Y_2 = 1$ and the odds of $Y_1 = 0$ given that $Y_2 = 0, \varphi = 1$ indicating independence between Y_1 and Y_2. The model is given by,

$$log\left(\frac{p_j}{1-p_j}\right) = \beta_j^T \boldsymbol{X}\ for\ j = 1,2$$

$$log\varphi = \beta_3^T \boldsymbol{X} \tag{5}$$

The covariate vector $\boldsymbol{X}$ may include block specific and subunit specific covariates. For covariates values associated with Y_1 but not with Y_2 the corresponding elements in β_2 are set to zero, similarly for β_1 (Palmgren, 1989). The joint probabilities P_{11} can be obtained in terms of P_1, P_2 and φ as,

$$P_{11} = \begin{cases} \frac{1}{2}(\varphi - 1)^{-1} & \{a - \sqrt{a^2 + b}\}\ for\ \varphi \neq 1 \\ P_1 P_2\ for\ \varphi = 1 & \end{cases} \tag{6}$$

where $a = 1(P_1 + P_2)(\varphi - 1)$ and $b = -4\varphi(\varphi - 1)P_1P_2$ (Dale, 1986). The three other joint probabilities P_{rs} can be recovered easily from the marginal $P_{10} = P_1 - P_{11}, P_{01} = P_2 - P_{11}\ and\ P_{00} = 1 - P_{10} - P_{01} - P_{11}$.

Bivariate Probit Model

An alternative to the bivariate logistic regression model is the bivariate probit model. A typical bivariate probit model have disturbances (or error) terms in the two equations that are assumed to be independent, identically distributed and follow the bivariate standard normal probability distribution with their correlation coefficient ρ. The two standard normally distributed joint pdf will be

$$\boldsymbol{\Phi}_2(\varepsilon_1, \varepsilon_2, \rho) = \frac{1}{2\pi\sqrt{1-\rho^2}} exp\left[\frac{-1}{2(1-\rho^2)}(\varepsilon_1^2 + \varepsilon_2^2 - 2\rho\varepsilon_1\varepsilon_2)\right]. \quad (7)$$

If $\rho = 0$, then the two variables (or errors) are independent, if $\rho \neq 0$, then the two variables (errors) are correlated and the probability of one variable will be dependent on the probability of the other. If $\rho = 1$, then the two variables are essentially the same. If $\rho = -1$, then the two variables are negatively related.

Estimation of Model Parameters

To estimate the unknown parameters in the binary bivariate regression model, a maximum likelihood approach is used. The log-likelihood estimation for the bivariate binary logit regression have the multinomial distribution with probability $P_{11i}, P_{10i}, P_{01i}, P_{00i}$. So the likelihood of a bivariate random variables is as follows:

$$L(\beta) = \prod_{i=0}^{1} P(Y_{11i} = y_{11i}, Y_{10i} = y_{10i}, Y_{01i} = y_{01i}, Y_{00i} = y_{00i}) = \prod_{i=1}^{n} P_{11i}^{y_{11i}} P_{10i}^{y_{10i}} P_{01i}^{y_{01i}} P_{00i}^{y_{00i}} \quad (8)$$

The parameter $\hat{\beta} = (\widehat{\beta_1}, \widehat{\beta_2}, \hat{\theta})$ is obtained by maximising the equation above by derive it to its parameters.

$$lnL(\beta) = ln\left(\prod_{i=1}^{n} P_{11i}^{y_{11i}} P_{10i}^{y_{10i}} P_{01i}^{y_{01i}} P_{00i}^{y_{00i}}\right) = \sum_{i=1}^{n} y_{11i} lnP_{11i} + y_{10i} lnP_{10i} + y_{01i} lnP_{01i} + y_{00i} lnP_{00i}$$

The completion of this parameters estimation can be done iteratively. Therefore, the maximum likelihood estimate is used in this study for fitting the bivariate binary regression model.

Data and Analysis

Data Source

This study used cross-sectional data from the 2015/16 Malawi Demographic and Health Survey (NSO, 2017). Full details of the study design, methodology and sampling method are described elsewhere (NSO, 2017). In brief, the 2015/16 MDHS sample was drawn using a two-stage stratified approach, with each district was stratified into urban and rural areas, yielding a total of 56 sampling strata. Then in the first stage, 850 enumeration areas or clusters (173 clusters in urban areas and 677 in rural areas) were selected with probability proportional to sample enumeration area (SEA) size. In the second stage, 30 households per urban cluster and 33 per rural cluster were selected using a systematic random sampling approach. All women of reproductive age (15–49 years) in the selected households were eligible to participate. A total of 27,516 households were selected for the sample, of which 26,564 were occupied at the time of data collection. Of the occupied households, 26,361 were successfully interviewed, for a response rate of 99%. In the interviewed households, 25,146 eligible women were identified for individual interviews. Interviews were completed with 24,562 women, for a response rate of 98%. The data set used in the analysis was child record data set, with a focus on children under the age of five years.

Data Management

Data management in terms of extracting and generation of variables from child record data set were done in STATA Version 14 (StataCorp, College Station, TX, USA). Several categories of certain variables were grouped, since some categories had few recorded cases, data cleaning were carried out to remove the missing data. The data was then exported to R Software (R Development Core Team, 2017) for further analysis. BLR were estimated using VGAM package (Yee, 2010) in R.

Ethical Consideration

According to the 2015-16 MDHS report, the 2015-16 MDHS study was ethically approved by Malawi Health Sciences Research Committee of the Malawi Ministry of Health. Informed consent for the survey was obtained from each respondent at the start of each individual interview. The MDHS data sets are publicly available, and permission to use the data was granted by MEASURE DHS.

Study Variables and Measurements

Outcome Variable

The outcome variables of interest were: (i) place first sought treatment for fever and (ii) number of days after fever began sought advice or treatment. Place first sought treatment for fever (or choice of first healthcare provider) was coded into two categories (public = 1 or not public = 0); and the number of days after fever began did seek advice or treatment (delay in seeking treatment) was coded into two categories (same day = 1 or not same day = 0).

Independent Variables

Independent variables used in this study were based on some variables identified in previous studies on health seeking behaviour (Kazembe and Neema, 2013; Mustapha and Arabi, 2016; and Oyekale, 2015). The independent variables in the generated data set were mother's education (0 = No education, 1 = Primary, 2 = Secondary, and 3 = Higher), family wealth index (0 = Poorest, 1 = Poorer, 2 = Middle, 3 = richer, and 4 = Richest), child's age in months (0 = <6, 1 = 6-11, 2 = 12-23, 3 = 24-35,4 = 36-47, and 5 = 48-59), mother's age in years (0 = <20, 1 = 20-29, 2 = 30-39, and 3 = 40-49), sex of the child (0 = Female and 1 = Male), place of residence (0 = Rural and 1 = Urban), region (0 = Northern Region, 1 = Central Region, and 2 = Southern Region), partner's education (0 = No education, 1 = Primary, 2 = Secondary, and 3 = Higher), sex of household head (0 = Female and 1 = Male), mothers currently working (0 = No and 1 = Yes), person who usually decides on mother's health care (0 = Not Mother alone and 1 = Mother alone), and getting medical help for self: distance to health facility (0 = Not a big problem and 1 = Big problem). Table 3.2 gives more details.

Statistical Model for Bivariate Binary Regression Model

The study observed mothers/caregivers who sought treatment for their children with fever, whether it is on same day or not and the second observation is that treatment was sought either in public health facilities or not. Mothers could decide to seek treatment for their under-five children with fever same day or not. Also, they can make decisions to seek treatment either in public facility or not. The different decision choices can be identified as seek treatment (Y_1 = same day or not), and seek treatment in (Y_2 = public facility or not), both taking a value of 1 and 0 respectively. When decisions by the same individuals are discrete and interrelated, bivariate probit or logit models are appropriate (Greene, 2003). The two equations bivariate binary model for Y_1 and Y_2 are presented in equation 1 and 2, respectively:

$$Delay\ in\ seeking\ treatment = Y_{1i}^{*} = X_1\beta_1 + \varepsilon_1;$$

such that

$$Y_{1i} = \begin{cases} 1 = & Same\ day\ if\ Y_{1i}^{*} > 0; \\ 0 = & otherwise \end{cases}$$

while

$$Choice\ of\ healthcare\ provider = Y_{2i}^* = X_2\beta_2 + \varepsilon_2;$$

with

$$Y_{2i} = \begin{cases} 1 = & Public\ facilities\ if\ Y_{2i}^* > 0; \\ 0 = & otherwise \end{cases}$$

where Y_{1i} (Y_{1i}^*), $Y_{2i}(Y_{2i}^*)$ in mathematical expression are the observed binary variables indicating seeking treatment same day or not and seeking treatment in public facility or not. The explanatory variables are given by the vectors X_1 and X_2 while β_1 and β_2 are the unknown parameters of the respective equations to be estimated in the model.

Analysis proceeded as follows: Firstly, two binomial logistic regression models were fitted separately for delay in seeking treatment and choice of health care provider for under-five children with fever in Malawi. Secondly, the study jointly models the two decisions, delay in seeking treatment and choice of healthcare provider.

RESULTS

Social Demographics Characteristics of the Study Participants

Table 3.2 summarises the characteristics of the study participants. In brief, it reveals that the average age of the children was 27.81 months with standard deviation of 16.26, while average age of the mothers was 27.60 years with standard deviation of 6.67. Close to half of the children (50.4%) were males, and the majority (87%) of the participants were rural dwellers. Similarly, nearly half (45.68%) of the participants were from southern region and 74.2% were male-headed households. Overall, 55.9% of the partner's participants had completed primary education, while 35. 6% of the partner's participants had secondary education, and only 4.3% had higher education. About 68% of the mothers have primary education, 11% had no education, 19% had secondary education and about 1.4% had higher education. On the other hand, 74.2% of the household heads were male and about 70.2% of the mothers were currently working.

Again from Table 3.2, about 58% acknowledged that the distance to the health facility is a big problem. Relative to the household wealth status, about 23% of the household were poorest, 22% in the poorer category, 21% were in the middle household, 19% were in the richer household and only 15% were in the richest household. Further, a total of 81.2% of the mothers does not decide on healthcare for themselves.

Table 3.2. Summaries for background characteristics

Variables	Categories	Mean (Std. dev)	Number (%)
Age of the mother in years		27.60 (6.70)	-
Age of the child in months		27.81 (16.30)	-
Age of the mother	<20		304 (8.17)
	20-29		2064 (55.50)
	30-39		1146 (30.81)
	40-49		205 (5.51)
Age of the child	<6		295 (7.93)
	6-11		464 (12.48)
	12-23		862 (23.18)
	24-35		833 (25.68)
	36-47		681 (15.03)
	48-60		584 (15.70)
Region	Northern		695 (18.69)
	Central		1325 (35.63)
	Southern		1699 (45.68)
Residence	Rural		3235 (86.99)
	Urban		484 (13.01)
Mothers education level	No Education		402 (10.81)
	Primary		2542 (68.35)
	Secondary		723 (19.44)
	Higher		52 (1.40)
Partners education level	No Education		224(7.30)
	Primary		1714 (55.87)
	Secondary		999 (32.56)
	Higher		131 (4.27)
Sex of household head	Female		960 (25.81)
	Male		2759 (74.19)
Sex of child	Female		1843 (49.56)
	Male		1876 (50.44)
Wealth Status	Poorest		856 (23.02)
	Poorer		815 (21.91)
	Medium		767 (20.62)
	Richer		709 (19.06)
	Richest		572 (15.38)
Currently working	No		1110 (29.85)
	Yes		2609 (70.15)
Decision on seeking health	Not Mother alone		252 (81.23)
	Mother alone		584 (18.77)
Distance	Not a big problem		1573 (42.30)
	Big problem		2146 (70.15)

Descriptive Results of Choice of First Healthcare Provider of Treatment for Fever by Various Characteristics Variables

Table 3.3 show that the mothers in age group 20-29 years formed the highest proportion (81.01%) of seeking treatment in the public health facilities and those of age group <20 years have the second highest with 80.26% in seeking treatment in public health facilities. The proportion of choice of first healthcare provider in mother's age group 30-39 years were 75.35% in public health facilities. Mothers aged 40-49 years tend to have a slighty lower percentage (75.61%) in using public health facilities. The proportion of choice of first healthcare provider in children age group 12-23 months was 79.58% in public health facilities, while those in age group 24-35 formed the highest proportion (81.15%) of seeking treatment in the public health facilities, whereas those in the age group <6 years have the second highest with 80.34% in seeking treatment in public health facilities. Children aged 36-47 tend to have a slightly lower percentage (76.65%) in using public health facilities.

The proportion of children aged 6-11 and 48-59 seeking treatment in public health facilities were 79.96% and 77.74% respectively. The proportion of urban mothers who sought treatment for their children in public health facilities was 82.85%, while rural mothers was 78.67% sought treatment for their under-five children with fever in public health facilities.

Table 3.3 also presents the percentage according to choice of first healthcare provider and mother's educational level and partner's educational level. The mothers with primary education have the highest proportion for seeking treatment in public health facilities with 80.21% for their under-five children with fever, while mothers with higher education had comparatively lower proportion of seeking treatment in public health facilities with 65.38%. The proportion of choice of first healthcare provider in mothers with secondary education have the second highest proportion with 78.56%, and mothers with no education have 75.87% in seeking treatment in public health facilities. The partners with primary education have the highest proportion for seeking treatment in public health facilities with 80.05% for their under-five children with fever, whereas partners with higher education have the lowest proportion of seeking treatment in public health facilities with 74.81%. The proportion of choice of first healthcare provider in partners with secondary education had the second highest proportion with 77.88%, and partners with no education had 76.79% in seeking treatment in public health facilities. It appeared that there was a consistent increase in the percentage of women and their partners with primary education who took their children to public health facilities with both having the highest proportion.

Table 3.3. Percent distribution of delay of seeking treatment and choice of first healthcare provider of treatment for fever by various characteristics variables

Covariates	Delay in seeking treatment		Choice of first healthcare provider	
	Not Same day = 0	Same day = 1	Not public = 0	Public facility = 1
Age of the mother in years	214 (70.39)	90 (29.61)	60 (19.74)	244 (80.26)
<20	1507 (73.01)	557 (26.99)	392(18.99)	1672(81.01)
20-29	814 (71.03)	332 28.97)	271(23.65)	875 (76.35)
30-39	139 (67.80)	66 (32.20)	50 (24.39)	155 (75.61)
40-49				
Age of the child in months				
<6	198 (67.12)	97 (32.88)	58 (19.66)	237 (80.34)
6-11	329 (70.91)	135 (29.09)	93 (20.04)	371 (79.96)
12-23	631 (73.2)	231 (26.8)	176(20.42)	686 (79.58)
24-35	590 (70.83)	243 (29.17)	157(18.85)	676 (81.15)
36-47	508 (74.6)	173 (25.4)	159(23.35)	522 (76.65)
48-59	418 (71.58)	166 (28.42)	130(22.26)	454 (77.74)
Region				
Northern	516 (74.24)	179 (25.76)	171 (24.6)	524 (75.4)
Central	930 (70.19)	395 (29.81)	285 (21.51)	1040 (78.49)
Southern	1228 (72.28)	471 (27.72)	317 (18.66)	1382(81.34)
Place of residence				
Rural	2344 (72.46)	891 (27.54)	690 (21.33)	2545 (78.67)
Urban	330 (68.18)	154 (31.82)	83 (17.15)	401 (82.85)
Mothers education level				
No Education	276 (68.66)	126 (31.34)	97 (24.13)	305 (75.87)
Primary	1839 (72.34)	703 (27.66)	503 (19.79)	2039 (80.21)
Secondary	527 (72.89)	196 (27.11)	155 (21.44)	568 (78.56)
Higher	32 (61.54)	20 (38.46)	18 (34.62)	34 (65.38)
Partners education level				
No Education	156(69.64)	68 (30.36)	52 (23.21)	172 (76.79)
Primary	1250 (72.93)	464 (27.07)	342 (19.95)	1372 (80.05)
Secondary	716 (71.67)	283(28.33)	221 (22.12)	778 (77.88)
Higher	89 (67.94)	42 (32.06)	33 (25.19)	98 (74.81)
Sex of household head				
Female	723 (75.31)	237 (24.69)	185 (19.27)	775 (80.73)
Male	1951 (70.71)	808 (29.29)	588 (21.31)	2171 (78.69)
Sex of child				
Female	1333(72.33)	510 (27.67)	377(20.46)	1466(79.54)
Male	1341(71.48)	535 (28.52)	396(21.11)	1480(78.89)
Wealth Status				
Poorest	638 (74.53)	218 (25.47)	144 (16.82)	712 (83.18)
Poor	607 (74.48)	208 (25.52)	171 (20.98)	644 (79.02)
Middle	545 (71.06)	222 (28.94)	146 (19.04)	621 (80.96)
Richer	504 (71.09)	205 (28.91)	154 (21.72)	555 (78.28)
Richest	380 (66.43)	192 (33.57)	158 (27.62)	414 (72.38)
Mothers currently working				
No	733 (66.04)	377 (33.96)	196 (17.66)	914 (82.34)
Yes	1941 (74.40)	668 (25.60)	577 (22.12)	2032 (77.88)

Table 3.3. (Continued)

Covariates	Delay in seeking treatment		Choice of first healthcare provider	
	Not Same day = 0	**Same day = 1**	**Not public = 0**	**Public facility = 1**
Decisions			545 (21.56)	1983 (78.44)
Mothers not alone	1799(71.16)	729(28.84)	110 (18.84)	474 (81.16)
Mothers alone	439(75.17)	145(24.83)		
Distance				
Not a big problem	1099 (69.87)	474 (30.13)	25 (20.66)	1248 (79.34)
Big problem	1575 (73.39)	571 (26.61)	448(20.88)	1698 (79.12)

Table 3.3 further shows that male-headed households, about 79% sought treatment for their children in public health facilities, compared to femaleheaded households with about 81%. Similarly about 79% male children and 80% female children were treated for fever in public health facilities. Results also show that mothers who lived in different regions had different status on the choice of first healthcare provider. The highest proportion was observed in southern region (81.34%) followed by central region (78.49%) and the least was observed in northern region (75.4%). The percentage of poorest mothers who used public health facilities for treatment of fever for their children health facilities was 83.18%, while the proportion of poorer, middle, richer and richest mothers who used public health facilities were 79.02%, 80.96%, 78.28% and 72.38% respectively. The proportion of currently working mothers who used public health facilities as the first choice healthcare provider was 77.88%, compared to mothers currently not working (with 82.34%). With respect to persons who usually decided on seeking healthcare, about 81.16% of mothers decided alone to seek treatment at public health facilities, while those that did not decide alone, about 78.44% sought treatment at public health facilities. Finally, the mothers who said that distance was a big problem, about 79.12% sought treatment in public health facilities, while those that said distance was not a big problem, about 79.34% sought treatment in public health facilities.

Descriptive Results of Delay of Seeking Treatment for Fever by Various Characteristics Variables

Table 3.3 presents results of association between explanatory variables and delay in seeking treatment for under-five children reported with fever in Malawi. The table shows that mothers in age group 40-49 formed the highest proportion (32.20%) in seeking treatment same day, followed by those in the age group <20 years and in 30-39 years (29.61% and 28.97% respectively). With regards to children's age, 26.8% of children in age group 12-23 months sought treatment the same day after presenting with fever. The highest proportion seeking treatment same day, however, were children in age group <6

(32.88%) of same day, followed by those in the age group 24-35 (29.17%). The proportion of urban women who sought treatment same day was 31.82%, while rural women was 27.54%.

Considering mother's educational level and partner's educational level, Table 3.3 shows that mothers with higher education have the highest proportion for seeking treatment same day with 38.46%, while thosee with secondary education, no education, or with primary education were 27.11%, 31.34%, and 27.66% respectively. Similarly, those with partners with higher education have the highest proportion for seeking treatment same day (32.06%). Those with partners with primary education have the lowest proportion of seeking treatment same day with 27.07%. For those with no education and secondary education, the percentages were 30.36% and 28.33% respectively. Household with male headship, about 29.29% sought treatment same day for their children, while female-headed households had 24.69% seeking treatment same day. About 28.52% of the male child and 27.67% of the female child were treated for fever same day.

The results also show differences across regions. The highest proportion seeking treatment same day was observed in central region (29.81%), followed by southern region (78.49%), and the northern region (25.76%). We also observed an association with wealth index. The percentage of poorest mothers who sought treatment same day was 25.47%, while the proportion of poorer, middle, richer and richest mothers who sought treatment same day for fever for their under-five children were 25.52%, 28.94%, 28.91% and 33.57% respectively. The proportion of seeking treatment same day in mothers who were currently working was 25.60%, while the women currently not working had proportion of 33.96%.

With respect to persons who usually decide health, about 25% of women who decided to seek care on their healthcare alone, while 29% did not decide on their own. Finally, the women who said that distance was a big problem, about 26.61% sought treatment same day, while those that said distance was not a big problem, about 30.13% sought treatment same day.

Statistical Model Results for Bivariate Binary Logistic Model

Table 3.4 shows a strong association between the two outcome variables (OR = 0.636, 95% CI 0.526, 0.768), and suggests that there is negative association between the choice of the first healthcare provider for treatment and delay in seeking treatment decision model. Same table shows that choice of the first healthcare provider for treatment of fever was associated with the residence of the mothers, mother's education, wealth status, currently working, and region.

Table 3.4. Bivariate binary logistic model results for delay in seeking treatment and choice of first healthcare provider for treatment of fever

	Delay in seeking treatment			**Choice of first healthcare provider**		
Covariates	**Odds ratio**	**95% CI**		**Odds ratio**	**95% CI**	
(Intercept)	0.621	0.343	1.123	5.144	2.612	10.131
Residence						
(ref)Rural	1.000			1.000		
Urban	0.968	0.739	1.267	1.756	1.278	2.413
Sex of household head						
(ref)female						
Male	1.286	1.024	1.615	0.917	0.716	1.176
Mothers education						
(ref)No education	1.000			1.000		
primary	0.934	0.713	1.223	1.382	1.033	1.848
secondary	0.874	0.621	1.231	1.342	0.928	1.940
higher	1.297	0.582	2.887	0.708	0.303	1.653
Partners education						
(ref)No education	1.000			1.000		
primary	0.769	0.574	1.029	1.090	0.785	1.516
secondary	0.782	0.565	1.082	1.027	0.713	1.479
higher	0.752	0.422	1.341	1.403	0.735	2.678
Wealth status						
(ref)poorest	1.000			1.000		
poorer	1.089	0.846	1.401	0.658	0.494	0.875
middle	1.246	0.964	1.609	0.768	0.570	1.033
richer	1.236	0.942	1.623	0.631	0.463	0.859
richest	1.535	1.102	2.138	0.372	0.258	0.535
Region						
(ref)Northern	1.000			1.000		
central	1.258	0.995	1.591	1.085	0.846	1.391
southern	1.179	0.939	1.479	1.296	1.016	1.654
Distance						
Big problem	0.859	0.728	1.015	0.964	0.801	1.160
(ref)Not a big problem	1.000			1.000		
Sex of the child						
(ref)female	1.000			1.000		
Male	1.029	0.878	1.206	0.967	0.811	1.153
Decisions						
(ref)not alone	1.000			1.000		
Alone	0.831	0.672	1.026	1.190	0.941	1.506
Age of the child						
(ref)<6	1.000			1.000		
6-11	0.712	0.502	1.009	0.874	0.580	1.318
12-23	0.703	0.513	0.962	0.886	0.610	1.289
24-35	0.789	0.579	1.075	0.914	0.630	1.326
36-47	0.655	0.465	0.921	0.785	0.529	1.167
48-59	0.736	0.527	1.028	0.873	0.588	1.295
Age of the mother						
(ref)<20	1.000			1.000		
20-29	0.932	0.667	1.303	1.118	0.761	1.644
30-39	0.993	0.698	1.413	0.854	0.572	1.273
40-49	1.229	0.774	1.954	0.895	0.529	1.511
Currently working						
(ref) no	1.000			1.000		
Yes	0.662	0.559	0.784	0.712	0.582	0.870
Association	(OR = 0.6355, 95% CI 0.526, 0.768)					

The parameters of a place of residence were statistically significant in the public health facilities (OR = 1.756; 95% CI = 1.278, 2.413). Those in urban areas were more likely to treat children infected with fever in public health facilities. With regards to mother's education holding all variables constant, the probability of seeking treatment in a public facility was found to be positively associated with mother's with primary education (OR = 1.382; 95% CI = 1.033, 1.848)

Currently working mothers were less likely to seek treatment in public facilities for their under-five children infected with fever compared with mothers who are not currently working (OR = 0.712; 95% CI = 0.582, 0.870). With regards to wealth status, mothers from poorer household (OR = 0.768; 95% CI = 0.494, 0.874), or richer (OR = 0.631; 95% CI = 0.463, 0.859), and richest households (OR = 0.372; 95% CI = 0.258, 0.535), were less likely to seek treatment in public health facilities for fever of under-five compared to the poorest mothers. With regards to region, we observed a positive association between use of public healthcare provider and region, with mothers in southern region more likely to seek treatment in public facility compared to those in northern region (OR = 1.296; 95% CI = 1.016,1.654).

Furthermore, Table 3.4 shows that four variables were associated with delay in treatment-seeking. These include sex of the household head, wealth status, age of the child and work status of the mother. In particular, male-headed households appeared to be a significant determinant for seeking early treatment (OR = 1.286; 95% CI = 1.102, 1.615), meaning that male-headed households are more likely to seek treatment for under-five children with fever. As expected, currently working mother (OR = 0.662; 95% CI = 0.559, 0.784) were less likely to seek timely treatment compared with those mothers who are not currently working.

The richest mothers (OR = 1.535; 95% CI = 1.102, 2.138) were more likely to seek timely treatment compared with poorest mothers. On the other hand, there was also a negative association between age of the child and timely treatment seeking, in particular, children aged 12-23 months (OR = 0.703; 95% CI = 0.513, 0.962), were less likely to seek timely treatment compared with children aged <6 months. Also, there was also a negative association between delay in seeking treatment and children aged 38-47 months (OR = 0.655; 95% CI = 0.465, 0.921).

Discussion

This study aimed atto jointly identifying factors that influence delay in seeking treatment and choice of healthcare provider for under-five children with malarial fever, as the outcome variable may be correlated. In part, the study sought to determine child, maternal, and household level characteristics that influence decision making on delay in seeking treatment and choice of first health care provider for treatment of fever in under-five children. While several studies on care-seeking behaviour exist (Oyekale, 2015; Kazembe et al., 2007), none of these have considered joint analysis of delay and choice of first healthcare provider using bivariate binary regression model.

The bivariate binary model result shows that mother's education, place of residence, household wealth status, mothers currently working and region were significantly associated with choice of first healthcare provider, while the decision to seek timely treatment care were influenced by sex of the household head, house hold wealth status, and mothers currently working were all significant.

The results of this study support the previously well-documented findings that children living in urban areas were more likely to be taken to a formal healthcare facility especially public health facilities as compared to children living in rural areas. Residents of urban areas have significantly higher probabilities of using public health facilities. Utilization of public facilities medical services by urban resident could have been motivated by concentration of such services in urban areas. These findings are similar to earlier studies (Gerald, 2015; Kazembe et al., 2007, Oyekale, 2015). In this study, mothers/caretakers with the primary levels of education were more likely to initiate and seek care in a public facility than at other health facilities, which is consistent with other studies. The education of the mothers was very significant in choice of first healthcare provider for treatment of fever, which is consistent with other study done in Malawi (Oyekale, 2015). Gerald (2015) has also emphasized the importance of maternal education for child health.

This study is not without limitation. Firstly, the study relies heavily on the mother's recall of the care and treatment sought for their under-five children with fever. This kind of information can be subject to recall bias, reporting bias and misclassification bias. In some instance a mother might not remember when she sought treatment for her child or where she sought treatment. But given that the survey was asking questions based on what happened two weeks before concerning their under-five children with fever, the bias is expected to be minimal. Secondly, some missing key observations that were missing are excluded from this study. Lastly, the sampling weight of the DHS survey was not accounted for and incorporated in the study, mainly due to the fact that the VGAM package in R currently does not have that capability. Despite these limitations, this research using maximum likelihood estimation provides additional information about treatment seeking behaviour for under-five children reported with fever using bivariate binary logistic model.

In summary, the results emphasize that adequate statistical modelling and analysis is of importance in health care seeking behaviour of women for their under-five children with fever. The findings further emphasized the need for the Malawian government to enhance welfare of people through capacity development and skill building activities. In Malawi, as in most least developed countries, the level of health care utilization in other health facilities is low, and when a decision has been taken to seek care is often delayed.

More research is needed to understand the factors influencing mother's treatment decisions in seeking healthcare for their under-five children with fever in Malawi. The study has highlighted some variables that are crucial in determining how parents would react to seek treatment for fever for their children. The study contribution to the studies may shed light on factors contributing to such human behaviour. Secondly, findings may help policy decision makers to pinpoint areas that need immediate health promotion campaigns to be designed that can ensure behaviour change. To conclude, it should be emphasized that a greater effort is needed to improve the quality of fever treatment or expand health facility utilization, at all levels of administration if the goal of reducing malaria/fever burden is to be realised in Malawi. Health promotion and education interventions should stress promptness of health facility visits, improved access to appropriate drugs and mothers/caregiver should seek timely and appropriate care for their under-five children.

POLICY IMPLICATIONS

The key point that we learn from this study, is that in addition to the efforts being made to reduce the frequency of childhood fever in general, specific attention should be given to reduce the delay in seeking treatment for fever and the choices made by mothers/caregivers in choosing first healthcare provider must be taking into consideration. The concerned bodies especially the mothers have to give top priorities to seek timely and adequate care for their under-five children with fever. Using the bivariate binary regression model, the factors influencing choice of first health care provider were place of residence, education of the mother, household wealth status, region and mothers currently working, while delay in seeking treatment were influenced by sex of the household head, household wealth status, age of the child and mothers currently working. Thus, the recommendations given below are along these significant predictors.

The educational level of mothers was found to be significant factors affecting choice of first healthcare provider, therefore, it is important to allow mothers attain their highest level of education and therefore improve the socioeconomic status of mothers and children. Give more attention to rural areas which have high rates of poverty, and problematic area in accessing health facilities. More health facilities should be built to enable them to seek health care services. Therefore, the most important issues to address in these areas are providing good health care facilities, and raising the educational level of parents. Mothers should be educated on the need to seek timely care from appropriate sources for children irrespective of age. The study also shown that sex of the household is very important in seeking timely treatment for children with fever, special attention should be given that the household head are able to seek timely treatment for their children. Educating heads of households and empowering women could be an effective policy drive towards using appropriate healthcare. The study recommends focusing on creating awareness at household level about health issues and effective ways to seek treatment early and appropriate treatment in good health facilities.

REFERENCES

Barofsky, J., Chase, C., Anekwe, T., Farzadfar, F. (2011). *The economic effects of malaria eradication: Evidence from an intervention in Uganda*. PGDA Working Paper No. 70; 2011.

Cessie, S. L., and Houwelingen, J. C. Van (1994). Logistic regression for correlated binary data. *Applied Statistics*, 43, 95–108.

Chukwuocha, U., Nwankwo, B., Amadi, A., Dozie, I., Ikegwuoha, A. (2009). Treatment seeking behaviour of mothers for febrile children in some rural parts of Imo State Nigeria: implications for home management of malaria in endemic areas. *Int J Trop Med*, 4, 132–135.

Collet, D. (1994). *Modelling Binary Data*. London, UK: Chapman & Hall.

Dale, J. R. (1986). Global Cross-ratio Models for Bivariate, Discrete, Ordered Responses. *Biometrics*, 42, 909- 917.

Darda, M. (2014). An Investigation of Agricultural Farm Dynamics using Bivariate Binary Regression. *Asian Journal of Research in Social Sciences and Humanities*, 4(9), 28-40.

Gerald, M. (2015). Assessing Factors Influencing Health Seeking Behaviour for Malaria Treatment in Children under Five Years in Rwimi Town Council Kabarole District. *Int J Sch Cog Psychol,* 2, 151.

Government of Malawi. (2010). *Health management information systems* (HMIS). Lilongwe: Ministry of Health.

Greene, W. H. (2003). *Econometric analysis* (5th ed.). Upper Saddle River, NJ: Prentice Hall.

Hosmer, D. W., and Lemeshow, S. (1989). *Applied Logistic Regression*. New York: Wiley.

Kazembe, L. N., Appleton, C. C., and Kleinschmidt, I. (2007). Choice of treatment for fever at household level in Malawi: examining spatial patterns. *Malaria Journal*; 6:40.

Kazembe, L. N., and Neema, I. (2013). Today, tomorrow, forever: A Bayesian ordered categories model for treatment seeking in febrile children. *Int. Sci. Technol. J. Namibia*, 1(1), 21-34.

Mackowiak, P. (1995). *Fever: basic mechanisms and management* (2nd ed.). Philadelphia: Lippincott Raven.

McCombie, S. C. (1996). Treatment seeking for malaria: a review of recent research. *Soc Sci Med*, 43, 933–945.

McCullagh, P., and Nelder, J. A. (1989). *Generalized Linear Models* (2nd ed.). London: Chapman & Hall.

Mustapha, I., and Arabi, U. (2016). What factors influence the choice of first healthcare provider for childhood fever or cough in Ghana? *International Journal of Medicine*, 4(2), 49-54.

National Statistical Office (NSO) and ICF Macro. 2017. *Malawi Demographic and Health Survey* 2015-16. Zomba, Malawi, and Calverton, Maryland, USA: NSO and ICF.

Olotu, A., Fegan, G., Williams, T. N., Sasi, P., Ogada, E., Bauni, E., Bejon, P. (2010). Defining clinical malaria: the specificity and incidence of endpoints from active and passive surveillance of children in rural Kenya. *PLoS One*, 5(12).

Oyekale, A. S. (2015). Assessment of Malawian Mothers' malaria knowledge, healthcare preferences and timeliness of seeking fever treatments for children under five. *Int J Environ Res Public Health.* 12(1), 521–540.

Palmgren, J. (1989). *Regression Models for Bivariate Binary Responses*. Technical Report no. 101, Department of Biostatistics, University of Washington, Seattle.

R Development Core Team (2017). R: A language and environment for statistical computing. R Foundation for Statistical Computing, Vienna, Austria. http: R-project.org

Shaikh, B. T., and Hatcher, J. (2005). Health seeking behaviour and health service utilization in Pakistan: challenging the policy makers. *J Public Health* (Oxf), 27(1), 49-54.

Sharma, V. R. (2008). When to seek health care: a duration analysis for malaria patients in Nepal. *Soc Sci Med*, 66, 2486–94.

Tabachnick, B. G., and Fidell, L. S. (1996). *Using Multivariate Statistics* (3rd Edn.). New York: Harper Collins College Publishers.

United Nations 2017. *Sustainable Development Knowledge Platform*. Retrieved September 26, 2017 from https://sustainabledevelopment.un.org/sdg3.

Upadhyay, R.P., Rai, S.K., Krishnan, A. (2013). Using three delays model to understand the social factors responsible for neonatal deaths in rural Haryana, India. *J Trop Pedriatrics* (Oxf), 59(2), 100-105.

Uzochukwu B. S. C., Onwujekwe, E. O., Onoka, C. A., & Ughasoro, M. D. (2008). Rural-urban differences in maternal responses to childhood fever in South East Nigeria. *PLoS One*, 3(3).

World Health *Organisation Africa* (2016). Retrieved September 25, 2017 from http://www.afro.who.int/sites/default/files/2017-07/afr-rc66-7-en-0411.pdf.

World Health Organisation Geneva (2016). *World Malaria Report*. Retrieved September 25, 2017 from http://www.who.int/malaria/media/world-malaria-report-2016/en/.

World Health Organisation Geneva (2016). *Malawi Neonatal and child health profile*. Retrieved October 4, 2017 from http://www.who.int/malaria/publications/country profiles/profile_mwi_en.pdf?ua=1.

Yee, T. W. (2010). The VGAM Package for Categorical Data Analysis. *Journal of Statistical Software*, 32 (10).

In: Statistical Modelling …
Editors: N-B. Kandala and L. N. Kazembe

ISBN: 978-1-53615-981-3

Chapter 4

MOVER-STAYER MODEL ON FUTURE CONTRACEPTIVE USE AMONG MARRIED WOMEN IN MALAWI

George T. Mwenye-Phiri*[*] *and Lawrence N. Kazembe
Malawi College of Health Sciences, Lilongwe,Malawi
Department of Statistics and Population Studies,
University of Namibia, Windhoek, Namibia

ABSTRACT

Analysis of future contraceptive use has usually assumed that all subjects would eventually adopt use of contraceptives, thereby ignoring unobserved heterogeneity that may arise from individual subject's failure to adopt their use. This study aimed at developing a Mover-stayer model to account for such heterogeneity. The study retrospectively followed a sample of 4716 married women contraceptive non-users of age 15 to 49, generated from the 2010 MDHS data. The study compared parameters in the mover-stayer model with those in a parametric clog-log model; a non-parametric clog-log model; and the pooled regression model. Results show that, in terms of model fit, the Mover-stayer model has supremacy over the other three models. The findings further suggest the importance of including a heterogeneity term in modelling future contraceptive use.

Keywords: population-averaged, mover-stayer, heterogeneity, models, contraception, log-logistic, hazard, non-random

---

[*] Corresponding Author's Email: georgemwenyephiri@gmail.com.

INTRODUCTION

The use of family planning services is an important issue for the developing world because it helps in reducing fertility levels, thereby contributing positively to their development (Clements & Madise 2004). According to WHO (2018), use of modern contraception worldwide has risen slightly in the years 1990 to 2015, particularly in Africa and Asia; with some stability observed in Latin America and the Caribbean. Although there has been a slight increase in the use of contraception in Africa, Malawi continues to experience a low contraceptive prevalence rate (CPR) of as low as 59% among all married women using any method (National Statistical Office (NSO) [Malawi] and ICF 2017). Such low CPR have been shown to be associated with some women experiencing unwanted pregnancies, resulting in a huge impact on the maternal mortality rate and lifetime risk of dying in childbirth or pregnancy (Collumbien et al. 2004, Bellizi et al., 2015).

As a means to ensure increased contraceptive uptake, several strategies have been employed, and are now available in many reproductive health programmes, including community-based interventions and antenatal contraceptive counseling. In community-based interventions, family planning outreach workers travel to rural and urban areas educating women regarding contraceptive alternatives (Hossain 2005). For such interventions to be successful, analysis of contraception data needs to be reliable. However, most of the modelling in contraception literature has usually applied standard regression models that assume that all women will eventually adopt use of contraceptives (See, for example, Mavromaras & Orme 2004; Nakosteen & Zimmer 1980; Alba-Ramirez, Arranz & Munoz-Bullon 2012; and Schmidt & Witte 1989). Consequently, such modelling may have ignored the problem of unobserved heterogeneity that arises due to some women changing preferences (states), leading to biased estimates and therefore, unreliable results.

The chapter, therefore, attempts to address such challenging methodological issues by proposing a Mover-stayer (MS) model to account for unobserved heterogeneity. The MS model has been applied in various fields, such as labour studies (see, for example, Simpson & Sproule 1998); finance and banking (see, for example, Altman 1998; Frydman, Kallberg & Kao 1995); smoking studies (see, for example, Cook, Kalbfleisch & Yi, 2002); cancer studies (see, for example, Chen, Duffy & Taber 1997), among others. However, there is paucity of information about its application in the field of contraception. It is against this background that this paper seeks to illustrate the robustness of the MS model over other duration models in modelling future contraceptive use among married women contraception nonusers. Thus, in this modelling framework, the performance of the MS modelling approach is compared with that of three other duration models that use the complementary log-logistic link: 1) Parametric Complementary Log-logistic (PCL) model; 2) Non-parametric Complementary Log-

logistic (NPCL) model; and 3) the Pooled regression (PR) model. Specifically, the paper seeks to: 1) determine whether time to adopt use of contraceptives varies by sub-groups of selected predictors; 2) estimate the probability that a married contraceptive non-user woman will use contraceptives in future, controlling for other factors; and 3) investigate the effect of selected predictors on future contraceptive use among married women who are currently non-users.

It is therefore anticipated that, the modelling explored in this paper, would be significant in several ways - first, the statistical modelling approach adopted may help reduce bias and or account for heterogeneity in identifying predictors for the probability of future contraceptive use. Secondly, it may also add on to an already existing statistical knowledge base in contraception. Furthermore, the resulting model may also create a platform for making sound reproductive health policy decisions, as through its findings, reproductive health outreach programmes may be targeted.

Previous Analyses on Contraceptive Behaviour

Before embarking on the modelling part of this study, this paper first highlights some approaches that previous analysts have used in their contraceptive studies.

Anasel & Mlinga (2014), for instance, in their study to determine effects of predictors on contraceptive use and the probability of using contraceptives, used a standard logistic regression model. Accordingly, a standard logistic regression method is suitable for data that have a dichotomous response variable. This model assumes independence of observations, as well as linearity, through log odds, between the explanatory and response variables. While the model has the advantage of the probability of the response lying within the bounds of 0 and 1, compared to some models like the probit, as well as converging fairly easily, it has a disadvantage that its assumption of independence may be violated in situations where data are clustered or in repeated measurements. This may lead to over-dispersion, and, consequently, biased parameter estimates.

To add more evidence to the problems inherent in the standard logistic regression model, Triparti (2008) compared the standard logistic regression model with the Cox Proportional Hazards (PH) model to analyse the hazard of married women to discontinue contraceptive use. The study found that there were more biases inherent in a logistic regression compared to the Cox PH model. He, therefore, recommended that the Cox proportional hazards model should be preferred over other techniques for analyzing factors associated with early discontinuation. However, when one is interested in observing the shape of the baseline hazard, the Cox PH model poses a challenge since it makes no assumption of the shape of the baseline hazard function (Jenkins, 2004). Another challenge with using the Cox PH model is its increased computational time when estimating the "exact" partial likelihood, particularly when the incidence of tied survival times is relatively high in one's data set (Jenkins, 2004).

To avert the problems inherent in the standard logistic regression and the Cox PH models, Bradley, Schwandt & Khan (2009) used a discrete-time hazard (pooled logistic regression) model to determine levels, trends and reasons for contraceptive discontinuation, using Demographic and Health Survey (DHS) data collected from eight countries within four regions from across the world. This model is used when the response variable is measured in countable units of time for example, in months. Specifically, the pooled logistic regression model assumes proportional odds, and was therefore used to compare analyses in the eight countries. The pooled logistic regression model assumes that the relative odds of making a transition at any given time is given by the product of the relative odds that is common to all individuals, and an individual-specific scaling factor (Jenkins 2004). One advantage of this model, over the logistic regression model, is that it gives better parameter estimates when data are dependent (For instance clustered or from repeated measurements). However, just like the Cox PH model, with longer spells and tied events, this method becomes computationally intensive (Allison 1982). More so, according to Jenkins (2004), use of this model depends on the specification of an underlying continuous time model in which the within-interval durations follow a log-logistic distribution.

In another study, Khan & Shaw (2011) used multilevel logistic regression model to investigate the selected factors affecting the regulation of fertility through contraception in the context of multilevel modelling in Bangladesh. The study compared standard with multilevel logistic regression models on the effects. The study demonstrated that the standard logistic model has a tendency to seriously bias the parameter estimates of observed covariates in multilevel data, compared to the multilevel logistic regression model that found significant multilevel effects of all explanatory variables on the response variable "current contraceptive use", and produced appropriate estimates and conclusions about parameters. This, according to Khan & Shaw (2011), is so because multilevel models have an advantage of accounting for dependencies between individual observations, due to non-random or cluster sampling from geographical areas. While this model has been shown to be effective in estimating cumulative probabilities over time (Hossain 2005), in events where many higher level units have response patterns of all zero or all one, imputation methods that are used to correct for bias may not be appropriate, as imputation methods have low power for detecting non-ignorable selection (Molenberghs & Fitzmaurice 2009).

The modelling that this study intends to explore assumes that the variable that captures future contraceptive use among married contraceptive non-user women is dichotomous; and that women are likely to change states (preferences) at any point in time. This paper therefore stipulates that the complexity of such contraception data may not be adequately captured by employing standard logistic or standard duration models available in existing contraception literature, as most of such models have been shown to fall short of: 1) producing appropriate estimates and conclusions about parameters; and 2) accounting for split and or unobserved heterogeneity that may arise due to some women changing states (preferences).

THE DATA

The data that were analyzed in this study were sourced from the 2010 Malawi Demographic and Health Survey (NSO & ICF Macro, 2011). This modelling retrospectively followed 4716 married women who were not using contraceptives at their first marriage among the randomly sampled 23, 020 women aged 15 – 49 who participated in the 2010 MDHS survey. We created a dependent variable: "*time to adopt use of contraceptives*", which was measured in number of years from year of first marriage to the year a woman started using contraceptives. The variable "*futuse*" (future use) was used as a censoring indicator =1 if the event occurred, and 0, otherwise. Then, variables that have been shown to be significant predictors for future contraceptive use in previous studies were selected from the data set. These independent variables included: *id* (individual woman identifier); *entry* (year of first marriage); *end* (year of interview); *exittime* (year a woman started using contraceptives or was censored); *uset* (time a woman used current contraceptive method); *obstime* (observation time, for instance *end* – *entry*); *td* (time a woman has not been using contraceptives since her first marriage, for instance *exittime* – *entry*); *age* (age at year of interview); *seqvar* (unique spell-year identifier); *cont* (contact dummy: 0 =no contact, 1=some contact) - this variable was created by creating a composite variable community modal contact at household and enumeration area level using variables: *visited by family planning worker last 12 months* (1 = yes; 2 = no), *visited health facility last month* (1 = yes; 2 = no); and *told of family planning at health facility* (1 = yes; 2 = no); *pov* (poverty dummy: 0 = poor; 1 = not poor); *edu* (highest education level dummy: 0 = never gone to school; 1 = ever gone to school"); *desire* (desire for more children dummy: 0 = wants, 1 = no more children); *tage* (woman's age in each spell-year – obtained by subsequent subtraction of 1 from current age, for instance age at year of interview to the year of first marriage); *logt* (logarithm of spell-year identifier); *urb* (type of place of residence: 1 = urban, 2 = rural); and *region* (de facto region of residence: 1 = north, 2 = central, 3 = south).

The full details of the data collection procedure and design used in the 2010 MDHS survey are published elsewhere (NSO & ICF Macro, 2011).

METHODS

For this methodological study, the paper investigated the performance of the **MS** model over other three duration models that use the complementary log-logistic link: 1) **PCL** model; 2) **NPCL** model; and 3) the **PR** model. Some of these models have already been investigated by other analysts in the context of contraception studies. A brief description of each model is provided below.

PCL Model

In the context of duration models, the PCL - also called the discrete-time proportional hazard (PH) model takes the form:

$$\ln(-\ln(1-h_{it})) = \beta'\mathbf{x}_{it} + \gamma_j \tag{1}$$

where h_{it} is the hazard, which is explained by predictors $\mathbf{x}_{it}$ with β as the corresponding regression coefficients, while the parameter γ_j includes *rlog*(*j*), a function that gives shape to the hazard function, and *j* is the function representing each spell (*j*=1, 2, ..., *J*). Thus, the shape of the hazard monotonically increases if r > 0, decreases if r < 0, or is constant if r = 0 (Jenkins 2004). One advantage of using this model over the Cox PH model is that the coefficient vector of the complementary log-log model is invariant to the length of the time intervals (Allison 1982). Estimation of the discrete-time PH (complementary log-logistic) model involves identifying the shape parameter, γ_j by using semi-parametric methods such as the Partial Likelihood proposed by Cox (1972), that do not impose any restrictions on how this parameter varies from interval to interval. However, with longer spells, this method becomes computationally intensive, especially if events are tied (Allison 1982). Therefore, MLE techniques have been proposed, that can maximize the likelihood function:

$$L = \prod_{i=1}^{i=n} \{[\Pr(T_i = t_i)]^{d_i} [\Pr(T_i > t_i)]^{1-d_i}\} \tag{2}$$

This can be done by first pooling the observations and maximizing the log-likelihood with respect to γ_j and $\boldsymbol{\beta}$. Estimation of parameters in parametric hazard models is done through Maximum Likelihood Estimation (MLE) technique. The maximum likelihood is found by computing the partial derivatives of the log-likelihood:

$$\sum_{i=1}^{i=n} d_i \ln[f(t_i, \beta, x_i)] + \sum_{i=1}^{i=n} \{(1-d_i)\ln[1-F(t_i, \beta, x_i)\}, \quad (3)$$

with respect to regression parameters, **β**, and equating the equations of the derivatives to 0, via iterative methods (Hosmer & Lemeshow 1999). This analysis requires a form of a parametric function, *f*, and a model for the dependence of its parameters on the explanatory variables.

NPCL Model

The NPCL modelis obtained by dividing the time scale into discrete time intervals, $t_i, i = 1,2,\ldots,k$, so as to derive estimates of the interval hazard rates,

$$h(t, \mathbf{x}) = \lambda_k \exp(\boldsymbol{\beta}'\mathbf{x}) \quad (4)$$

Again $h(t, \mathbf{x})$ is the hazard function at time t, explained by a set of independent variables $\mathbf{x}$ with a corresponding vector of regression coeeficients **β**. The parameter λ_k, is the baseline hazard at time *k*. This ensures that each interval has a separate constant hazard rate (Agresti, 2002). The piece-wise constant exponential approach attains a model with such specification. Estimation in this model uses non-parametric estimators that make no prior assumptions about the shape of the relevant hazard function. Alternatively, one can use the Nelson-Aalen estimator for the cumulative hazard function. This estimator describes the accumulated risk of failure up to a specified time, t. See, for example, Jenkins (2004), for more details about this estimator.

PR Model

The PR model, also called the population-averaged model for panel data, focuses on modelling the mean response across population observations at each time point (Davidian & Giltinan, 1995). It mainly describes the relationship between averages across population of responses at different time points over time. Thus, at any time point t_{ij}, for individual *i*, the model describes the mean response as a function of fixed parameters ***β***, time t_{ij}, and sometimes including covariates. The model makes two alternative assumptions: 1) response variables are normally distributed; 2) data are not normally distributed. When one makes the first assumption, it is necessary to specify a mean response model and a model for the covariance matrix that may help in coming up with a multivariate normal distribution. On the other hand, when one makes the second assumption, it may not be sufficient to specify a mean response model to come up with an appropriate multivariate probability distribution. Therefore additional assumptions about more important observations taken several times at a time are also required to develop extensions to the multivariate probability distribution for such data. Eventually, an analytical challenge ...occurs to... overcome this challenge, Liang & Zeger (1986) proposed an idea of fitting mean-covariance models for panel data that just requires the mean and covariance matrix. The starting point is to take generalized linear models for each observation at the beginning by representing the distribution of possible values across all observations by a Bernoulli probability distribution with an appropriate link, such that each observation will have its own mean and variance model. Thus, for a model with a complementary log-logistic link: the mean response of $\boldsymbol{Y}_{ij}$ can be modeled by the function:

$$h_i(j) = 1 - \exp(-\exp(\mathbf{X}_i'\boldsymbol{\beta} + \alpha)) \tag{5}$$

and is given by:

$$E(\boldsymbol{Y}_i) = f_i(\boldsymbol{\beta}) \tag{6}$$

where, $E(\boldsymbol{Y}_i)$ is the vector matrix with elements $f(X_{ij}'\beta), j = 1,2,...,n$. The variance of each element in the mean and variance model can be modeled by the same Bernoulli probability distribution, as a function of the mean. Thus the variance of $\boldsymbol{Y}_{ij}$ can be modeled by some function:

$$\mathrm{var}(\boldsymbol{Y}_i) = \boldsymbol{\Phi}\boldsymbol{\Psi}_i \tag{7}$$

where,

$$\Psi_i = D_i^{1/2} W_i D_i^{1/2} \tag{8}$$

and$D_i^{1/2}$ is the standard deviation matrix; W_i is working correlation assumption matrix: compound symmetry; autoregressive (1); one-dependent; or unstructured. The dispersion parameter, $\boldsymbol{\Phi}$ modifies the model to allow for greater variation, both within and among individual observations. The PR model of this type assumes that individual data vectors are independent across all individual observations. Estimation of the PR model can be done using generalized estimating equations proposed by Liang & Zeger (1986). However, the proposed estimation strategy has a draw – back in that it is not possible to compare different working correlation matrices and isolate one whose assumptions are most suitable using quantities like the Akaike Information Criteria (AIC) and Bayesian

Information Criteria (BIC). Another draw-back is that, sometimes, heterogeneity may differ across observations over time. Nevertheless, Green (2012) offers a robust estimator as a solution to this draw-back.

MS Model

Mover-Stayer (MS) model assumes that a population comprises a group of "movers" who change states according to a Markov chain, and another group of "stayers" who never change states at any time. Suppose v_k and c_k are proportions of movers and stayers, respectively, in state k at the first time point; then the MS model is defined by the equations (9) to (10). Eshima (1991) describes the MS model as follows: let K denote the number of states, and S = {1, …,k} be the sample space of states.

$$\sum_{k=1}^{K} v_k + \sum_{k=1}^{K} c_k = 1 \tag{9}$$

$$v_k \geq 0,\ c_k \geq 0 \tag{10}$$

Suppose a dichotomous response $\boldsymbol{Y}(t)$ is measured at T occasions, where t denotes a particular occasion, $1 \leq t \leq T$. Suppose also, a mover changes states according to a time-homogeneous markov chain. Let p_{ij} be the transitional probability to state j at occasion t+1, given in state i at occasion t, t=1, 2,…, T-1, then

$$\sum_{j=1}^{K} p_{ij} = 1 \tag{11}$$

$$p_{ij} \geq 0 \tag{12}$$

Suppose $\pi(s_1,\ldots,s_T)$ is the probability with which an individual changes states, $s_1 \rightarrow s_2 \rightarrow \ldots \rightarrow s_T$, then:

$$\pi(s_1,\ldots,s_T) = v_{s_1} \prod_{t=1}^{T-1} p_{s_t s_{t+1}} + c_{s_1} \prod_{t=1}^{T-1} \delta_{s_t s_{t+1}} \tag{13}$$

where the indicator variable, $\delta_{ij} = \begin{cases} 1 & (i = j) \\ 0 & (i \neq j) \end{cases}$. The relationship between the MS model and Markov chain is that when $c_{s_1} = 0$ (for all k=1,...,K), then equation (13) reduces to the Markov chain model:

$$\pi(s_1,\ldots,s_T) = v_{s_1} \prod_{t=1}^{T-1} p_{s_t s_{t+1}} \tag{14}$$

According to Vermunt, Tran & Magidson (2008), the MS model is well identified if T>=3. Estimation of the MS model is usually done using Bayesian methods (Fourgere & Kamionka 2003) or likelihood Methods (Frydman 2012). Alternatively, one can use the Newton-Raphson method to estimate the parameters (see, for example, Agresti 2002, for details about the method).

MS Model Formulation, Estimation and Diagnostics

Standard duration models assume that every subject would eventually fail. In this modeling framework, this assumption is relaxed, by assuming that some women will change states by adopting use of contraceptives at some time instant (Schmidt & Witte 1989). With this assumption, the study adopted Schmidt & Witte (1989) to specify the Spilt-Population (SP) survival model. According to Jenkins (1995), the SP survival model is a form of an MS model. This model is therefore expected to account for heterogeneity that may arise due to some women changing states (adopting use of contraceptives). Let F be an unobserved "latent" variable, such that F=1, if a woman eventually uses contraceptives, and 0, otherwise. Also, let:

$$\Pr(F = 1) = \upsilon \tag{15}$$

$$\Pr(F = 0) = 1 - \upsilon \tag{16}$$

where, υ is the rate at which women non-users eventually use contraceptives.

According to Jenkins (1995), the spsurv Stata routine treats v as 1 – c, where c is the cure probability (the probability of "never failing"). If it happens that $\upsilon=1$,then all women will eventually use contraceptives, and there will be no split, implying that a standard duration model would suffice. Otherwise, an SP model that accounts for unobserved heterogeneity should be used. The study used the complementary log-logistic model with fixed covariates to model the incidence part of the SP survival model. Thus, conditional on F=1, assume $M(t|F=1)$ is a cdf, with density, $m(t|F=1)$, t=1, 2, 3, …. Let Y_t be an observable variable (*future use*), which is 1 if a woman is observed adopting use of contraceptives in future, and 0, otherwise. If a woman starts using contraceptives, she contributes to the likelihood, through:

$$\Pr(Y=1) = \Pr(F=1)m(t|F=1) = \upsilon m(t|F=1) \tag{17}$$

A woman who never uses contraceptives throughout the observation period contributes to the likelihood: through

$$\Pr(Y=0) = \Pr(F=0)m(t|F=1) = (1-\upsilon)m(t|F=1) \tag{18}$$

To model the duration part of the SP survival model, acomplementary log-logistic model with fixed and time-varying covariates in equation (19) is used:

$$h_{it} = 1 - \exp[-\exp(\beta' X_{it} + \gamma_t)] \tag{19}$$

This model estimates the hazard of adopting use of contraceptives, given the probability of eventually using them and vector $\boldsymbol{X}_{it}$, of fixed covariates: highest educational level, poverty dummy, desire for more children dummy, community modal contact, de facto region of residence, type of place of residence, and time-varying covariates – woman's age in each spell year and logarithm of spell-year identifier [log (t)], for an individual, i at time, t. In the model, γ_t is a parameter that includes log (t) - the term that summarizes duration dependence in the hazard common to each individual, i. The estimated coefficients, $\boldsymbol{\beta}$, on the predictors measure the effect of the predictors on a woman's duration of non-use until when the start of contraceptive use is observed. The model also assumes that the hazards are proportional. Therefore, using PH interpretation, given that a woman will eventually use contraceptives, a positive coefficient implies that the covariate is associated with increased hazard of failure or decreased duration of survival (an average woman takes too short to adopt to use contraceptives).

This practice of modelling the incidence and duration parts of the SP survival model follows a suggestion by Douglas & Hariharan (1994). The SP model is also used to estimate the split parameter that determines the estimated mean probability of married women adopting use of contraceptives. To determine the effect of each predictor on the probability of adopting use of contraceptives, partial effects of each predictor were used. As these partial effects may not be directly calculated in the model, the modelling followed a strategy proposed by Jenkins (2004). This strategy compares predicted hazard rates for each predictor by comparing hazard rates in three individuals: 1) an individual with its predictors at the sample mean; 2) an individual with a predictor value other than the mean of each predictor; and 3) an individual with a predictor value other than values for individuals 1 and 2. Then, the density function;

$$z(t) = \beta' X_{it} + D(t) + u \tag{20}$$

is generated, where *D* (t) = (*q*-1) *log* (*t*) is the baseline hazard function, with: (*q*-1) = *r* as the coefficient in logarithm of time; *u* = log (*v*) as the term that includes the heterogeneity term, *v*, and $\beta' X_{it}$ includes the intercept term, α in the model. To account for unobserved heterogeneity in the population, the strategy assumes that the heterogeneity term, *v* = 1(i.e., attains its expectation), such that *u* = log (*v*) = 0 in its linear representation. This generates z_0, z_1 and z_2 as densities for each of the three individuals. Using the hazard function, $h(t) = 1 - \exp(-\exp(z(t))$ predicted hazard rates for each individual are calculated, and then plotted against time to event. Using:

$$m = \frac{-\log(2)}{\log(1 - h(t))} \tag{21}$$

where *h(t)* is the predicted hazard rate [10], the median survival time, *m*, is calculated. Estimation in this model is based on maximization of the log-likelihood. To determine this log-likelihood; the hazard function in equation (19) is used to estimate the discrete survivor function:

$$S_{it} = \prod_{j=1}^{j=t} (1 - h_{ij}) \tag{22}$$

Suppose d_i is an indicator variable that is 1 if a woman starts using contraceptives, and 0, otherwise. Since the *n* observations are independent, the likelihood function is, therefore calculated as:

$$\prod_{i=1}^{i=n} \{[(1-c)(h_{it})(S_{i(t-1)})]^{d_i} .[c + (1-c)S_{it}]^{1-d_i}\} \tag{23}$$

From this likelihood, the (log) likelihood contribution for woman *i* with a survival time of *t* ‘years’ is derived as:

$$\ln L_i = d_i \ln[(1-c)(h_{it})(S_{it-1})] + (1-d_i)\ln[c + (1-c)S_{it}] \tag{24}$$

Maximization of this log-likelihood is, however, a daunting task. Nevertheless, some statistical software like Stata uses the Newton-Raphson method to facilitate this process. Finally, three assumptions of the model are then assessed: 1) the PH assumption, under the null hypothesis of non-proportional hazards. This assumption is confirmed using a statistical test approach - the global test for non-proportional hazards; 2) the assumption of no heterogeneity – under the null hypothesis that the probability of never using contraceptives (staying) is equal to zero, the likelihood ratio test is used; 3) the assumption of model fit - under the null hypothesis that the proportions of movers and stayers are not significantly different, a plot of the non-parametric estimates of the cumulative hazard function is used.

RESULTS

Descriptive Statistics

Table 4.1 presents descriptive statistics for women contraceptive non-users among failed and censored observations. There were 4716 women who were eligible for the study, among which 2990 (63.40%) started using contraceptives, while 1726 were censored. Women who adopted use of contraceptives had an average age of 22.002 years (SE= 0.096). In addition, the contraceptive use outcome was concentrated on certain individuals: about 86.6% of those who adopted use of contraceptives were those who had ever gone to school. In addition, among women who adopted use of contraceptives are those who were not poor (84.0%); those who desired to have no children (48.7%); those who had some contact with family planning workers (72.3%); those who came from the central region of Malawi (30.3%); those who came from the southern region of Malawi (50.6%), and those who resided in the rural areas (85.3%).

Variation of Survival Times by Subgroups of Predictors

Figures 4.1(a) – (f) show variation of survival times by subgroups of selected predictors. It would appear that non-users who have no education would take more time before adopting use of contraceptives, compared to those with some education (Figure 4.1a); the non-poor adopted to use of contraceptives earlier than those who were poor (Figure 4.1c); those who had no or less contact with family planning workers adopted to use contraceptives earlier than those who had some contact (Figure 4.1b); Women who wanted more children adopted to use contraceptives earlier than those who did not want any more children (Figure 4.1f). This is counter-intuitive to the expectation that those who want more children will take more time to adopting contraceptive use. On the other hand, Figure 1d show that women who lived in the urban adopted to use contraceptives earlier than those who lived in the rural; and women who live in the northern region of Malawi adopt to use contraceptives earlier than those who live in the southern region (Figure 4.1e), although this is not clear from the graph. The log-rank test for the equality of survivor functions showed that, except for de facto region of residence (p = 0.863), the chi-squared values associated with the other variables were all sufficiently large, with their associated p-values sufficiently small ($p<0.05$) (Table 4.2).

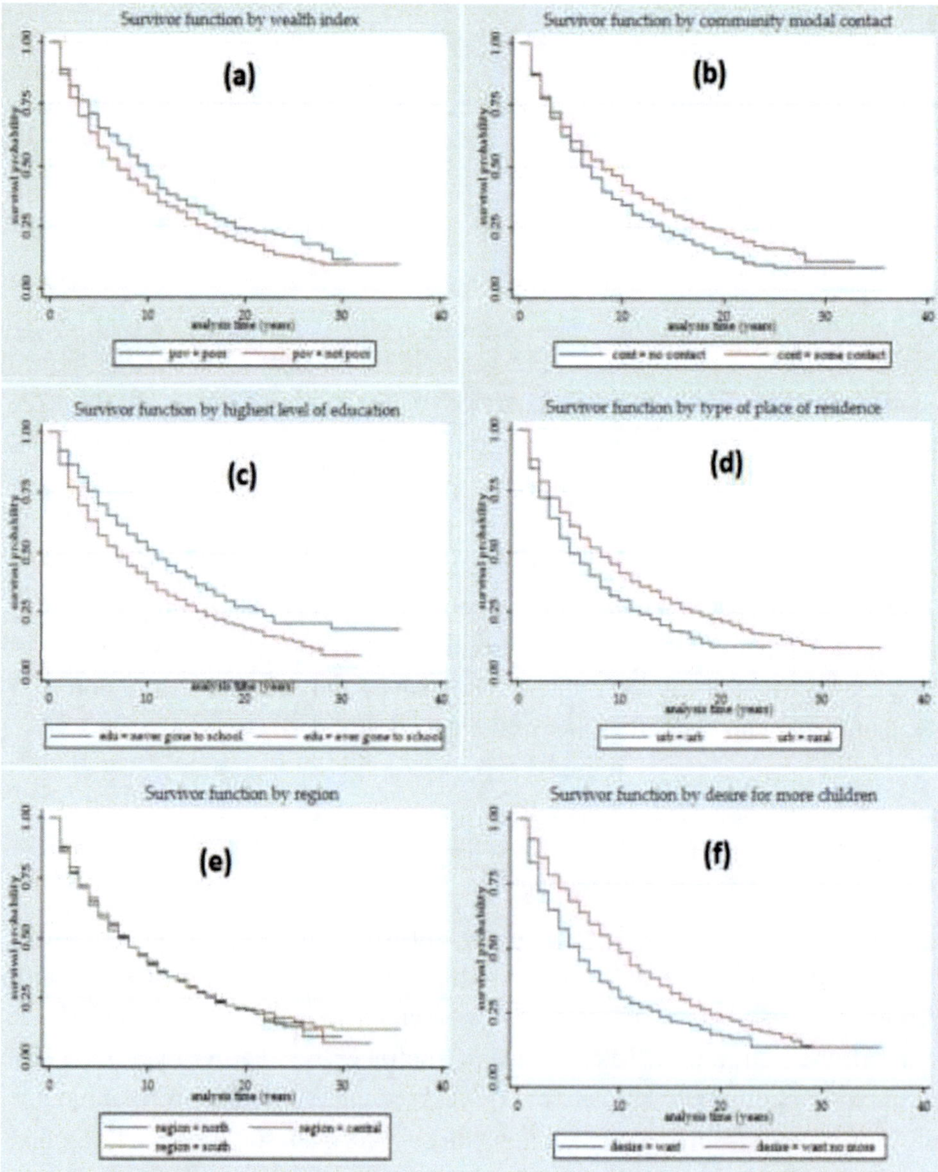

Figure 4.1. Variation in survival times by background characteristics.

Table 4.1. Descriptive statistics of married women contraceptive non-users

Variable	Adopted use		Never used	
	Mean	SE	Mean	SE
Age	22.002	0.096	21.748	0.126
Education				
ever gone to school	0.866	0.006	0.870	0.008
Poverty status				
not poor	0.840	0.007	0.795	0.010
Desire for more children				
want no more	0.487	0.009	0.414	0.012
Community modal contact				
some contact	0.723	0.009	0.781	0.011
Region of residence				
Central	0.303	0.008	0.338	0.011
South	0.506	0.009	0.463	0.012
Type of place of residence				
Rural	0.853	0.006	0.882	0.008
Sample size	**n = 2990**		**n = 1726**	

Table 4.2. Log-rank test for the equality of survivor functions

Variable	Total Observed	χ^2	p-value
Education (ever gone to school)	2990	42.12	<0.0001
Poverty status	2990	18.06	<0.0001
Desire for more children	2990	141.82	<0.0001
Community modal contact	2504	19.96	<0.0001
Region of residence	2990	0.30	0.863
Type of place of residence (rural/urb)	2990	39.31	<0.0001

Comparison of Models

Table 4.3 presents a comparative analysis of the four models. It is clear from the results in Table 3 that: estimates for age; woman's education; desire for more children are highly significant predictors for the hazard of adopting use of contraceptives. Except in the model with duration dummies where age has been excluded from the regression, PCL; the PR model that ignores the heterogeneity term, and the SP survival model that accounts for the heterogeneity term, indicate that age is a significant predictor (*P*-value<0.05). Clearly also, having gone to school significantly increases the hazard of adopting use of contraceptives in all models, compared to having not gone to school. Women who do not want any more children are less likely to adopt to use contraceptives earlier than those who want more children. As for type of place of residence, this predictor was significant in two models – PR and NPCL, but not significant in the SP model. Also, the interaction between desire for more children dummy "*no more children*" and log (*time*) has also been found to be significant in two models – PR and PCL, and not in the SP. This interaction has been shown to increase the odds of adopting use of contraceptives among married women. In addition the shape parameter, *r*, was also highly significant in the SP; the PR; as well as the PCL.

Table 4.3. Comparison of SP, PR, PCL and NPCL models

Variable	SP*	PR	NPCL	PCL
(Intercept)	- 4.490	- 4.990		- 5.482
	(0.443)	**(0.364)**		**(0.272)**
	< 0.001	< 0.001		< 0.001
Age	0.145	0.107		0.109
	(0.013)	**(0.012)**		**(0.012)**
	< 0.001	< 0.001		< 0.001
ever gone to school	0.991	0.785	0.276	0.807
	(0.200)	**(0.161)**	**(0.120)**	**(0.160)**
	< 0.001	< 0.001	0.022	< 0.001
no more children	- 1.684	- 1.811	-1.495	- 1.798
	(0.138)	**(0.181)**	**(0.095)**	**(0.180)**
	< 0.001	< 0.001	< 0.001	< 0.001
Rural	- 0.280	- 0.240	- 1.071	
	(0.154)	**(0.117)**	**(0.067)**	
	0.069	0.040	< 0.001	
*no more children*log(time)*		0.443		0.437
		(0.104)		**(0.103)**
		< 0.001		< 0.001
Shape parameter	- 0.876	- 1.166		- 1.178
	(0.084	**0.072)**		**(0.072)**
	< 0.001	< 0.001		< 0.001
Split parameter	0.957			
	(0.138)			
	< 0.001			
Statistics				
Sample size (weighted)	47595	47595	47747	47595
Log-likelihood	-2729.225		-2809.676	-2736.082
AIC	5472.450		5669.352	5484.165
BIC	5533.843		5888.694	5536.788

Note:b/(se)/ p.

Concerning the size of the estimates, it can be seen that estimates in the **SP** model are generally larger compared to those in the other three models; with generally larger standard errors and highly significant *P*-values for predictors that are significant. Based on this comparison, we can conclude that the **SP** model, which is a form of the **MS** model, could be a better fit. Furthermore, the hazard rates for all predictors have been shown to decrease monotonically with time. This implies that the rate at which women will adopt to use contraceptive is likely to fall with time. In addition, the mean probability of adopting use of contraceptives, conditional on not using them in the past, has been found to be about 0.28, implying that the probability of staying is about 0.72 (Table 4.3).

Interpretation of Parameter Estimates in the MS Model

Table 4.4 presents results of both the incidence and duration part of the **SP** model in an exponentiated form. On one hand the incidence part of the model suggests that the odds of using contraceptives are almost 2.5 times more among married women who have ever gone to school, than among those who have never gone to school (OR=2.542, *P*<0.001)). In addition, women who do not want any more children are 76% less likely to use contraceptives, compared to those who want more children (OR=0.244, *P*<0.001).

Table 4.4. Risk Factors of Future Contraceptive Use in the SP model

Future use	Duration part			Incidence part		
	HR	**SE**	**P-value**	**OR**	**SE**	**P-value**
(Intercept)	0.011	0.005	<0.001	0.018	0.006	<0.001
Age	1.156	0.015	<0.001			
ever gone to school	2.693	0.539	<0.001	2.542	0.454	<0.001
no more children	0.186	0.026	<0.001	0.244	0.025	<0.001
Rural	0.756	0.116	0.069	0.709	0.090	0.007
Shape parameter	0.416	0.035	<0.001			
Split parameter	2.604	0.359	<0.001			
Statistics						
Sample size (weighted)	47595			40130		
Log-likelihood	2729.23			2435.0		
AIC	5472.45			4878.0		
BIC	5533.84			4912.4		
Mean probability of staying	72.3%			(2.8%)		P<0.001
LRχ^2		29.02				

Table 4.4 also show that women who live in rural areas are 29% less likely to use contraceptives than those who live in urban areas (OR=0.709, *P*=0.007). On the other hand, the duration part of the model suggests that, conditional on eventually using contraceptives, on average after every passing year in age, increases the odds of adopting use of contraceptives earlier by 16% (*HR*=1.156, *P*<0.001). Also, wanting no more children among married women non-users reduces the odds of adopting use of contraceptives earlier by 81%, compared to wanting more children (*HR*=0.186, *P*<0.001). Furthermore, living in rural areas reduces the odds of starting contraceptives earlier by 24% (*HR*=0.756, *P*=0.069). However, the effect of type of residence is not statistically significant in the model. Thus, the shape parameter, $\boldsymbol{r} = q\text{-}1 = ln\ (0.416) = -\ 0.876$ is less than 1; implying that the hazard function is decreasing monotonically with time. The split parameter, $p = ln\ (2.604) = 0.957$ is also significant at 95% confidence level (*P*<0.001).

Probability of Adopting Use of Contraceptives

Using results presented in Table 4.4, the mean probability of staying among married women contraceptive non-users is about 0.72 (0.028). The likelihood ratio chi square statistic for the null hypothesis that the probability is zero, is significant at 5% significance level ($\chi^2(1) = 29.02$; $P<0.001$).

Effects of Predictors on the Probability of Adopting Use of Contraceptives

Figures 4.2(a)-(d) present the results of the effects of each predictor (age and education) on the hazard rates of adopting use of contraceptives. Based on the results, the hazard rates of adopting use of contraceptives are higher in older women compared to younger women. The difference in the hazard rates between a 31-year old woman and a 15-year old woman, ceteris paribus, is about 13% (0.149-0.019) The median survival time for a 31-year old woman is about 4.3 years (Figures 4.2a and 4.2b).

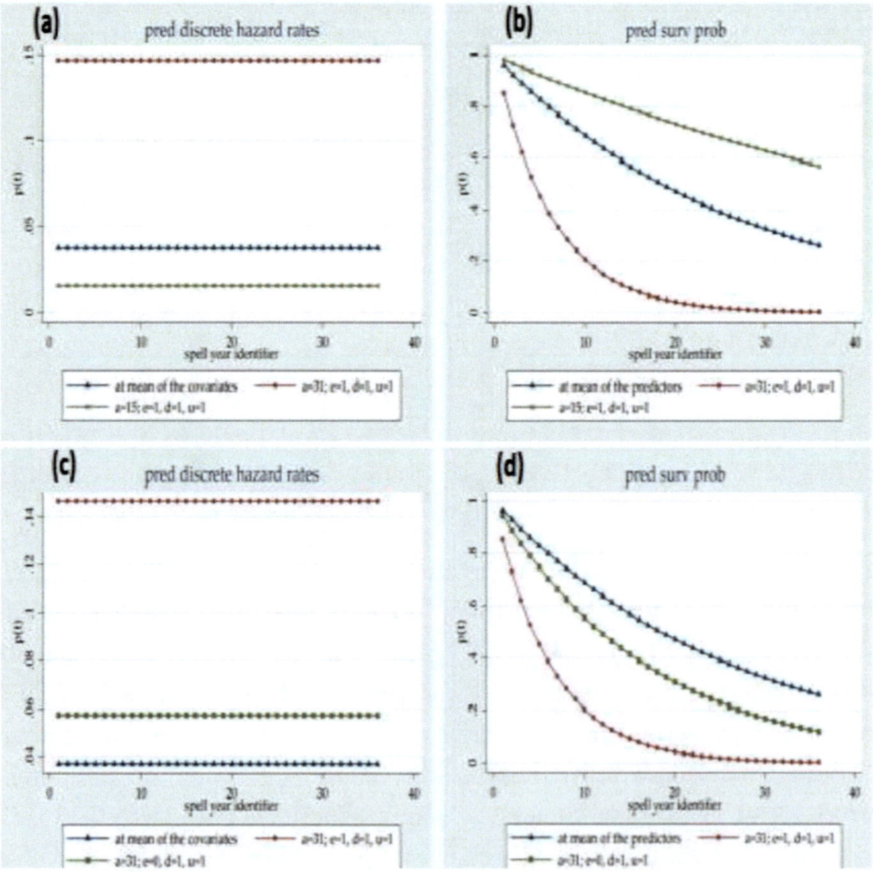

(Note: a = age; e = education level; d = desire; and u = type of place of residence).

Figure 4.2. Predicted adjusted hazard rates and survival probability for: (a) Age-adjusted Discrete Hazard Rates; (b) Age-Adjusted Survival Probability; (c) Education-adjusted Discrete Hazard Rates; (d) Education-adjusted Survival Probabilities.

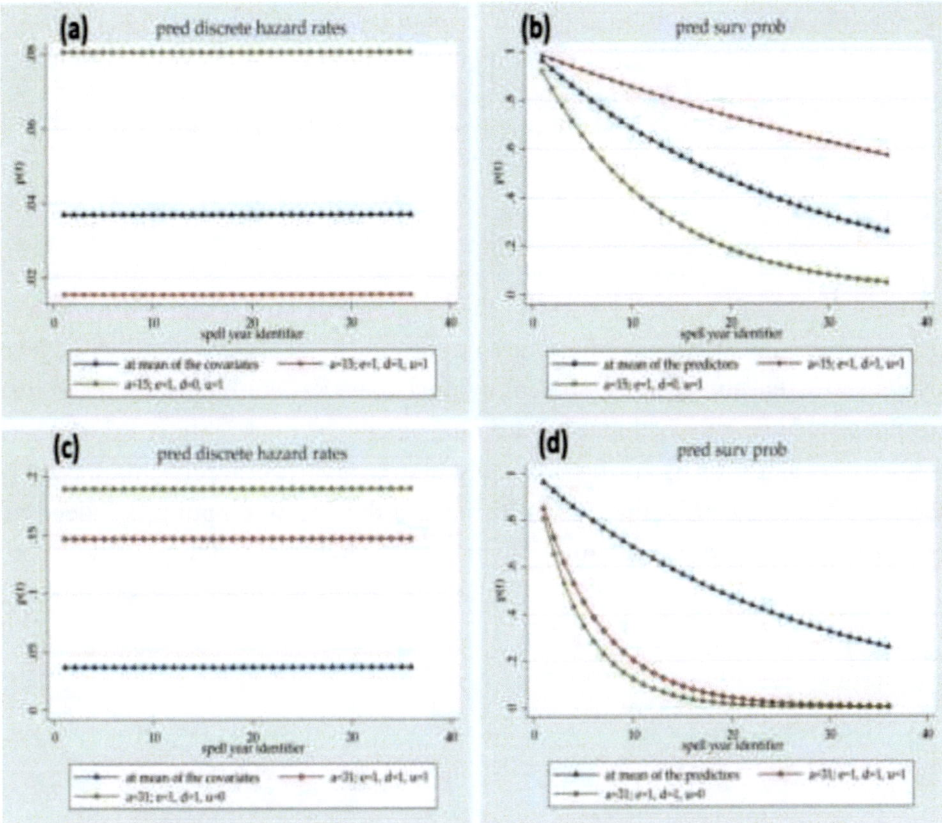

(Note: a = age; e = education level; d = desire; and u = type of place of residence).

Figure 4.3. Predicted adjusted hazard rates and survival probability for: (a) Hazard rates for desire; (b) Survival Probability for desire; (c) Hazard Rates for residence; (d) Survival Probabilities for residence.

In addition, the hazard rates of adopting use of contraceptives are higher in women who have ever gone to school, compared to those who have never gone to school. The difference in the hazard rates between a woman who has ever gone to school and a woman who has never gone to school, all other variables constant, is about 8.6% (0.144-0.058). The median survival time for a woman who has ever gone to school is 4.46 years, against 11.6 years for one who has never gone to school (Figures 4.2c and 4.2d). Also, the hazard rates of adopting use of contraceptives are higher in women who want more children, compared to those who do not want any more children. The difference in the hazard rates between a woman who wants no more children, and a woman who wants more children, all other variables constant, is about 4.5% (0.081-0.036).

Figure 4.3 shows the predicted hazard rates and survival probability for desire for more children and place of residence. The median survival time for a woman who wants more children is 8.21 years, compared to a median of 18.91 for a woman who wants no more children (Figures 4.3a and 4.3b). Furthermore, the hazard rates of adopting use of contraceptives are higher among married women who live in the urban areas, compared to those who live in the rural areas. The difference in the hazard rates between these women, holding other variables constant, is about 1.2% (0.157 - 0.145). The median survival time for women who live in urban areas is about 4.06 years, compared to a median of 4.42 years for those who live in rural areas (Figures 4.3c and 4.3d).

Model Diagnostics

Figures 4.4 present results of a test of the PH assumption for the null hypothesis that the hazards for groups of each predictor are not proportional at 95% confidence level.

Results show that there is insufficient evidence at 95% confidence level to reject the null hypothesis of non-proportional hazards for desire for more children dummy (Figure 4.4c). When this assertion is confirmed using a statistical test approach - the global test for non-proportional hazards, the test confirms that the PH assumption is indeed violated for desire for more children dummy "No more children" with the p-value of <0.0001 (Table 4.5).

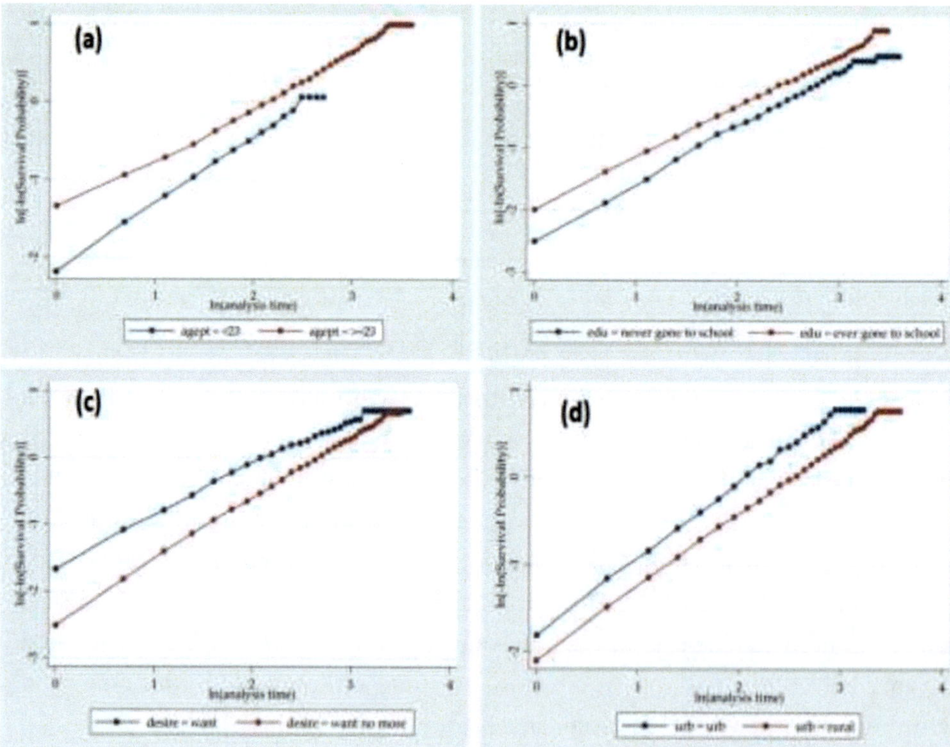

Figure 4.4. Graphical test of the PH assumption for: (a) age; (b) education; (c) desire for more children (d) type of place of residence.

When the desire for more children dummy "No more children" is interacted with time (_*t*> = 18 years), it is found that the interaction effect is not significant at 95% confidence level, suggesting that the hazards for the dummy are proportional, conditional on stratification (Table 4.6).

Table 4.5. Test for proportional hazards assumption

	Rank: _t			
Variable	**Correlation**	χ^2	**df**	**P-value**
age	-0.02987	2.46	1	0.117
ever gone to school	-0.02822	2.07	1	0.150
no more children	0.10719	28.3	1	<0.001
rural	-0.01274	0.43	1	0.514
Global test		33.75	4	<0.001

Table 4.6. Cox Stratification for desire for more children interacted with time

_t	Estimate	SE	95% CI	P-value
Main				
age	0.248	0.073	(0.105, 0.392)	0.001
ever gone to school	0.279	0.058	(0.166, 0.393)	< 0.001
no more children	- 0.441	0.041	(- 0.522, - 0.360)	< 0.001
rural	- 0.329	0.056	(- 0.438, - 0.220)	< 0.001
Tvc				
no more children	0.333	0.327	(- 0.308, 0.973)	0.309

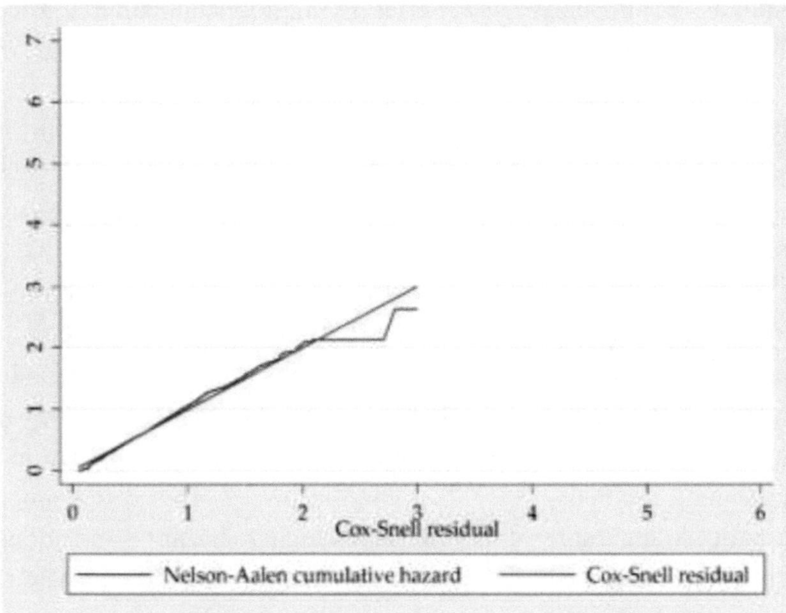

Figure 4.5. Plot of the non-parametric estimate of the cumulative hazard function.

The likelihood ratio test for the null hypothesis that probability of never using contraceptives (staying) is equal to zero is found to be significant (χ^2 (1) = 49.02; $p<0.001$). This means there is insufficient evidence to reject the null hypothesis of no heterogeneity at 95% confidence level (Table 4.3). This also implies that the proportion of movers is significantly different from that of stayers. Finally, the test for the null hypothesis that the model is not a better fit shows that the plot of the non-parametric estimate of the cumulative hazard function generally follows the 45-degree line very closely, jumps for few events, for larger values of time - implying that, overall; the SP survival duration model used could indeed be a better fit to the data (Figure 4.5)

DISCUSSION

The study has found that time to adopt use of contraceptives varies by sub-groups of woman's education; poverty status; desire for more children; community modal contact, and type of place of residence. These results were expected and are, therefore, consistent with findings from other studies (Stephenson & Tsui 2002; Gage 1995; and Nazar-Beutelspacher et al., 1999). However, the findings that time to adopt to use contraceptives does not significantly vary by sub-groups of de facto region of residence are somewhat inconsistent with findings in other studies. One would expect time to adopt to use contraceptives in the three regions to vary as these regions have different cultural orientations. As Kaggwa, Diop & Storey (2008) argue, even in low (contraceptive) prevalence settings, contraceptive use varies by region and culture. Therefore, what makes time not to vary by sub-groups of de facto region of residence in Malawi is not known.

The study has also found that, controlling for other factors, the mean probability of adopting use of contraceptives, conditional on not using them in the past, has significantly been found to be about 0.28, implying that the probability of staying is about 0.72. The 2010 Malawi Demographic and Health Survey found that 74% of married women contraceptive non-users intend to use any family planning method in future (NSO & ICF Macro 2011), yet this study has found that about 28% are likely to adopt to use contraceptives. These findings, collectively, could be an alarm to the reproductive health system in Malawi as they may imply that the remaining percentage (46%) of married women non-users who intend to use any family planning method in future may opt for other family planning methods, other than contraceptives. This could probably explain why women who want more children have higher hazards of adopting use of contraceptives earlier, compared to those who do not want any more children.

Nevertheless, the finding that, conditional on eventually using contraceptives; the hazard rates of adopting use of contraceptives are higher in older women compared to younger women was expected since as the age of a woman increases, she is likely to have more children (Kaggwa, Diop & Storey, 2008). Consequently, this is likely to lead to increased uptake of contraceptives with a view of spacing children. Furthermore, the finding that the hazard rates of adopting use of contraceptives earlier are higher among married women who have ever gone to school, compared to those who have never gone to school is also consistent with that of Nazar-Beutelspacher, et al. (1999) and Hogan, Berhanu & Hailemariam (1999), who found that women who are more educated are more likely to adopt family planning, compared to those who are not. However, the finding that women who want more children have higher hazards of adopting use of contraceptives earlier, compared to those who do not want any more children, contradicts what would naturally be expected. It would be logical to expect that if a woman desires to have no more children, she is likely to start or intensify use of contraceptives, compared to a woman who has desire to have more children since some studies have shown that desire for no more children is associated with higher odds of contraceptive use (See, for example, Kaggwa, Diop & Storey 2008). On the other hand, it would sound illogical to assume that all women who desire to have no more children would adopt use of contraceptives, as contraceptives may not be the only family planning methods available within most reproductive health systems. Or it could be possible that they use flexible contraceptive methods rather than those which permanently ensures zero fertility. In addition, findings of this study on the effect of type of place of residence on the hazard of adopting use of contraceptives show that living in urban or rural areas seem not to have a significant effect on the hazard of adopting use of contraceptives. This result differs from that of other studies who have found a significant effect of type of place of residence on the probability of using contraceptives. For instance, studies by Gage (1995) and Jato, et al. (1999) have found that women in urban areas were more likely to adopt contraceptives than women in rural areas. It is possible that in the rural areas, accessibility could be a challenge, and so outreach activities could not be successful or effective. Consequently contact of family planning workers with women contraceptive non-users could be low.

CONCLUSION

The study has attempted to show that, in terms of model fit, the **MS** model has supremacy over the PCL, NPCL and the PR models. Furthermore, the findings further suggest the importance of including a heterogeneity term in modelling future contraceptive use. However, how the results found in this study compare with simulation studies in literature still remains scanty. There is therefore need to consider possibilities of carrying out simulation studies to compare results from the MS model with those obtained in the simulationstudy.

ACKNOWLEDGMENT

On a very important note, the researchers acknowledge the permission granted by Measure DHS to download and use their data for this study. No funding has been sourced from any institution that may have led to the results of this study.

REFERENCES

Agresti, A. (2002).*Categorical Data Analysis*, 2nd Ed., John Wiley & Sons, Inc., Hoboken, New Jersey.

Alba-Ramirez, A., Arranz, J. M. & Munoz-Bullon, F. (2012). 'Reemployment probabilities of unemployment benefit recipients', *Applied Economics*, no. 4428, pp. 3645-3664.

Allison, P. D. (1982). 'Discrete-Time Methods for the Analysis of Event Histories', *Sociological Methodology*, no. 13, pp.61-98.

Anasel, M. G. &Mlinga, U. J. (2014). 'Determinants of contraceptive use among married women in Tanzania: Policy implication', *African Population Studies*, vol. *28*, no. 2.

Bellizi, S., Sobel, H., Obara, H. &Temmerman, M. (2015). 'Underuse of modern methods of contraception: underlying causes and consequent undesired pregnancies in 35 low- and middle-income countries', *Human Reproduction*, retrieved 14 May 2018, <doi:10.1093/humrep/deu348>.

Bradley, S. E. K., Schwandt, H. M. & Khan, S, (2009).*Levels, Trends, and Reasons for Contraceptive Discontinuation*, DHS Analytical Studies No. 20, Available at: https://pdf.usaid.gov/pdf_docs/PNADQ639.pdf (Accessed: 12 May 2018).

Chen, H. H., Duffy, S. W. & Taber, L. (1997). 'A Mover-stayer mixture of Markov chain models for the assessment of dedifferentiation and tumour progression in breast cancer', *Journal of Applied Statistics*, vol. *24*, no. 3, pp.265-278.

Clements, S. &Madise, N. (2004). 'Who is being Served Least by Family Planning Workers?',*African Journal of Reproductive Health*, vol. *8*, no. 2, pp.124-136.

Collumbien, M., Gerresu, M. &Clerand, J. (2004). Non-use of ineffective methods of contraception. Available at http://www.who.int/publications/cra/chapters/volume2/1255-1320.pdf. Accessed: 7 September 2015.

Cook, R. J., Kalbfleisch, J. D. & Yi, Y. (2002). 'A Generalised Mover-stayer Model for Panel Data', *Biostatistics*, no. 3, pp.407-420.

Cox, D. R. (1972).'Regression models and life-tables', *Journal of the Royal Statistical Society*, no. 34, pp.187-202.

Davidian, M. & Giltinan, D. M. (1995).*Non-linear Models for Repeated Measurement Data*, Chapman & Hall, London.

Douglas, S. & Hariharan, G. (1994). 'The hazard of starting smoking: estimates from a Split-population survival model', *Journal of Health Economics*, no. 13, pp. 213- 230.

Eshima, N. (1991). 'The Mover-stayer Model and a Procedure for the Maximum Likelihood Estimation of the Parameters', *Journal of the Japan Statistical Society*, vol. *21*, no. 1, pp.73-83.

Fourgere, D. & Kamionka, T. (2003).'Bayesian inference for the Mover-Stayer model in continuous time with an application to labour market transition data', *Journal of Applied Econometrics*, vol. *18*, no. 6, pp.697-723 <doi: 10.1002/jae.727>.

Frydman, H. (2012).'Maximum Likelihood Estimation in the Mover-Stayer Model', *Journal of the American Statistical Association*, vol. *79*, no. 387, pp.632-638 <doi: 10.1080/01621459.1984.10478090>.

Frydman, H., Kallberg, J. G. & Kao, D. L. (1985). 'Testing the adequacy of Markov chains and Mover-stayer models as Representations of Credit Behaviour', *Operations Research*, no. 33, pp.1203-1214.

Gage, A. J. (1995). 'Women's socioeconomic position and contraceptive behaviour in Togo', *Studies in Family Planning*, vol. *26*, no. 5, pp. 264–277.

Green, W. H. (2012).*Econometric Analysis*, 7th Ed., Pearson Education Ltd., Edinburgh Gate, Harlow.

Hosmer, W. H. &Lemeshow, S. (1999).*Applied Survival Analysis: Regression Modeling of Time to Event Data*, John Wiley & Sons, Inc., NY.

Hossain, M. B. (2005). 'Analysing the relationship between family planning workers' contact and contraceptive switching in rural Bangladesh using multilevel modeling', *Journal of Biosocial Science*, vol. *37*, no. 5, pp.60-67.

Jenkins, S. P. (2004).Survival Analysis, in possession of the Institute for Social and Economic Research, University of Essex, Colchester, UK.

Jenkins, S.P. (1995). 'Easy estimation methods for discrete-time duration models', *Oxford Bulletin of Economics and Statistics*,no. 57, pp.129-138.

Kaggwa, B., Diop, N. &Storey, J. D. (2008).'The Role of Individual and Community Normative Factors: A Multilevel Analysis of Contraceptive Use among women in union in Mali',*International Family Planning Perspectives*, vol. *34*, no. 2, pp.79–88.

Khan, H. R. & Shaw, E. H. (2011).'Multilevel Logistic Regression Analysis Applied to Binary Contraceptive Prevalence Data', *Journal of Data Science*, no. 9, pp.93-110.

Liang, K. Y. &Zeger, S. L. (1986). 'Longitudinal data analysis using generalized linear models', *Biometrica*, no. 73, pp.13-22.

Mavromaras, K. G. & Orme, C. D. (2004). 'Temporary layoffs and Split-population models', *Journal of Applied Econometrics*, no. 19, pp.49-67.

Molenberghs, G. & Fitzmaurice, G. (2009). 'Incomplete data: Introduction and overview', in *Handbooks of modern statistical methods*, Eds G. Fitzmaurice, M. Davidian, G. Verbeke& G. Molenberghs, CRC Press, USA.

Nakosteen, R. & Zimmer, M. (1980). 'Migration and Income: The Question of Self-Selection', *Southern Economic Journal*, no. 46, pp. 840-851.

National Statistical Office (NSO) & ICF Macro. (2011).*Malawi Demographic and Health Survey 2010*Available at: http://www.nsomalawi.mw. (Accessed: 12 May 2018).

National Statistical Office (NSO) [Malawi] and ICF. (2017).*Malawi Demographic and Health Survey 2015-16*Available at: http://www.nsomalawi.mw. (Accessed: 12 May 2018).

Nazar-Beutelspacher, A., et al. (1999). 'Education and nonuse of contraceptives among poor women in Chiapas, Mexico', *International Family PlanningPerspectives*, vol. *25*, no. 3, pp.132–138.

Schmidt, P. & Witte, A. D. 1989, 'Predicting recidivism using 'split population', survival time models', *Journal of Econometrics*, no. 401, pp. 141-159.

Simpson, W. & Sproule, R. (1998).*Economic Analysis of Canadian Self-Employment Using SLIDIncome and Labour Dynamics* Statistics Canada Working Paper Series No. 75F0002M.

Stephenson, R. &Tsui, A.O. (2002).'Contextual influences on reproductive health service use in Uttar Pradesh, India', *Studies in Family Planning*, vol. *33*, no. 4, pp.309–320.

Tripathi, V. (2008). 'Statistical Approach for Evaluation of Contraceptive Data', *African Reproductive Health*, vol. *12*, no. 1, pp.17-21.

World Health Organisation.(2018). 'Family Planning Fact-Sheet', weblog post, 8 February, retrieved 14 May 2018, http://www.who.int/en/news-room/fact-sheets/detail/family-planning-contraception.

In: Statistical Modelling …
Editors: N-B. Kandala and L. N. Kazembe
ISBN: 978-1-53615-981-3

Chapter 5

INVESTIGATING CAUSAL AND MEDIATING RISK FACTORS FOR STUNTING IN UNDER FIVE CHILDREN IN MALAWI USING STRUCTURAL EQUATION MODELLING TECHNIQUES

Emmanuel Banda*[*]*, Mavuto Mukaka and Lawrence N. Kazembe
Somalia/Somaliland Save the Children Country Office, Nairobi, Kenya
Mathematical Sciences Department, University of Malawi,
Chancellor College, Zomba, Malawi
Mahidol-Oxford Tropical Medicine Research Unit,
Faculty of Tropical Medicine, Mahidol University, Bangkok, Thailand
Oxford Centre for Tropical Medicine and Global Health, Oxford, UK
Department of Statistics and Population Sciences,
University of Namibia, Windhoek, Namibia

ABSTRACT

Most studies that have modelled and identified risk factors for stunting in Malawi have used ordinary least squares (OLS) regression such as linear and multilevel models. These might have limitations when dealing with outcomes like malnutrition which are: i) multi-faceted, ii) where predictors are correlated, iii) there is a possibility of measurement errors, iv) data might not follow multivariate normal distribution, and v) when there is a path relationship between different predictor latent factors. This study used Structural Equation Model (SEM) techniques to investigate causal and mediation relationships between dependent observable variable stunting (HAZ) and latent factors as defined in the UNICEF malnutrition causal framework. Specifically, the study assessed causal

[*] Corresponding Author's Email: emmamhanya@yahoo.com.

relationships between stunting and its immediate, underlying and basic factors in the UNICEF framework and explored the mediation relationships between immediate factors, underlying and basic factors in predicting stunting as given by the same framework. The study further evaluated whether the Malawi Demographic and Health Survey for 2015-16 data fits the UNICEF framework causal pathways for stunting or not. The results showed that all three latent factors (Immediate, Underlying and Basic causes) were important predictors of stunting, thereby confirming the UNICEF malnutrition causal framework. The findings showed that the underlying factors only predicted stunting through the immediate factors (mediator), also confirming the UNICEF causal framework. The study further demonstrated that underlying factors were a significant mediator of causal relationship between basic factors and stunting. In essence the study confirmed that Malawi DHS 2015/16 data fits the UNICEF malnutrition causal framework. The SEM demonstrated to be an effective model for analyzing stunting data because it can assess DHS data conformity to causal framework by applying both regression and path analysis techniques, with flexibility in assumptions when data does not always follow multivariate normality and has complex pathways. The study findings can help policy makers to have confidence in adopting the essential nutrition actions (ENA) which were based on a global UNICEF causal framework as it is applicable to Malawi data.

Keywords: Structural Equation Models, stunting, under-nutrition, Malawi, Demographic and Health Survey, UNICEF malnutrition causal framework

INTRODUCTION

Malnutrition refers to deficiencies, excesses or imbalances in a person's intake of energy and/or nutrients. The term malnutrition covers two broad groups of conditions. One is 'undernutrition' — which includes stunting (low height for age), wasting (low weight for height), underweight (low weight for age) and micronutrient deficiencies or insufficiencies (a lack of important vitamins and minerals) (Nagati et al., 2003; UNICEF, 2006; Bailey et al., 2015). The other group is overweight, obesity and diet-related non-communicable diseases such as heart disease, stroke, diabetes and cancer (WHO Health Topic, July 2016). Undernutrition is a consequence of consuming too few essential nutrients, or using or excreting them more rapidly than they can be replaced. In children, the outcome is growth (weight or height) faltering and/or specific symptoms and signs of micronutrient deficiency disorders (Office of Development Effectiveness, 2014). According to UNICEF, nearly half of all deaths in children under five are attributable to undernutrition, translating into the loss of about 3 million young lives world-wide in a year. Undernutrition puts children at a greater risk of dying from common infections, increases the frequency and severity of such infections, and contributes to delayed recovery. The interaction between undernutrition and infection can create a potentially lethal cycle of worsening illness and deteriorating nutritional status. Poor nutrition in the first 1,000 days of a child's life can also lead to stunted growth, which is associated with impaired cognitive ability and reduced school and work performance (UNICEF, 2006).

Undernourished children are more likely to die from common childhood ailments, and for those who survive, have recurring sicknesses and faltering growth. The sustainable development goals (SDGs) target to end all forms of malnutrition, including achieving, by 2025, the internationally agreed targets on stunting and wasting in children under 5 years of age, and address the nutritional needs of adolescent girls, pregnant and lactating women and older persons by 2030 (UNDP, 2016).

A reduction in the prevalence of malnutrition would also assist in the goal to reduce child mortality (Scaling Up Nutrition, April 2011). In early childhood, under nutrition has negative life-long and intergenerational consequences; undernourished children are more likely to require medical care as a result of under nutrition-related diseases and deficiencies. This increases the burden on public social services and health costs incurred by the government and the affected families. Without proper care, underweight and wasting in children results in higher risk of mortality. During schooling years, stunted children are more likely to repeat grades and drop out of school, thus reducing their income-earning capability later in life. Furthermore, adults who are stunted as children are less likely to achieve their expected physical and cognitive development, thereby impacting on their productivity (MGDS III, 2017). Stunting is the biggest problem among the undernutrition indicators in Malawi and SADC region as whole. Almost 37% of children aged less than five years are stunted in the SADC region, (UNICEF, WHO, WB, 2015) and therefore not able to fully achieve their mental or physical potential. The prevalence of stunting in the region ranges from 7.9% in the Seychelles to 49.2% in Madagascar (UNICEF, WHO, WB, 2015). The prevalence of stunting among young children is above 30% in more than half of the SADC countries (UNCHO, May 2016).

The problem of child malnutrition in Malawi is still significant despite efforts to eradicate it. Between 2015 and 2016, 37% of children under the age of five years were reported to be stunted or too short for their age, compared to 47% in 2010 and 48% in 1992 (National Statistical Office & ICF Macro, 2011 and 2015). The proportion of children wasted remained at about 4% between 1992 and 2015. Despite numerous nutrition policies and programs in various countries, very little is known about why child malnutrition remains a challenge affecting Malawi and many African countries (Kumchulesi, 2017). Figure 5.1 shows the trends of key under-nutrition indicators in Malawi between 1990 to 2015. The graph shows that despites a notable progress in decrease of malnutrition over the period, stunting has remained relatively very high compared to the other indicators. For example, stunting; underweight; and wasting reduced from 55.8% to 37.1% , 24.4% to 11.7%, and 6.6% to 2.7% respectively of children under five years over the period 1990 to 2015. The stunting prevalence still remains amongst the highest in the region.

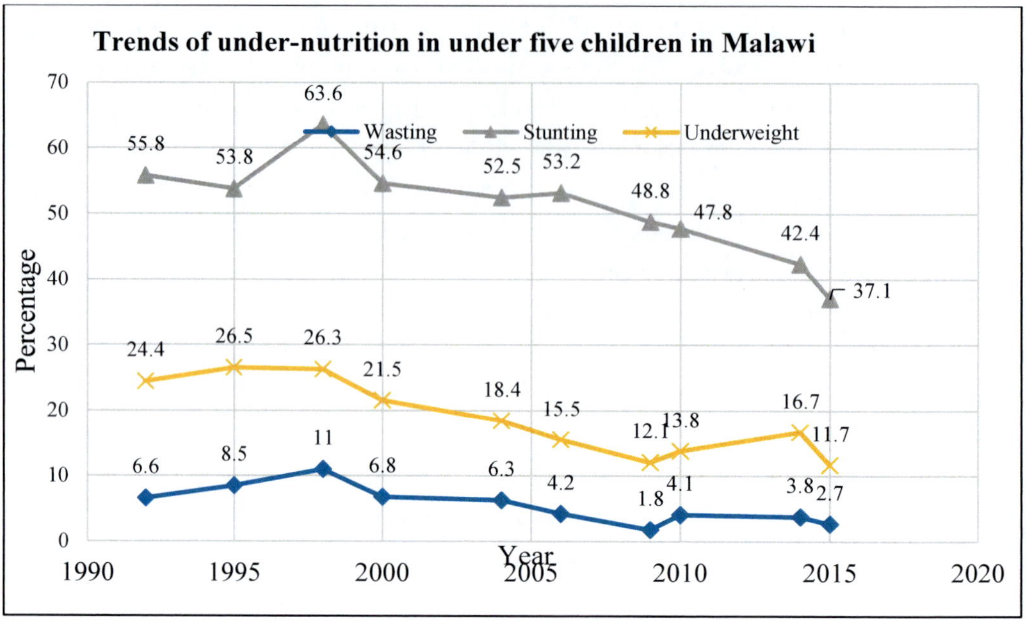

Figure 5.1. Trends of malnutrition indicators in Malawi (Source: Raw data from UNICEF, Worldbank 2018).

UNICEF Causal Framework for Child Nutrition Status

A number of frameworks seek to describe the complex range of factors that influence child nutrition. One that is widely cited, is the United Nations Children's Fund (UNICEF) framework for improving child nutrition which forms the integral part of Scaling Up Nutrition global movement. At the core of this framework are a number of direct determinants of nutrition, called 'immediate' causes, followed by a further group called 'underlying' causes and, at the periphery, a group of 'basic' causes. Basic causes include political, ideological, economic, environmental, resource and technology factors. The UNICEF framework describes 'short-route' interventions that address the immediate causes and 'long-route' interventions that address underlying and basic causes. This framework is a theoretical framework that guided our structural mediation modelling in this chapter. Figure 5.2 presents the UNICEF conceptual framework for child nutritional status shows causal relation of factors in for undernutrition.

Under the UNICEF Conceptual framework in Figure 5.2, child undernutrition is described as caused not just by the lack of adequate, nutritious food, but by frequent illness, poor care practices and lack of access to health and other social services. These multifactorial determinants were first outlined in UNICEF's conceptual framework of child undernutrition more than two decades ago (UNICEF, 1990). Identifying immediate, underlying and basic causes of undernutrition, the framework has evolved to incorporate new knowledge and evidence on the causes, consequences and impacts of undernutrition. Immediate causes of undernutrition are inadequate dietary intake and disease. A child's dietary intake and exposure to disease are affected by underlying factors, including household food insecurity (lack of availability of, access to, and/or utilization of a diverse diet), inadequate care and feeding practices for children, unhealthy household and surrounding environments, and inaccessible and often inadequate health care. Basic causes of poor nutrition encompass the societal structures and processes that neglect human rights and perpetuate poverty, limiting or denying the access of vulnerable populations to essential resources. Social, economic, and political factors can have a long-term influence on maternal and childhood undernutrition. Moreover, chronic undernutrition can lead to poverty, creating a vicious cycle (World Bank, 2006).

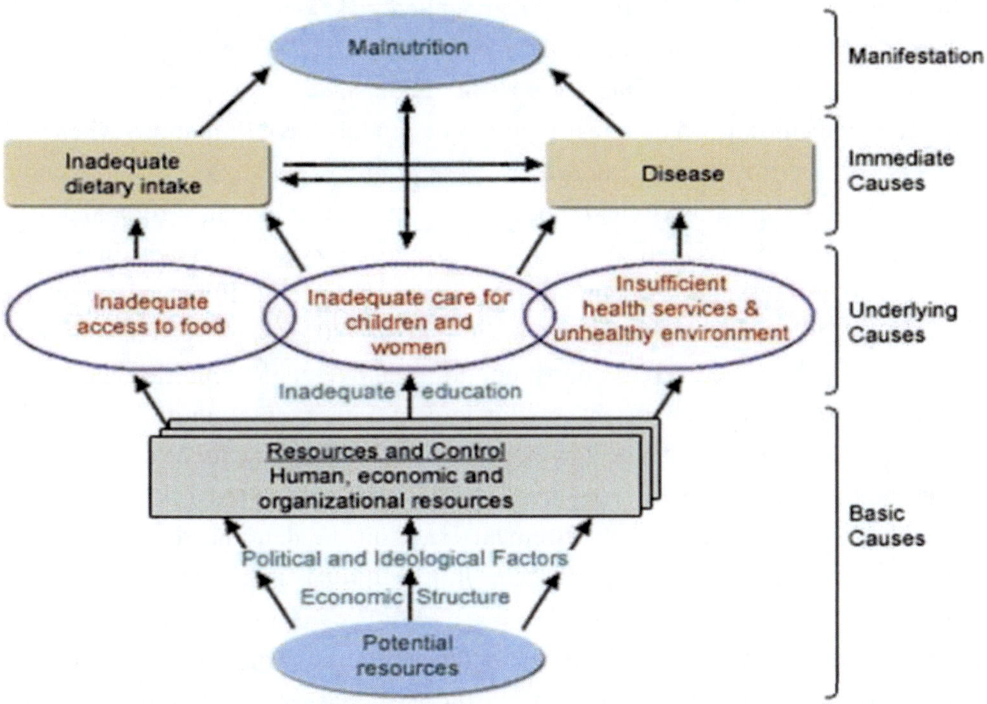

Figure 5.2. UNICEF Conceptual Framework for child nutrition status. (Adapted from UNICEF's Approach to Scaling Up Nutrition, 2015).

Defining the Problem

Public Health Problem

The high stunting levels for under-five children in Malawi led to the Malawi Government putting more emphasis on undertaking programs that should tackle stunting in Malawi. In 2011, Malawi Government launched a program called Scaling Up Nutrition (SUN) 1000 special days which is a broad sensitization program calling all development partners to unite to end stunting by targeting resources to mothers and children in the first 1000 days of life beginning from pregnancy. The SUN program used the UNICEF malnutrition causal framework to design Essential Nutrition Actions (ENA) as strategies to reduce the prevalence stunting amongst under five children. Since the causal framework is a global tool developed from data across the world, it is therefore important applicability of the framework to Malawi by assessing it against national data on nutrition. Continued study of factors associated with stunting and how the factors relate to each other in Malawi is therefore pivotal if significant progress is to be made in reducing the levels of child stunting in Malawi.

Statistical Problem

Most studies that have identified and modelled risk factors for stunting in Malawi have used Ordinary Least Squares (OLS) or linear regression models and generalised linear models like logistic, log-linear regression models and a few lately have used multi-level modelling techniques. The linear regressions models have limitations when you deal with outcomes like under nutrition which are a) multi-faceted, b) most predictors are correlated c), there is a reasonable possibility of measurement errors and d) most importantly when the causal frameworks of the problem which policy makers use to design preventive interventions is a path relationship between different predictor factors. Lately, others (Chirwa and Ngalawa, 2008; Chikhungu, 2013; Madise and Mpoma, 1997; Makoka, 2013) modelled child nutrition in Malawi using multi-level regression techniques to investigate important predicators at either community, household and individual levels. While multi-level models can to certain extend deal with correlated predictors and can cluster predictors at different levels like community, household and individual; they still assume no measurement errors in the data, need to construct a few models with individual goodness of fits and thirdly though there are hierarchal they do not analyze paths (mediation or moderation) of the causal relationships. This study attempted to model stunting in Malawi using Structural Equation Modelling (SEM) techniques which used latent factors to cluster observed variables and simultaneously model mediation pathways assumed in the theoretical frameworks like UNICEF Causal Framework for malnutrition.

The main objective of this study was to investigate the causal and mediation pathways (structural relationship) for stunting in under five children in Malawi using SEM on the UNICEF framework.

METHODS

Participants

The study used 2015-16 Malawi Demographic and Health Survey (2015-16 MDHS) which was conducted between October 2015 and February 2016 by the National

Statistical Office (NSO) of Malawi in joint collaboration with the Ministry of Health (MoH) and the Community Health Services Unit (CHSU). The survey was based on a nationally representative sample that provided estimates at the national and regional levels and for urban and rural areas with key indicator estimates at the district level. The survey included 26,361 households, 24,562 female respondents, and 7,478 male respondents. The 2015-16 MDHS included household and respondent characteristics, fertility and family planning, infant and child health and mortality, maternal health and maternal and adult mortality, child and adult nutrition, malaria, HIV/AIDS, domestic violence, orphans, and vulnerable children. A total of 6,033 children under age 5 were eligible for height and weight measurements. The height-for-age index included 94% (5,671) of eligible children with complete and valid height measurement and age data. The study used 3,047 individual and households level records of under five children mostly because they did not have significant missing values for variable of interest which was a precondition of running SEM.

Variable Selection

The selection of variable was, firstly, influenced by the UNICEF causal framework and the Essential Nutrition Action (ENA) for child nutrition. Secondly, the variable selection was also influenced by the statistical approaches that the study selected. SEM with maximum likelihood (ML) estimation method mostly makes useful estimates with continuous or highly ordinal variables as its works on the correlation matrix between variables. Variables like morbidity were generated as a scale value from number of recent cases of diarrhoea, respiratory infection and fever and were turned into ordinal value. Thirdly, the selection was influenced by availability of the variables in the 2015-16 Malawi Demographic and Health Survey. Only the variables which were collected in the survey were considered in the model. Lastly, variables were selected based on what literature on predictors of child nutrition has published before. Table 5.1 presents the observable (manifest) variables used in the study gotten from MDHS data, their labels, their respective names given in the study and lastly the latent factors that the variables describe in the study.

Structural Equation Model

The main technique that was applied to develop models under this study to model risk factors for stunting in Malawi was the Structural Equation Modelling for the reasons stated in the above sections. Structural Equation Modeling can be defined as a class of methodologies that seeks to represent hypotheses about the means, variances, and covariance of observed data in terms of a smaller number of 'structural' parameters defined by a hypothesized underlying conceptual or theoretical model (Social and Behaviour Sciences, 2001).

Table 5.1. Description of study latent and observable variables

Latent Factor	**Label**	**Indicators**	**DHS Variable name**	**Study Variable name**
Malnutrition	Stunting	Z-score HAZ	*HAZ*	*HAZ*
Immediate causes	Inadequate Dietary intake	Months of breast feeding	*m5*	*month_bf*
		Number of times a child ate semi-solid foods	*m39*	*num_semfood*
	Disease	Diarrhea in the last 2 weeks	*morb* (created a scale variable from frequency disease occurance) original variables *h22, h31, h11*	*Morb*
		Respiratory infection in the last 2 weeks		
		Fever in the last 2 weeks		
Underlying causes	Inadequate care for children and women	Mother's age	*v012*	*m_bmi*
		Mothers education	*v149*	*m_educa*
		Birth Order of the child	*bord*	*Bord*
Basic causes	Resources and control	Number of HH members	*v136*	*HH_members*
		Number of Children <5 years of age	*v137*	*N_under5*
		Wealth index	*v190*	*w_index*

The purpose of Structural Equation Modeling is to define a theoretical causal model consisting of a set of predicted covariances between variables and then test whether it is plausible when compared to the observed data (Jöreskog, 1970; Wright, 1934). A Structural Equation Model is a system of linear equations among several unobservable variables (constructs) and observed variables. An SEM is composed of two parts: a structural part, linking the constructs to each other (usually, this part expresses the endogenous or dependent constructs as linear functions of the exogenous or independent constructs), and a measurement part, linking the constructs to observed measurements. The second part resembles a confirmatory factor analysis model and often referred to as a Measurement models.

Hypothesized Structural Equation

In the hypothesized structural model (Figure 5.3), immediate and underlying causes have a direct causal relationship with malnutrition, while basic causes of malnutrition act through the underlying causes. Further, the underlying causes of malnutrition acts through two path ways: directly through basic and indirectly through immediate causes (Figure 5.3).

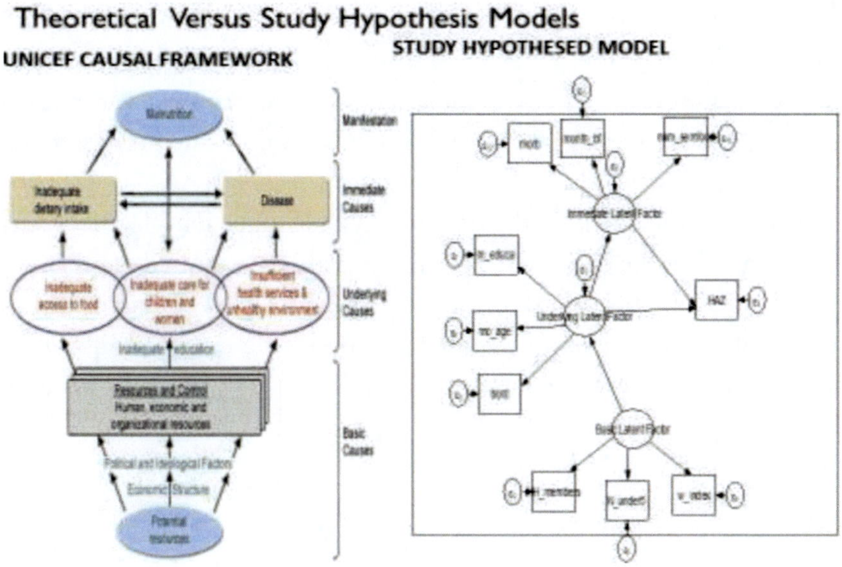

Figure 5.3. Theoratical and Study hypothesised models.

Therefore, the conceptual model to be tested can be described as follows. Stunting, as a measure of malnutrition, was measured by the *Height for Age Z-score (HAZ),* while latent immediate causes were measured by observable variables: *morbidity, months of breast feeding and number of times the child eat semi-solid food.* On the other hand, latent underlying causes were measured by observable variables including: *mother's education, preceding birth order and mothers age,* while the latent basic causes is measured by observable variables like: *number of under five children, number of household members and wealth index score of the household* (Table 5.1). So the relationship between the latent factors and their observed variables is called the measurement model. There are three measurement models in the hypothesized model represented by the three latent constructs and their respective measurement models (Figure 5.3).

Estimation of SEM Models

A Structural Equation Model is a broad class of statistical models that consists of two parts: the structural model and the latent variable model. The structural model consists of the regression-like relationships among the variables. The latent variable model (LVM) forms the latent variables (LVs) used in the structural model. When a LVM is analyzed without a structural model, it is sometimes called a confirmatory factor analysis (CFA). If there was not a hypothesized structure for the latent variable model, then it would be an exploratory factor analysis (EFA). In this study, UNICEF malnutrition causal framework forms our theoretical model from which we have developed our hypothesis model. This then allows us to use CFA methods of developing our measurement models. The purpose of a LVM is to understand the underlying structure that produced relationships among multiple manifest variables (i.e., covariance matrix). The manifest variables (MVs) that are directly influenced by the LV are the indicator variables. The proposed causal agents in these underling structures are LVs.

At the heart of SEM model identification is the question: Is there enough non-redundant information in the data to be able to estimate the required parameters uniquely. With regression, identification is never an issue because they are always just-identified models, meaning that the number of parameters to estimate exactly equals the amount of non-redundant information in the dataset. This can be easily calculated using the formula in Equation (1).

$$q = \frac{p(p+1)}{2} \qquad (1)$$

where p is the number of manifest (observable) variables.

With an LVM, model identification is more complex than with standard regression, as the models can be just-identified, under-identified, or over-identified. If there are more parameters to estimate than there are pieces of non-redundant information, then the model is under-identified. Conversely, if there are more pieces of non-redundant information than there are parameters to estimate, then the model is usually over-identified. If a model is under-identified, the model's parameters cannot be estimated uniquely and *lavaan* (or any other LVM program) will return an error message. If a model is over- or just-identified, there should be unique estimates for each parameter. For over-identified models, not only do they provide unique parameter estimates, but they can also provide measures of model fit. This is because they have degrees of freedom (df). In LVMs, df can be thought of as the number of non-redundant pieces of information in the data minus the number of parameters to estimate. Only over-identified models have df > 0. Because LVs are not directly observed, there are no inherent units by which to measure them. Consequently, a LVM is not identified unless some parameter estimates are constrained to set the latent variable's scale. There are three common ways to set this scale.

1. *Standardized latent variable:* This method constrains the latent variable's variance to 1.0. This, in effect, makes the latent variable a standardized variable (i.e., on a Z score scale). If the indicator variables are standardized as well, the loadings can be interpreted the same as a standardized regression coefficient: the number of standard deviations the MV changes as the LV increases one standard deviation. In addition, if there is more than one LV, then the covariance among

the LVs becomes a correlation. This is the method that we used in these models but others are;

2. *Marker variable:* This method requires a single factor loading for each LV be constrained to an arbitrary value (usually 1.0). The indicator variable whose loading is constrained is called the marker variable. This method uses the marker variable to define the LV's variance.
3. *Effects-coding:* This method estimates all the loadings, but constrains that the loadings for a given LV average 1. 0 or equivalently, that their sum is equal to the number of unique indicator variables.

Estimation of Mediation in SEM

Much of the output from an SEM was similar to that from a CFA. For example, estimates of the factor loading, intercept, error variance, and factor variances were the same in either approach. For SEM we were not as interested in those parameters as we would be for a standard CFA when producing measurement models. Given that the first step was to estimate the measurement models for the factors, and know that they fit the data well, the second step would generally not focus on the factor parameters themselves unless we noted problems in the structural part of the model. Our primary focussed on the structural parameters, including the direct and indirect effects that were specified in our model statements. A test statistic with p-value of less than 0.05 would conclude that there was a direct relationship between latent factors or with stunting. The indirect (ab) and total effects (c) appear at the very end of the output as *Defined parameters*. The indirect effect was considered statistically significant if $p = <0.05$, and on the other hand, the total effect need not to be significant ($p = >0.05$) for the mediation path (ab) to be significant. These are elaborated under results.

Overall Model Evaluation

The commonly used fit indices in the literature include comparative fit index (CFI), Tukey-Lewis index (TLI), Root Mean Square Error of Approximation (RMSEA), and Standardized Root Mean Square Residual (SRMR). According to Garson (2017), CFI and TLI measures equal to or greater than 0.9 signify good fit indices. Also, RMSEA less than 0.06 and SRMR $\leq .08$ (Hu & Bentler, 1999) displays the most acceptable fit index. For evaluation of the models (structural model and confirmatory factor analysis), other literature have suggested CFI and TLI should be greater than 0.95 and RMSEA should be less than 0.05. In this study, a conservative approach was taken, where TLI and CFI greater than 0.95 were good fit and values greater than 0.90 were adequate fit and

RMSEA less than 0.05 were good fit but less 0.06 were adequate fit. SRMR equal to or less than 0.08 were good fit. A model was considered good fit for the data if it fitted any of the four criteria.

Data Processing

Data cleaning and descriptive statistics were conducted in STATA 13.1. Structural models were conducted in *R version 3.50* using lavaan.survey package. In addition, semPlot package in *R* was used for plotting structural diagrams.

RESULTS

Univariate Analysis and Normality Assumption Checks

Table 5.2. Summary statistics for the variables used for modelling in the study

Variables	No of Obs	Mean	Min	Max
HAZ	3047	-136.1	-589	597
Number of semi-food eaten	2008	1. 9	0	8
Months of breast feeding	1904	12.1	0	35
Morbidity	3047	1.4	0	5
Mother's Age in years	3047	27.0	15	48
Preceding birth order	3047	3.1	1	13
Mother's Education	3047	1.1	0	3
Number HH members	3047	5.4	2	17
Number of under-five children	3047	1.6	0	7
Wealth index	3047	2.9	1	5

A total of 3047 children under five with complete height and age have been used in the study. Table 5.2 shows descriptive statistics for variables of interest in the study. From Figure 5.4, the study also checked for normality assumption in the study data because some estimation techniques for models used in this study assumes normal distribution or multivariate normal distribution. From the histograms, mix picture was gotten but mother's age and HAZ showed close to a normal distribution. Secondly, the study conducted the multivariate normal test and all tests rejected the null hypothesis of multivariate normality. Test for multivariate normality Mardia mSkewness $p < 0.001$, Mardia mKurtosis $p < 0.001$, Henze-Zirkler $p < 0.001$ and Doornik-Hansen $p < 0.001$ This implies that the study will have to use robust maximum likelihood methods in estimating its models.

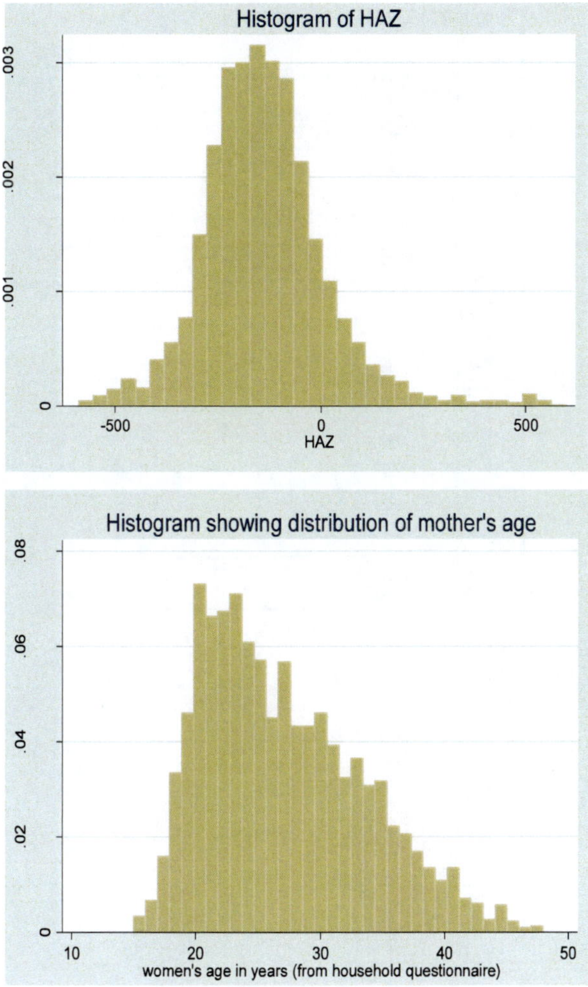

Figure 5.4. Histogram of stunting (HAZ) and maternal age.

Bivariate Analysis

The bivariate analysis of covariance between the study variables was run to evaluate the relationship between the all variables of interest in the study. This was important because SEM assumed there was a significant covariance relationship between variables and it helps to make necessary adjustments for residual correlation to improve model fit in SEM. Table 5.3 shows that there were few significant correlations that should be considered when adjusting the residual correlations in the SEM equation. For example, number of semi-solid food and month of breast feeding (r = 0.41), birth order and mother's age (r = 0.85), and number of under five children with birth order (r = 0.19) were significantly correlated.

Table 5.3. Correlation matrix of all obversable variable of interest in the study

Variables	HAZ	month_bf	morb	num_se~d	m_educa	mo_age	bord	N_under5	HH_mem~s	w_index
HAZ	1.00									
month of breast feeding	-0.214	1.00								
Morbidity	-0.039	0.128	1.00							
number times of semi solid food	-0.069	0.412	0.116	1.00						
mother's education	0.099	-0.041	0.0151	0.022	1.00					
mothers age	0.012	0.056	-0.065	0.002	-0.162	1.00				
Birth border	0.016	0.015	-0.043	-0.033	-0.32	0.851	1.00			
Number of under5	0.044	-0.103	-0.032	-0.081	-0.154	0.127	0.291	1.00		
Number of HH members	0.057	-0.033	-0.012	-0.051	-0.108	0.364	0.453	0.523	1.00	
Wealth index	0.096	-0.031	0.0006	0.037	0.388	0.062	-0.158	-0.052	0.166	1.00

Model Fitting and Interpretation

Six models were fitted in the process to respond to the study objectives. At first stage, three measurement models for latent factors: Immediate, underlying and basic causes were fit using CFA approach to assess how the observable variables explained the latent factors. In the second stage, a full structural hypothesized model was fit to assess significance of each of the three latent constructs: Basic, underlying and immediate causes in predicting stunting. The third stage involved construction of two mediation models; one model assessed the mediation relationship between immediate causes and underlying causes in predicting stunting. Another was constructed to assess the mediation relationship of underlying and basic causes in predicting stunting. All the six models outlined above were assessed for good of fit using the criteria described in the literature, where CFI and TLI >= 0.90 would indicate adequate fit and values of >= 0.95 suggest good fit, while RMSEA $\leq .06$ and SRMR $\leq .08$ was used to determine good fit. In this study, a model was of good fit if it achieved at least adequate fit for CFI or TLI and good fit for RMSEA or SRMR.

Measurement Models for Latent Factors

First process in constructing a SEM was to construct measurement models that define the latent factors. The study had three latent factors; immediate, underlying and basic factors as defined in the framework. Since the study already had a theoretical framework, CFA was used to confirm if the observable variable defined their respective latent and whether separately measured their goodness of fit.

Immediate Causes Latent Factor

A confirmatory factor analysis was run to construct a latent factor for immediate causes (Im_) that was explained by observable variable morbidity (mrb), month of breast feeding (mn_) and number of times the child eat semi-solid foods (nm_). The path model is given in Figure 5.5. This was the measurement model for immediate causes factor in the UNICEF framework. Figure 5.5 shows the standardized estimates for the model. The study observed that month of breast feeding and number of semi-solid foods explained the latent 14 and 3 times respectively more than morbidity. Both month of breast feeding and number of semi-solid foods were significant observable variable in the latent with p-value less than 0.001. With CFI and TLI = 1.00 and RMSEA and SRMR less than 0.001, it was concluded that the estimated model showed good fit.

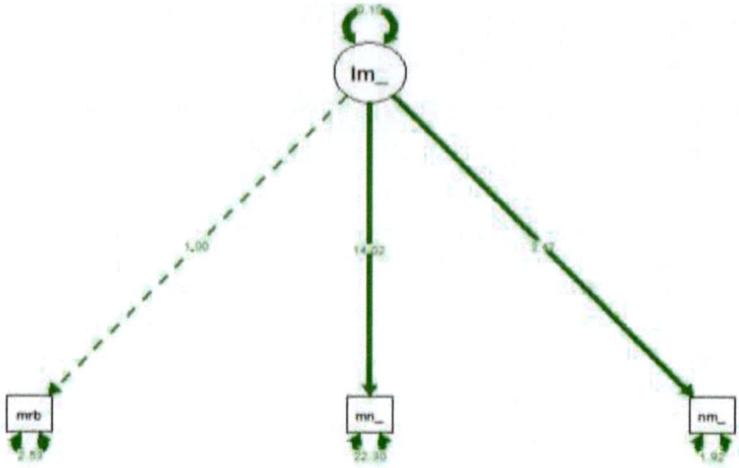

Figure 5.5. Measurement model for latent immediate factor (Im_).

Underlying Causes Latent Factor

A second CFA (Figure 5.6) was run to construct a latent factor that is explains underlying factors (Un_) through observable variables mother's education (m_d), mother's age (m_g) and preceding birth order (brd).

Fixing mothers education to 1, mother's age was estimated to influence underlying causes factor minus (-) 29 times compared to minus (-) 19 times the preceding birth order does. With p-value of less than 0.001 both were significant observable variables in the latent factor underlying causes. Table 5.4 presents all fit criteria and demonstrated that measurement model fitted the data well.

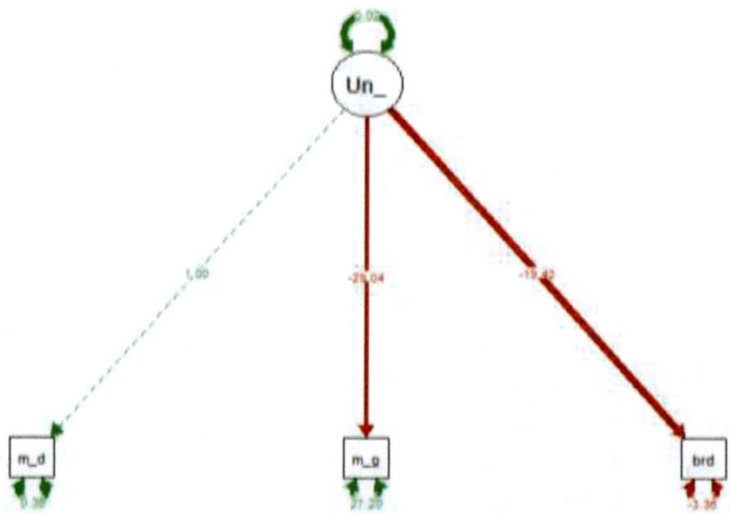

Figure 5.6. Measurement model for Underlying causes latent factor (Un_).

Table 5.4. Model evaluation for underlying latent measurement model

Criteria	**CFI**	**TLI**	**SRMR**	**RMSEA**
Value	1	1	<0.001	<0.001
Cut-off	>=0.90 adequate	>=0.90 adequate	<=0.08 good	=<0.06 adequate
	>=0.95 good	>=0.95 good		=<0.5 good

Basic Causes Latent Factor

A third CFA was run to construct a latent factor that explains basic factors through observable variables, number of under-five children, number of household members and wealth index. The p-values for both number of household members and wealth index were less than 0.05 indicating both variables were significant in the latent basic causes when under number of under children was restricted to constant 1. Table 5.5 indicate that CFI and TLI suggest good fit (1.00) was greater than 0.95 while SMRM (<0.01) and RMSEA (<0.01) were less than 0.8 and 0.5 respectively.

Table 5.5. Observable variables explaining latent factor basic causes latent

Basic causes latent		Estimate	Std Error	Z-value	P(>|z|)
Observable variables	N_under5	1.00			
	HH_members	-9.37	3.7	-2.53	0.011
	w_index	0.72	0.08	8.95	<0.001

A Full Structural Equation Model of the Hypothesis Model

As presented above, Structural Equation Modeling can be defined as a class of methodologies that seeks to represent hypotheses about the means, variances, and covariance of observed data in terms of a smaller number of 'structural' parameters defined by a hypothesized underlying conceptual or theoretical model. The purpose of Structural Equation Modeling is to define a theoretical causal model consisting of a set of predicted covariance between variables and then test whether it is plausible when compared to the observed data (Jöreskog, 1970; Wright, 1934). A Structural Equation Model is a system of linear equations among several unobservable variables (constructs) and observed variables. An SEM is composed of two parts: a structural part, linking the constructs to each other (usually, this part expresses the endogenous or dependent constructs as linear functions of the exogenous or independent constructs), and a measurement part, linking the constructs to observed measurements. In this study, we assessed the significance of three latent constructs in the regression equation of predicting stunting. Figure 5.7 presents the SEM diagram showing how the three latent factors: Immediate causes (Im_), Underlying causes (Un_) and Basic causes (Bs_) and their respective observable variables are related to Stunting (HAZ). In Figure 5.7, Underlying and Basic factors have a positive relationship with HAZ, therefore we expect less stunting when better underlying and basic factors are in place. Overall immediate factors have negative relationship so we expect more stunting when Immediate factors increase.

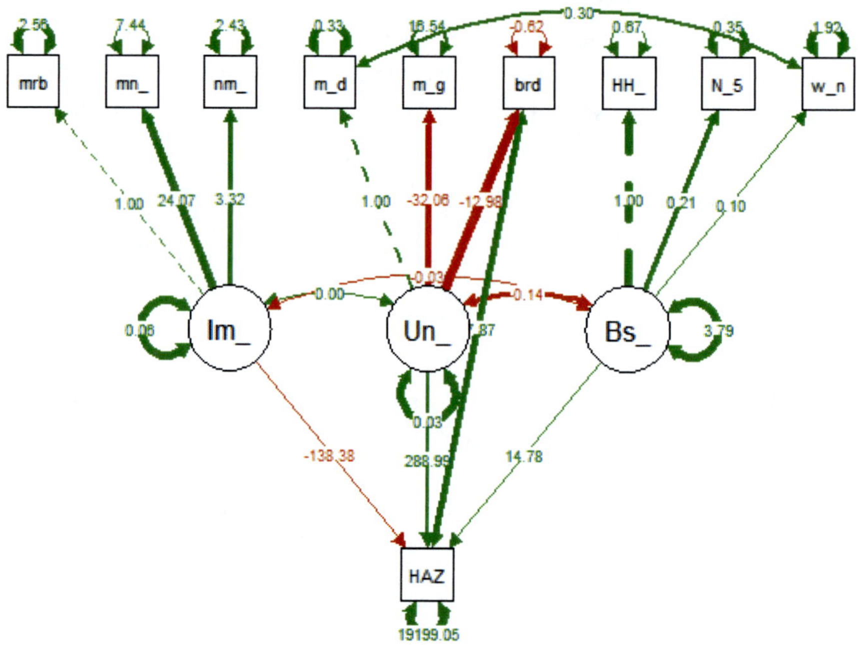

Figure 5.7. Structural models showing relationship between the three latents and stunting.

Table 5.6. Regression coefficients of latent factors in predicting Stunting

Regression		Estimate	Std Err	Z-value	P(>|z|)
HAZ~					
Latent variables	Immediate	-138.4	27.9	-4.9	0.001
	Underlying	289	100	2.9	0.004
	Basic	14.8	4.3	3.4	0.001

Table 5.7. Model evaluation of a full structural model

Criteria	CFI	TLI	SRMR	RMSEA
Value	0.927	0.879	0.042	0.084
Cut-off	>=0.90 adequate >=0.95 good	>=0.90 adequate >=0.95 good	<=0.08 good	=<0.06 adequate =<0.05 good

From the outputs in Tables 5.6 and 5.7, all the observable variables were significant at 95% confidence level in their respective latent constructs (p-value of β is less than 0.05). Secondly, all the three latent constructs were important predictors of HAZ (stunting) in the nested regression model. The p-values estimates for Immediate (<0.000), Underlying causes (0.029) and Basic causes (0.006) were all less than 0.05 therefore the three constructs were significant predictors of HAZ (stunting) in the model (Table 5.6). On the overall fit of data to the model Comparative Fit Index (CFI) was 0.913, Tucker-Lewis Index (TLI) was 0.849, RMSEA was 0.059, SRMR was 0.042 (Table 5.7). Both RMSEA and SRMR showed adequate and good fit of the data in the model. Similarly, CFI was above 0.9 but less than 0.95 which translated to adequate fit. With at least two tests showing good fit, we can conclude that this model fitted the data well. From the model (Figure 5.7), stunting can be predicted by

HAZi = β_0 - 138.4*(Immediate causes)i + 289*(Underlying causes)i + 14.8*(Basic causes)i

Mediation Structural Equation Models

Three necessary conditions must be met in order to claim that mediation is occurring (Kenny, Kashy, & Bolger, 1998; MacKinnon, Lockwood, Hoffman, West, & Sheets, 2002).

1. X is significantly related to M.
2. M is significantly related to Y.
3. The relationship of X to Y diminishes when M is in the model.

Figure 5.8 presents the mediating effect of immediate causes on the relationship between underlying causes and stunting. In this mediation path analysis, the study tested the significance of the causal indirect route of Underlying causes to Immediate Causes to Stunting (ab) and the Underlying Causes direct to Stunting total (c), which diminishes if (ab) is significant.

Table 5.8 shows a p-value of 0.04 for the path (ab) was less than 0.05 showing its significance, while the total path (c) had p-value of 0.85, demonstrating it was non-significance path. This therefore showed that Immediate causes were important mediators of Underlying factors in causing Stunting. This confirmed the UNICEF causal framework path way in our theoretical model.

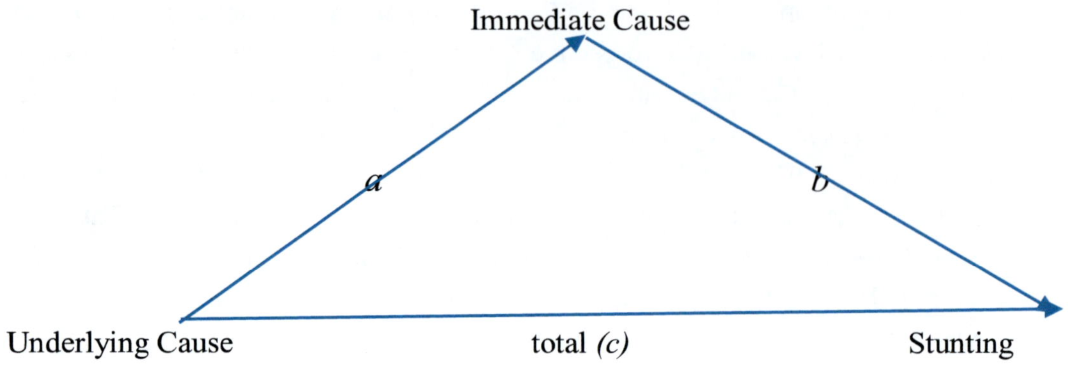

Figure 5.8. Immediate causes mediating relationship between underlying causes and Stunting.

Table 5.8. Showing significance of direct and meditation path

Defined Parameters		Estimate	Std Err	Z-value	P(>|z|)
Paths	*Ab*	-8.3	4.2	-1.9	0.04
	total *(c)*	-6.9	36.3	-0.2	0.85

Table 5.9 presents results of model validation. All the four criteria (CFI, TLI, SRMR, RMSEA) showed the models were of very good fit.

Table 5.9. Model evaluation of immediate latent factor as a mediator of underlying factor in casual relationship with stunting

Criteria	CFI	TLI	SRMR	RMSEA
Value	0.984	0.970	0.033	0.049
Cut-off	>=0.90 adequate >=0.95 good	>=0.90 adequate >=0.95 good	<=0.08 good	=<0.06 adequate =<0.05 good

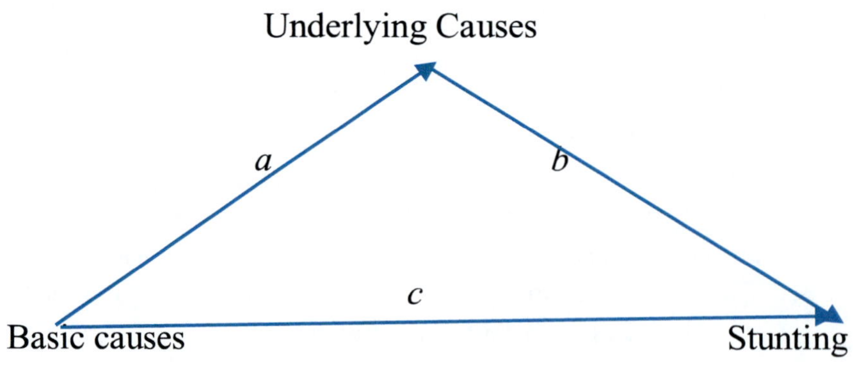

Figure 5.9. Underlying causes mediating the causal relationship between basic causes and Stunting.

Next in Figure 5.9 we investigated the mediation model for underlying causes. Here we tested whether Underlying Causes latent factor was a significant mediator for Basic Causes latent factor in causal relationship with Stunting. So the model tested if path (ab) was significant and path total (c) diminished its significance when (ab) was in the model. Below is the part of the mediation output.

Table 5.10 shows a p-value of <0.01 for path (ab) was significant (<0.05), while total path (c) Basic factors to Stunting p-value (0.932) was insignificant path in the model. The output indicated that the Underlying Causes latent factor was mediating the causal relationship between Basic Causes latent factor and Stunting. This therefore demonstrated that UNICEF theoretical frameworks fitted the data in the study.

Table 5.10. Model path parameters

<table>
<tr><th colspan="2">Defined Parameters</th><th>Estimate</th><th>Std Err</th><th>Z-value</th><th>P(>|z|)</th></tr>
<tr><td rowspan="2">Paths</td><td>Ab</td><td>-4.49</td><td>1.52</td><td>-2.96</td><td>0.003</td></tr>
<tr><td>total (c)</td><td>0.24</td><td>2.86</td><td>0.08</td><td>0.932</td></tr>
</table>

Table 5.11. Model evaluation of underlying latent factor as mediator for basics latent factor

Criteria	CFI	TLI	SRMR	RMSEA
Value	0.85	0.61	0.08	0.19
Cut-off	>=0.90 adequate >=0.95 good	>=0.90 adequate >=0.95 good	<=0.08 good	=<0.06 adequate =<0.5 good

Table 5.11 shows the model had good fit with SRMR of equal to 0.08. The CFI (0.85), TLI (0.61) and RMSEA (0.19) did not show good fit. Since the study used survey data, most literature accept the use of SRMR as the most consistent criteria with the kind of data and therefore the study accepted the model to be just good fit.

Discussion and Conclusion

Discussion

This work aimed at demonstrating and testing how well the Malawi Demographic and Health survey for 2015-16 fits a UNICEF malnutrition causal framework which guides development of strategies on combating malnutrition globally and in Malawi. In order to establish this, the study constructed Structural Equation Models to test significance of causal and mediation relationships between latent factors: Immediate, Underlying and Basic causes, and Stunting as defined in the UNICEF malnutrition causal framework. Since the study adopted the UNICEF framework as its theoretical model, a hypothesized model was developed based on the literature, and available observable variables as guided by the framework.

The results showed that all the three latent factors were significant predictors of the stunting in the model. Basic causes latent factor (number of household members, number of under 5 children and wealth index) which mostly describes access to resources by the children in the households showed a positive ($\hat{\beta}$ =14.8) causal relationship with stunting indicating as resources at the household increase the more likely you are going to have a higher HAZ score and therefore less chances of stunting. A household's access to resources is also pertinent to enable household members' access to medical/health services which are crucial for the maintenance of good health status. It is therefore not surprising that several studies have reported that wealthier households experience less child under-nutrition compared to poorer households (Janevic et al., 2010; Hong et al., 2006; Hong and Mishra, 2006). This study has therefore confirmed what other studies have found and also assumptions made in the UNICEF framework.

Similarly, Underlying causes latent factor (birth order, mother's age and mother's education) depicted a positive causal relationship ($\hat{\beta}$=289.0) with Stunting in the model. This implies that a child whose is born later (at higher order) and whose mother's age is older and education is better is more likely to have a better HAZ score and therefore less chances of being stunted. A study on the role of maternal education in child health and nutritional status found that maternal education as measured by the number of years of schooling is a consistent positive associate of child's health status in general and nutritional status in particular in Ecuador. Thus our SEM approach is confirmed through similar results (Larrea and Kawachi, 2005).

Lastly, Immediate causes latent factor (morbidity, number of breasting months and number of times child eat semi-solid food) showed a negative association ($\hat{\beta}$= -138.3), suggesting, overall that as the latent decrease there is an increase in HAZ and therefore less chances of stunting. This can be explained by the influence of morbidity on the overall outcome of the influence. Even though we expect increased breast feeding period and frequency of semi-solid food to have a positive outcome in HAZ, morbidity of a child takes out the gains and a child who is frequently sick is more likely to be stunted. A dual relationship exists between nutrient intake and childhood illnesses as shown in the conceptual framework for child nutritional status given above. Children with poor nutrient intake either due to lack of access to nutritious food or poor absorption are likely to have a reduced immune system and have an increased chance of catching illnesses as a result.

On the other hand, childhood illnesses reduce children's nutrient uptake by reducing children's appetite, reducing the absorption of nutrients and increasing the loss of nutrients in their bowel movements. Children that do not recover from their illnesses get caught up in the vicious circle of poor health and poor nutrition and are therefore very weak to fight any new infections that they catch decreasing their chances of surviving from any new illnesses. Undernourished children are reported to have had longer durations of diarrhea compared to well- nourished children in a study conducted in northern Nigeria (Tomkins, 1981). In the study the duration of diarrhea was 33% longer for underweight children, 39% longer for stunted children and 79% longer for wasted children. This study therefore confirms this relationship using structural modeling techniques.

Further the results shows presence of mediating factors on causes of stunting, be it Immediate or Underlying causes. With regard to the Immediate causes, evidence of the influence of maternal education on child wellbeing measured by child immunization and stunting is reported by Abuya et al. (2010) in their study conducted in Kenya. Children born to mothers with primary education were 2.17 times more likely to be fully immunized and less likely to get sick compared to those whose mothers lacked formal education and the unadjusted results for nutrition revealed that children born to mothers with primary education had 94% lower odds of having stunted growth compared to children whose mothers had no primary education. The study above generated evidence that the better underlying factors are the less likely the occurrence of disease incidences (immediate factors) which resulted in less stunting.

Considering the mediating effect of Underlying causes, results elsewhere show the importance of mother's economic status in children's nutritional status (Chirwa and Ngalawa, 2008), where children from households where the mother or female head of household was in a salaried employment had better nutritional status than children whose mothers were not. On another note, mother's education is reported to be a major influence on child feeding practices in Accra Ghana, more than household food availability and economic resources (Armar-Klemesu et al., 2000). This shows that access to resources and awareness of good child care practices go hand in hand in influencing a child's nutritional status, and in circumstances where one factor appears more important than the other, it could be the case that there is a major imbalance of the two factors. Using the SEM techniques, this study has validated the findings of the previous studies.

Limitations of Study

The UNICEF causal framework describes malnutrition with all three indicators stunting, wasting and underweight but in this study only used stunting to represent malnutrition. Our focus on stunting hinges on the fact that in Malawi stunting is stubbornly high, and it has been a major focus of efforts towards reducing stunting. Moreover, wasting is mostly seasonal in nature, and the validity of relation may vary, making it difficult to use the results for programming interventions. Secondly, study could not use all records for children under five years in the analysis because high numbers of missing variables in some of our variables of interest. Nevertheless, over 3,000 children were used for analysis, which was good enough using rule of thumb, to provide enough power corresponding to the number of parameters estimated in the model.

Conclusion

Based on the findings and discussion above, the study concludes that Malawi Demographic and Health Survey data well fits UNICEF malnutrition causal framework and is able to explain all causal relationships and mediation pathways as defined by the framework. In addition, the study has demonstrated the efficiency of using structural modelling techniques in modelling stunting data. Unlike in previous studies where stunting or malnutrition was modelled using ordinary least regression methods and most recently multi-level regression techniques, SEM techniques has not just found predictors of stunting but also the mediation paths that those predictors utilize to cause malnutrition. Secondly, SEM has utilized latent factors which are defined in the frameworks used by policy makers to design preventive interventions of stunting in modelling risk factors therefore its results are more user friendly than other models. Thirdly, SEM has utilized data that has significant correlation within the predictors and whose multivariate normally assumption did not hold to still make useful models of predicting malnutrition. Overall SEM has demonstrated to be a better technique of modelling risk factors for stunting.

Recommendations

Structural Equation Modelling is becoming a useful technique in biostatistics in most recent times having previously been used in psychology mostly. This study has demonstrated how this technique can be used in modelling data with complex relationships, non-multivariate normal distribution utilizing what are weaknesses of the other common approaches. Further studies with national data and even subnational data on malnutrition applying these methods are recommended to validate the findings.

REFERENCES

Abuya, B., Onsomu, E., Kimani, J. & Moore, D. 2010. Influence of Maternal Education on Child Immunization and Stunting in Kenya. *Maternal and Child Health Journal,* 1-11.

Armar-Klemesu, M., Ruel, M. T., Maxwell, D. G., Levin, C. E. & Morris, S. S. 2000. Poor Maternal Schooling Is the Main Constraint to Good Child Care Practices in Accra. *The Journal of Nutrition,* 130, 1597-1607.

Australian Government Office of Development Effectiveness. 2014; *Addressing child under nutrition: evidence based review.*

Bailey R. L., West K. P. Jr, Black R. E. (2015). The Epidemiology of Global Micronutrient Deficiencies. *Ann Nutr Metab.* 66(suppl 2):22–33.

Black, R. E., Victora, C. G., Walker, S. P., Bhutta, Z. A., et al. 2013. Maternal and Child Nutrition Study Group. Maternal and child undernutrition and overweight in low-income and middle-income countries. *The Lancet,* 382(9890), 427-45.

Bollen, K. A. (1989). Structural equations with latent variables. New York: Wiley.

Boomsma, A., & Hoogland, J. J. (2001). The robustness of LISREL modeling revisited In R. Cudeck, S. du Toit & D. Sörbom (Eds.), *Structural equation models: Present and future. A Festschrift in honor of Karl Jöreskog* (pp. 139-168). Chicago: Scientific Software International.

Byrne B. M. 2001. *Structural equation modelling with AMOS: basic concepts, applications and programming*. New Jersey, NY: Lawrence Erlbaum Associate.

Chikhungu L. C. 2013, *Modelling Child Nutrition in Malawi,* University of Southampton, PhD thesis

Chirwa, E. W. & Ngalawa, H. P. 2008. Determinants of Child Nutrition in Malawi. *South African Journal of Economics,* 76, 628-640.

Cochrane, S. H., Leslie, J. & O'hara, D. J. 1982. Parental education and child health: Intracountry evidence. *Health Policy and Education,* 2, 213-250.

Engle, P. L. 1991. Maternal Work and Child-Care Strategies in Peri-Urban Guatemala: Nutritional Effects. *Child Development,* 62, 954-965.

Finch, W. Holmes, Jr., French Brian, F. 2015. *Latent Variable Modelling in R: Step by Step Guide:* Routledge 711 Third Avenue, New York, NY 10017 and Routledge 27 Church Road, Hove, East Sussex BN3 2FA

Garson G., Structural Equation Modeling: PA 765, 2017.

Gerbing D. and Anderson J. C. 1988. An updated paradigm for scale development incorporating unidimensionality and its assessment: *Journal of Marketing Research,* vol. 25, no. 2, pp.186-199.

Gibson, R. S., Abebe, Y., Hambidge, K. M., Arbide, I., Teshome, A. & Stoecker, B. J. 2009. Inadequate feeding practices and impaired growth among children from subsistence farming households in Sidama, Southern Ethiopia. *Maternal & Child Nutrition,* 5, 260-275.

Hoddinott, J., Maluccio, J. A., Behrman, J. R., Flores, R. and Martorell, R. (2008). Effect of a nutrition intervention during early childhood on economic productivity in Guatemalan adults. *The Lancet* 371(9610): 411-416.

Jöreskog, K. G., & Sörbom, D. (1981). *LISREL V: Analysis of linear structural relationships by maximum likelihood and least squares methods:* University of Uppsala, Department of Statistics.

Kendall, M. G., Stuart, A., & Ord, J. K. (1987). *Kendall's Advanced Theory of Statistics Distribution theory* (5th ed. Vol. 1). New York, NY: Oxford University Press.

Kline, R. B. (2010). *Principles and practice of structural equation modeling* (3rd ed.). New York, NY: The Guilford Press.

Kumchulesi G. 2017. *Persistence of Child Malnutrition in Malawi*: Explanations from Demographic and Health Surveys.

Larrea, C. & Kawachi, I. 2005. Does economic inequality affect child malnutrition? The case of Ecuador. *Social Science & Medicine,* 60, 165-178.

Madise, N. & Mpoma, M. 1997. Child Malnutrition and feeding practices in Malawi. *Food and Nutrition Bulletin,* 18, 190- 201.

Makoka, D. (2013). Adult education and child nutrition: the role of family and community. *DHS Working Papers*, pp. 153–159.

Malawi Government. *Malawi Growth and Development Strategy* III, 2017.

Malcom, L. A. 1970. Growth retardation in a New Guinea boarding school and its response to supplementary feeding. *British Journal of Nutrition,* 24, 297-305.

Mukherjee, R., Chaturvedi, S. & Bhalwar, R. 2008. Determinants of Nutritional Status of School Children. *Medical Journal Armed Forces India,* 64, 27-231.

Nagati K, Ben Mansour, A., Alouane, L. (2003). Multiple approaches to micronutrient deficiency in developing countries. *Forum Nutr*. 56:282-3.

Nahar, B., Ahmed, T., Brown, K. H. & Hossain, M. I. 2010. *Risk Factors Associated with Severe Underweight among Young Children Reporting to a Diarrhoea Treatment Facility in Bangladesh.*

Scaling Up Nutrition 2011. *A Framework for Action*; Reprint April 2011

Smolin, L. A. & Grosvenor, M. B. 2003. *Nutrition: science and applications.*

Tomkins, A. 1981. Nutritional status and severity of diarrhea among pre-school children in rural Nigeria. *The Lancet,* 317, 860-862.

Tzioumis, E., & Adair, L. S. (2014). Childhood dual burden of under-and over-nutrition in low-and middle-income countries: A critical review. *Food & Nutrition Bulletin,* 35(2), 230-243.

UNICEF, 2006. *Monitoring the situation of children and women*: https://data.unicef.org/topic/nutrition/malnutrition/.

UNOCHA, 2016. *Regional Outlook for Southern Africa, Recommendations for Humanitarian Action and Resilience Response.*

Wamani, H., Tylleskar, T., Astrom, A., Tumwine, J. & Peterson, S. 2004. Mothers' education but not fathers' education, household assets or land ownership is the best predictor of child health inequalities in rural Uganda. *International Journal for Equity in Health,* 3, 9.

WHO 2016. *WHO Health Topic,* http://www.who.int/features/qa/malnutrition/en/

World Bank 2006. *Repositioning Nutrition as Central for Development.* Washington, DC: World Bank

Xuan Xiao, B.S 2013. *Structural Equation Modeling Compared with Ordinary Least Squares in Simulations and Life Insurers' Data,* The University of Texas at Austin.

Zhu Z, Suzukr, K. et al. 2009. Confirmatory factor analysis of the Child Feeding Questionnaire (CFQ) in Japanese elementary school children; *Appetite,* vol. 52, no. 1, pp. 8-1.

In: Statistical Modelling …
Editors: N-B. Kandala and L. Kazembe
ISBN: 978-1-53615-981-3

Chapter 6

Linking Food Insecurity to Quality of Life Using Structural Equation Models

Tobias W. Shinyemba**, Ndeyapo M. Nickanor and Lawrence N. Kazembe
Department of Statistics and Population Studies, University of Namibia, Windhoek, Namibia

Abstract

Namibia is experiencing an increase in its population living in urban areas due to rapid urbanisation, with the biggest growth rate experienced in Windhoek, where about 16% of the total population lives. With rapid population growth, Windhoek faces a number of challenges including the development of informal settlements, which manifest compromised well-being. The objective of the study is to investigate the relationship between food insecurity and quality of life in Windhoek informal settlements, since going without food does have an impact on people's well-being. We used Structural Equation Modelling (SEM) approach on secondary data obtained from a quantitative cross-sectional study in informal settlements of Windhoek, Namibia. The study measured food insecurity using HFIAS indicators, while WHOQOL-BREF and EQ-5D instruments were used to measure quality of life. Multivariate statistical analysis using exploratory factor analysis (EFA), principal component analysis (PCA), and confirmatory factor analysis (CFA) methods were applied to reduce dimensionality in the data and link food insecurity to quality of life. The results confirmed that food insecurity was a significant predictor of quality of life outcomes in Windhoek informal settlements. The study further revealed that lack of access to food affect social well-being, environmental well-being and physical well-being in a negative way. Similarly, the study found that pain/discomfort, anxiety/depression, self-care, and usual activities were positively associated with food insecurity.

---

* Corresponding Author's Email: floshinyemba@gmail.com.

Keywords: food insecurity, quality of life, informal settlements, Windhoek, structural equation modelling

INTRODUCTION

Over the past 20 years, the world has experienced a massive growth in its population living in urban areas from 43% (2.3 billion) in the mid-1990s to 54% (4 billion) inhabitants in mid-2015, and it is projected to reach 60% (4.9 billion) by 2030 and 66% (6.5 billion) in 2050 (UN-DESA, 2015; UN-Habitat, 2016). Urban population has been increasing rapidly due economic expansion and development of better infrastructure in urban areas compared to rural areas. As a result of high poverty rate in rural areas most people have been migrating to urban areas in search for better standard of living, such as better health care services, employment opportunities, and better education (UN-Habitat, 2016). Presently, Asia has the highest per cent of the population living in urban areas followed by Europe, Latin America, Africa, North America and Oceania (UN-Habitat, 2016).

Urbanisation growth rates have not been the same in all parts of the world. Currently the highest rate is experienced in Africa due to high rural-urban migration, high fertility rate, and reclassification of rural areas (Buhaug & Urdal, 2013). Similarly, Namibia has experienced an increase in its urban population between 2001 and 2011, from 33% (9.08 hundreds) to 53% (1.34 million) inhabitants (NSA, 2013). Consequently, the growth rate has not been the same across Namibia, with Windhoek experiencing the highest. Currently, 16% of the country's total population live in Windhoek, making it the most populated urban centre in the country (NSA, 2013). Subsequently, throughout the years, Windhoek has faced a number challenges including proliferation of informal settlements as a result of unsustainable number of people migrating to the city in search for better livelihoods (Indongo, Angombe, & Nickanor, 2013).

An informal settlement is defined as a residential area where a cluster of dwelling units are built on land where individuals have no legal claim, and such housing facilities do not meet municipal planning requirements and building regulations (Wekesa, Steyn, & Fred-Otieno, 2011). These settlements are mostly found located on the outskirts of cities where there is minimal or no access to basic municipal services and infrastructure such as clean water, waste disposal, poor sewage system, and tarred roads and they are mostly overcrowded (UN-Habitat, 2015). Verma (2017) indicated that informal settlements mainly develop when the city's land and administration divisions fail to deliver housing services for its people. Further, Fox (2014); Tacoli, McGranahan, and Satterthwaite (2015) pointed out that informal settlements develop as a result of poor migrants and poor city-born who try to integrate themselves into a city's economic lifestyle, given that most of them cannot afford to build, buy or rent a decent housing unit that will keep them closer to employment opportunities. Fox (2014) argued that, informal settlements exist because of urban poverty, and their continuous growth is linked to the city's population growth.

Over time, overcrowding may cause various social problems in informal settlements such as poor health conditions, unsustainable livelihoods, and food insecurity among other things (Indongo et al., 2013; Jha & Tripathi, 2014; Kimani-Murage et al., 2014; Tacoli et al., 2015). Social problems negatively affect the quality of life of people living in these areas in a negative way, resulting in poor livelihood outcomes.

Quality of life is a term used to describe how individuals or communities experience health, happiness and comfort. It has to do with how individuals live, how they feel and how they comprehend their daily lives. It is measured by examining individuals' psychological, physical, social, environmental and material well-being and its outcomes are determined by social, economic, psychological, physical, and environmental factors (Cella, 1994; Felce & Perry, 1995; Verma, 2017). Social factors include promotion of social justice and availability of affordable houses. Whereas, economic aspects include availability of jobs and business opportunities, and psychological factors contain creating a quiet enjoyable and pleasurable environment by improving the current social environment. Physical factors consist of access to public services and friendly neighbourhood, while environmental factors include access to green spaces, hygienic air, and regulation of solid waste and waste-water disposal (Verma, 2017). Unavailability of the above mentioned aspects has led to residents of informal settlements to be dissatisfied with their living conditions, and thus resulting in poor quality of life outcomes (Wokekoro & Owei, 2014).

Apart from developmental challenges faced by people residing in informal settlements, they also encounter the problem of food insecurity. FAO (2003) defined food security as a state whereby "all people, at all times, have physical, social and economic access to sufficient, safe and nutritious food which meets their dietary needs and food preferences for an active and healthy life" (p. 29). Studies have identified that the state of food insecurity occurs when food systems collapses and it is more prevalent in areas with rapid urbanisation and limited economic growth (Friel & Ford, 2015; Gregory, Ingram, & Brklacich, 2005). In most countries, food systems have collapsed due to ever increasing number of residents that are living in poverty, and this has also broaden the food gap between urban households (Crush & Frayne, 2010). Consequently, it has restricted poor households from purchasing high nutritious food, and their inability to afford refrigerators to store such food has limited them to buying low nutritious and less preferred food (Garrett, 2000; Garvelink & Wedding, 2013).

Additionally, poverty has also exposed most urban poor to make decisions on whether to spend more on food or on other necessities due to limited resources, resulting in most households sacrificing on food diversity and food quality over food quantity, to ensure that their households have enough food (Kimani-Murage et al., 2014; Pendleton,

Crush, & Nickanor, 2014). Whereas, other households tend to skip meals, reduce quality and variety of food, eating street food, buying food on credit, stealing, and scavenging as part of their coping strategies. Such decisions mostly affect women and dependent children, making them to be frustrated, feeling helpless, and depressed (Garrett, 2000; Ivers & Cullen, 2011; Weigel, Armijos, Racines, Cevallos, & Castro, 2016). Even though these coping strategies reduce psychological hunger in the short period, they have long-term effects on human lives. Lack of inadequate intake of nutrients can lead to mental or physical illness, because nutrients are essential for maintaining good mental and physical health (Weigel et al., 2016, p. 8). Similarly, FAO, IFAD, UNICEF, WFP, and WHO (2017) reported that 33% of women in the reproductive age group suffer from anaemia globally as a result of food insecurity and this puts the health and nutrition of many infants at risk. Hunger and under-nutrition are both outcomes of food insecurity and they have negative effect on health outcomes that affect the quality of life of people and at times manifested through chronic diseases, obesity and mental health disorders (Altman, Hart, & Jacobs, 2009; Atuoye & Luginaah, 2017; Naicker, Mathee, & Teare, 2015).

Over the last decade, most scholars have focused more on addressing the issue of rural food insecurity and overlooked the prevalence of food insecurity in urban areas (Crush & Frayne, 2010). The analysis done by Crush, Frayne, and Pendleton (2012) found that food insecurity has spread more widely in poor urban neighbourhoods across the fast-growing cities in Southern African due to high unemployment rates. Other studies have pointed out that, food insecurity is an ongoing public health concern that has implications on individual's life span and it contributes to poor outcomes of quality of life (Chung et al., 2016; Gundersen, 2013; Naicker et al., 2015). Not having enough access to nutritious food has implications on human health and it is highly important to understand the consequences faced by those affected. Even though there are number of studies undertaken on urban food insecurity in Namibia, there is scanty research on the impact of food insecurity of quality of life. Hence, there is a need to assess the impact of food insecurity on quality of life, since going without food does have an impact on human's well-being.

The paper aims to quantify the relationship between food insecurity and quality of life in the informal settlements of Windhoek using structural equation modelling (SEM) approach.

METHODS

Population and Study Setting

The study followed a quantitative cross-sectional study design using secondary data collected for a project on Urban Livelihood and Quality of Life in Windhoek informal settlements. A cluster sampling design was used to select 420 households following a two-stage sample design, where enumeration areas (EAs) were first selected as the Primary Sampling Units (PSU) from the 2011 Namibia Population and Housing Census sampling frame, and then a fixed number of households were systematically selected in each EAs for interviews. The sample was drawn from three constituencies that have informal settlements: Samora Machel, Tobias Hainyeko, and Moses //Garoeb.

Measurements

The data was collected using a face-to-face questionnaire, that captured information on Household Food Insecurity Scale (HFIAS) indicators (based on ten variables) to measure food insecurity. Whereas, quality of life was measured using 21 variables based on the World Health Organisation (WHO) Quality-of-Life Scale (WHOQOL-BREF) and 36 variables based on European Quality of life (EQ-5D) instrument.

The HFIAS indicators were developed by the United States Agency for International Development (USAID)'s Food and Nutrition Technical Assistance (FANTA) group to estimate the prevalence of food insecurity (access). The indicators have three domains namely; "anxiety and uncertainty about household food supply", "insufficient quality", and "insufficient intake and its physical consequences" (Coates, Swindale, & Bilinsky, 2007, p. 5). This approach is more useful, because the set of questions used distinguishes food secure households from food insecure households. Further, the indicators are also useful to assess food insecurity prevalence and detect how it has changed over time.

On the other hand, the WHOQOL-BREF was developed by WHO with data from fifteen culturally diverse international field centres to legitimately measure quality of life outcomes internationally and it has four domain of QoL, namely: physical, environmental, psychological, and social health (WHO, 1996). Further, European quality of life (EuroQol) group developed EQ-5D to measure individuals' health status. The instrument has five dimensions: mobility, self-care, usual activities, pain/discomfort and anxiety/depression, with each dimension having five levels of increasing severity (no problems, slight problems, moderate problems, severe problems, and extreme problems). This instrument has been widely used in clinical trials, observational studies, population surveys and other health surveys (van Reenen & Janssen, 2015).

Variables listed under each instrument were reduced into domains of food insecurity and quality of life respectively. Food insecurity construct was used as a predictor variable for outcomes variables derived from quality of life constructs.

STATISTICAL ANALYSIS

Multivariate statistical analyses were carried out using SEM to identify the structural relationships between food insecurity and quality of life. SEM is a complex modelling statistical technique that allow modelling multivariate data using unobserved variables (Sanchez, Budtz-Jorgensen, Ryan, & Howard, 2005). It is used to test the hypothesis about the association between manifest (observed) and latent (unobserved) variables that are defined in terms of manifest variables (Teo, Tsai, & Yang, 2013). SEMs are essential when dealing with the complexity that exists hidden in social phenomenon as it permits casual analysis of numerous variables. In this study, SEM was carried out in two stages; in the first stage, we run principal component analysis (PCA) and exploratory factor analysis (EFA) to reduce dimensionality in the data into constructs of food insecurity and quality of life. In the second stage, we run confirmatory factor analysis (CFA) to confirm the preconceived relationship established using EFA and to link the two constructs of quality of life to food insecurity construct in a process that completes SEM.

Fitted Models

Principal Component Analysis

PCA is one of the widely used old multivariate statistical techniques, it was first introduced by Pearson in 1901, and later on independently developed by Hotelling in 1933 (Dunteman, 1989; Everitt & Dunn 2001). The goal of PCA is to reduce a large number of multivariate variables into relatively smaller number of linear combinations that can be used to account for much variability in the data, such that the first principal component have the largest variability in the data, followed by the second component chosen to account for as much as possible variation in the data, and it should be uncorrelated to the first principal component. Similarly, the third component is selected to account for much variation in the data, and it is uncorrelated to both the first and second component, and so on (Everitt & Dunn 2001; Zelterman, 2015).

In this analysis, since there were ten variables on food insecurity and 57 (21 on WHOQOL-BREF and 36 on EQ-5D) variables on quality of life, PCA was the most suitable step in SEM to begin with by reducing the number of variables under each aspects into the main constructs of food insecurity and quality of life respectively.

The study begins with a multivariate random vector $x = (x_1, x_2, \ldots, x_q)$ that has a mean μ and covariance Σ. The sample principal components are then calculated as follows. The first principal component for this vector will be a linear combination of:

$$y_1 = w_1^T x = w_{11}x_1 + w_{12}x_2 + \ldots + w_{1q}x_q \tag{1}$$

whose sample variance is greatest among all linear combinations, and since the variance of the first component y_1 can be increased without limits, by simply increasing the coefficients $w_1^T = (w_{11}, w_{12},..., w_{1q})$. A restriction is placed on these coefficients, to ensure that the sum of squares of these coefficients takes the value of one ($w_1^T w_1 = 1$). Further, the second principal component will be a linear combination of:

$$y_2 = w_2^T x = w_{21}x_1 + w_{22}x_2 + \ldots + w_{2q}x_q \tag{2}$$

that has a greatest variance subject to the following conditions $\begin{cases} w_2^T w_2 = 1 \\ w_2^T w_1 = 0 \ . \end{cases}$ (3)

These conditions (equation 3) ensure that y_1 and y_2 are uncorrelated. Finally, the jth principal component will be a linear combination of:

$$y_j = w_q^T x = w_{q1}x_1 + w_{q2}x_2 + \ldots + w_{qq}x_q \tag{4}$$

that has a greatest variance subject to the conditions $\begin{cases} w_p^T w_p = 1 \\ w_j^T w_i = 0 \quad (i < j) \ . \end{cases}$ (5)

In the above equations $y_1, y_2,..., y_p$ denote the principal components of x, and w_{ij} denote the weights or loadings that explain how each of the original observation contribute to each of the principal components (Zelterman, 2015, p. 208).

The variances for principal components are estimated using the following equation expressed as:

$$Var\ y_i = w_i^T \Sigma w_i \tag{6}$$

The value of variances can be made arbitrarily large by multiplying w_i by a larger scalar, and it is avoided by setting a restriction on the loadings w_i to have a unit length of $w_i^T w_i = 1$, and making y_i to be mutually uncorrelated by setting:

$$Cov\ (y_i, y_j) = w_i^T \Sigma w_i = 0\ \text{for all}\ i \neq j\ \text{both between 1 and}\ q. \tag{7}$$

Since, the study wish to make the variance in equation (6) to be as large as possible and have the covariance in equation (7) to be zero, this is attained by estimating the covariance Σ using the sample variance matrix S, by examining the pairs of eigenvalues (λ_i) and eigenvectors (e_i). If the weights w_i are replaced by respective eigenvectors e_i, this maximizes the estimated variances of principal component y_i with respect to the estimated covariance matrix S, leading to the principal component y_i to be uncorrelated (Hair, Anderson, Babin, & Black, 2009; Johnson & Wichern, 2007; Zelterman, 2015). Thus, the estimated variance will then be given by:

$$\overline{Var}\, y_i = e_i^T S e_i = \lambda_i \tag{8}$$

Finally, the total variance of the q principal component will be equal to the total variance of the original variable so that $\sum_{i=1}^{q} \lambda_i = s_1^2 + s_2^2 + \ldots + s_q^2$, where s_i^2 is the sample variance of x_i. This can then be rewritten as $\sum_{i=1}^{q} \lambda_i = trace(S)$ (Everitt & Hothorn, 2011; Hair et al., 2009). Then the jth principal component accounts for a proportion P_j of the total variability in the original data, where:

$$P_j = \frac{\lambda_i}{trace(S)} \tag{9}$$

To decide on the number of components to be retained, Rencher (2002, p. 397) suggested that the following proposed guidelines maybe used:

1. Retain components that account for a specified percentage of the total variance.
2. Retain the components whose eigenvalues are greater than the average of the eigenvalues. For a correlation matrix, the average is one.
3. Use the scree plot, and look for a natural break down between the large eigenvalues and the small eigenvalues.
4. Test the significance of the larger components, that is, the components corresponding to the larger eigenvalues.

Even though there are number of studies that have used the above stated subjective guidelines to decide on the number of components to be retained, the methods sometimes lead to under extraction or over extraction of components (Franklin, Gibson, Robertson, Pohlmann, & Fralish, 1995). Hayton, Allen, and Scarpello (2004) suggested that Parallel Analysis is the most accurate method used to determine the number of significant components. This technique work by comparing eigenvalues from the real data correlation matrix before they are rotated, to those of the random correlation matrix of the same dimensionality, this method retain components that have actual eigenvalues greater than the parallel random average eigenvalues (Franklin et al., 1995; Hayton et al., 2004).

Exploratory Factor Analysis

The study also fitted EFA to identify the relationship between the assumed latent variables and manifest variables without making any assumption about how manifest variables are related to the factors (Everitt & Hothorn, 2011, p. 135). EFA is more like a classical regression model, they differ in a way that in factor analysis manifest variables are linked to a set of unobserved variables, compared to the regression model where dependent variables are linked to independent variables (Everitt & Hothorn, 2011). In the last stage of the analysis, CFA will be used to test if whether the specified factor models found using EFA provide adequate fit for the covariance or correlation between manifest variables. This is how the two method of factor analysis differs (Everitt & Hothorn, 2011, p. 135).

At this point, we wish to mention that, even though both principal component analysis and factor analysis aim to reduce dimensionality in the data, their difference is that: (1) in factor analysis variables are expressed as linear combinations of the factors, as opposed to principal component analysis where principal components are linear functions of the variables. (2) In principal component analysis, we emphasize on explaining the total variance, whereas in factor analysis we emphasize on explaining the covariance (Rencher, 2002).

In EFA, we assumed that the observed relationship between manifest variables are as a results of the relationship between latent variables (Everitt & Hothorn, 2011, p. 137). Then the model of the kth-factor analysis is assumed to have a set of manifest variables $x^T = (x_1, x_2, \ldots, x_q)$ that are linked to k latent variables or common factors $f_1, f_2, .., f_k$, where $k < q$ written mathematically as follows:

$$
\begin{aligned}
x_1 &= \lambda_{11} f_1 + \lambda_{12} f_2 + \ldots + \lambda_{1m} f_m + \varepsilon_1 \\
x_2 &= \lambda_{21} f_1 + \lambda_{22} f_2 + \ldots + \lambda_{2m} f_m + \varepsilon_2 \\
&\vdots \\
x_q &= \lambda_{q1} f_1 + \lambda_{q2} f_2 + \ldots + \lambda_{qm} f_m + \varepsilon_q .
\end{aligned}
$$

This equation can be rewritten in short as $x = \Lambda f + \varepsilon$

$$\text{such that,} \quad \Lambda = \begin{pmatrix} \lambda_{11} & . & . & \lambda_{1k} \\ . & & & . \\ . & & & . \\ \lambda_{q1} & . & . & \lambda_{qk} \end{pmatrix}, \; f = \begin{pmatrix} f_1 \\ . \\ . \\ f_q \end{pmatrix}, \; \varepsilon = \begin{pmatrix} \varepsilon_1 \\ . \\ . \\ \varepsilon_q \end{pmatrix}. \tag{10}$$

In this equation, we assumed that the random disturbance terms $\varepsilon_1, \varepsilon_2 ..., \varepsilon_q$ are uncorrelated with each other and with the factors $f_1, f_2, .., f_k$. These assumptions implies that manifest variables are independent given any values of the factors (Everitt & Hothorn, 2011).

In addition, the variance of x_i ,σ_i^2 , in factor analysis model is given by:

$$\sigma_i^2 = \sum_{j=1}^{k} \lambda_{ij}^2 + \psi_i \tag{11}$$

where ψ_i is the variance of ε_i

From equation (11), the study assumed that the variance of each observed variable can be split into two parts. The first part is known as the communality of the variable that represents the variance shared with other variables through a common factor given by: $h_i^2 = \sum_{j=1}^{k} \lambda_{ij}^2$. The second part ψ_i is known as a specific or unique variance, that is related to the variability in x_i not shared with other variables (Everitt & Hothorn, 2011, p. 138). Similarly, the covariance of variables x_i *and* x_j in the factor model can be written as:

$$\sigma_{ij} = \sum_{l=1}^{k} \lambda_{il} \lambda_{jl} \tag{12}$$

From equation (12), we assumed that the covariance does not depend on any specific variate. Hence, it is only the common factors that intend to account for the relationships between manifest variables (Everitt & Hothorn, 2011, p. 138). Further, we noted that the covariance matrix Σ of the observed variable is of the form:

$$\Sigma = \Lambda\Lambda^T + \psi \tag{13}$$

where $\psi = diag(\psi_i)$

However, in practice the study will estimate Σ by the sample covariance matrix S, by obtaining the estimate of Λ *and* ψ so that the observed covariance matrix takes the required form of the model. To apply the above stated equations, we need to estimate the factor loadings and their specific variances. Hence, the study need to find the factor-loading matrix $\hat{\Lambda}$ and the diagonal matrix $\hat{\psi}$ that contains the specific variances. The specific variances are given by:

$$\hat{\psi}_i = s_i^2 - \sum_{j=1}^{k} \hat{\lambda}_{ij}^2 , \; i = 1,\ldots,q \tag{14}$$

Equation (14) helps to estimate the diagonal terms in the covariance matrix S.

Further, to estimate parameters in the model, the study uses maximum likelihood methods. This method is believed to be a respectable method in factor analysis parameter estimation (Everitt & Hothorn, 2011, p. 142). It assumes that the data follows a multivariate normal distribution, and that the model holds a likelihood function L shown to be $-\frac{1}{2}nF$, where F is given by:

$$F = \ln\left|\Lambda\Lambda^T + \psi\right| + trace\left(S\left|\Lambda\Lambda^T + \psi\right|^{-1}\right) - \ln\left|S\right| - q \tag{15}$$

The function takes on the value of zero if $\Lambda\Lambda^T$ is equals to the covariance matrix S. The loading estimates are estimated by minimizing the F function with respect to these parameters (Everitt & Hothorn, 2011, p. 142).

Finally, the study estimates the factor scores that are needed in the interpretation of the fitted model. To estimate the factor scores the study introduces an orthogonal matrix M of order $k \times k$, since the variables in equation (15) are expressed in terms of factors. The calculation of factor scores requires variables to have the relationship that is in an opposite direction. Equation (15) can then be rewritten as:

$$x = (\Lambda M)(M^T f) + \varepsilon \tag{16}$$

This model is now associated with new factors $f^* = Mf$ and new factor loadings ΛM. The covariance matrix of observed variables in this model is then given by:

$$\Sigma = (\Lambda M)\,(\Lambda M)^T + \psi_i \tag{17}$$

Therefore, the factor scores are calculated using

$$\hat{f} = \hat{\Lambda}\, S^T x \tag{18}$$

Confirmatory Factor Analysis

CFA models enable us to either confirm or reject our preconceived theory established using EFA (Hair et al., 2009). CFA forms part of SEMs that aims to explain how both explanatory and latent variables are linked in series of linear combinations.

In CFA, there are three steps involved in model building. The first step, begin by estimating the values of model parameters that minimises the discrepancy function between the elements of the observed covariance matrix **S** of manifest variables and the covariance matrix $\Sigma(\theta)$ implied by the fitted model, contained in the vector $\theta = (\theta_1, \theta_2, \ldots, \theta_t)^T$. Everitt and Hothorn (2011, p. 202) stated that maximum likelihood estimation method, is the most commonly used method for estimating parameters in CFA. The method assumes that the observed data follows a multivariate normal distribution. The maximum likelihood for minimizing the discrepancy function (FML) is given by:

$$FML = \left(S, \Sigma(\theta)\right) = \log\left|\Sigma(\theta)\right| - \log|S| + trace\left(S\Sigma(\theta)^{-1}\right) - q \tag{19}$$

The second step, deals with parameter identification in the model. In CFA, model identifiability depends on the choice of the model, specification of fixed constraints, and number of free parameters in the model (Everitt & Hothorn, 2011, p. 204). They further stated that model identification in CFA can be difficult times, since there are no universally applicable rules for evaluating model identification. However, there is a simple but not sufficient condition for identification, that states that the number of free parameters in a model *t,* need to be less than $q(q+1)/2$.

The third step, assess how the model predicts well the covariance matrix of manifest variables using the likelihood ratio statistic given by:

$$\chi^2 = (N-1)FML_{\min} \tag{20}$$

where N is a sample size and FML_{min} is the minimized value of the maximum likelihood discrepancy function given in equation (19). Everitt and Hothorn (2011, p. 204) noted that even though the likelihood ratio statistic is the only measure used to assess the model goodness of fit, sometimes it is impractical to use this method when the sample size is large, and when manifest variables have a non-normal distribution. Hence, the likelihood ratio statistics should be used alongside any of the following procedures: (1) visual inspection of residual covariance, (2) examination of the standard errors of the parameters and the correlations between estimates, and (3) inspecting the estimated values outside their range. For instance, inspecting results that are showing negative variances indicate that the fitted model is wrong (Everitt & Hothorn, 2011).

CFA forms part of SEMs, the technique uses standardised regression scores to link latent variables to manifest variables and this relationship is visualised using path diagrams. The method aims to test the null hypothesis that, there is no difference between the patterns of manifest variables in the data set and the specified model.

To test if the study sample size is adequate for factor analysis, the study uses Kaiser-Meyer-Olkin (KMO) and Bartlett's Test to examine if the sample size is acceptable and if the data is appropriate for factor analysis (Rencher, 2002, p. 443).

KMO is given by: $$\frac{\sum_{i}\sum_{j\neq i} r_{ij}^2}{\sum_{i}\sum_{j\neq i} r_{ij}^2 + \sum_{i}\sum_{j\neq i} u_{ij}^2} \quad (21)$$

where r_{ij} is the correlation matrix, and u_{ij} is the partial covariance matrix

Bartlett's Test is given by: $$\chi^2 = -\left(n - 1 - \frac{2p+5}{6}\right) \times \ln|R| \quad (22)$$

such that n is the number of observation, p is the number of variables, and R is the correlation matrix. For KMO, a correlation around 0.60 to 0.70 is deemed to be sufficient for analysing EFA and Bartlett's Test is significant when the $p < 0.05$. We also need to assess the relationship between variables using a correlation matrix, it is recommended to inspect variables with correlation coefficients over 3.0.

Model identification and goodness of fit the model used included the Chi-square statistics (CMIN), Standardised Root Mean Square Error Residuals (SRMR), Root Mean Square Error of Approximation (RMSEA), Tucker-Lewis Index (TLI), Comparative Fit Index (CFI), and the Goodness of Fit Index (GFI). Hu and Bentler (1999) suggested that the value of SRMR and RMSEA should be close to 0.08 and 0.06 respectively, whereas the TLI, CFI, and the GFI should be equivalent or larger than 0.9.

The data was analysed in R version 3.3.1. The sample codes used to produce the study results are given in the Appendix A.

RESULTS

Model Fit and Validation

Food Insecurity

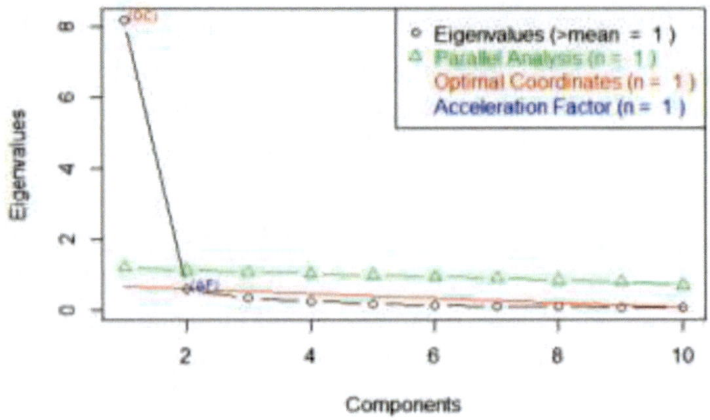

Figure 1. Scree plot for food insecurity components.

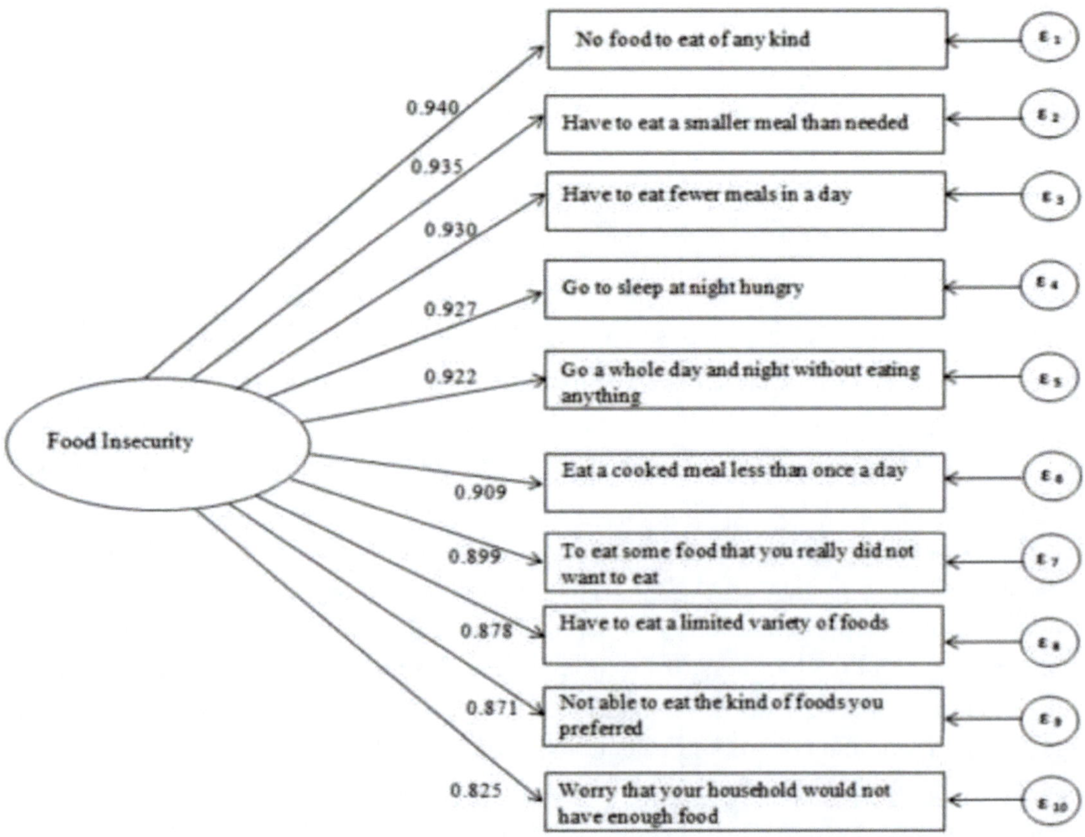

Figure 2. Summary of food insecurity results.

Table 1. Model fit analysis for food insecurity

Indicator	Primary model	Ideal Magnitude	
Chi.sq.	679.4	> 200	Yes
RMSEA	0.22	≤ 0.05	No
SRMR	0.05	< 0.1	Yes
TLI	0.85	> 0.9	No
CFI	0.89	> 0.9	No
GFI	0.7	> 0.9	No

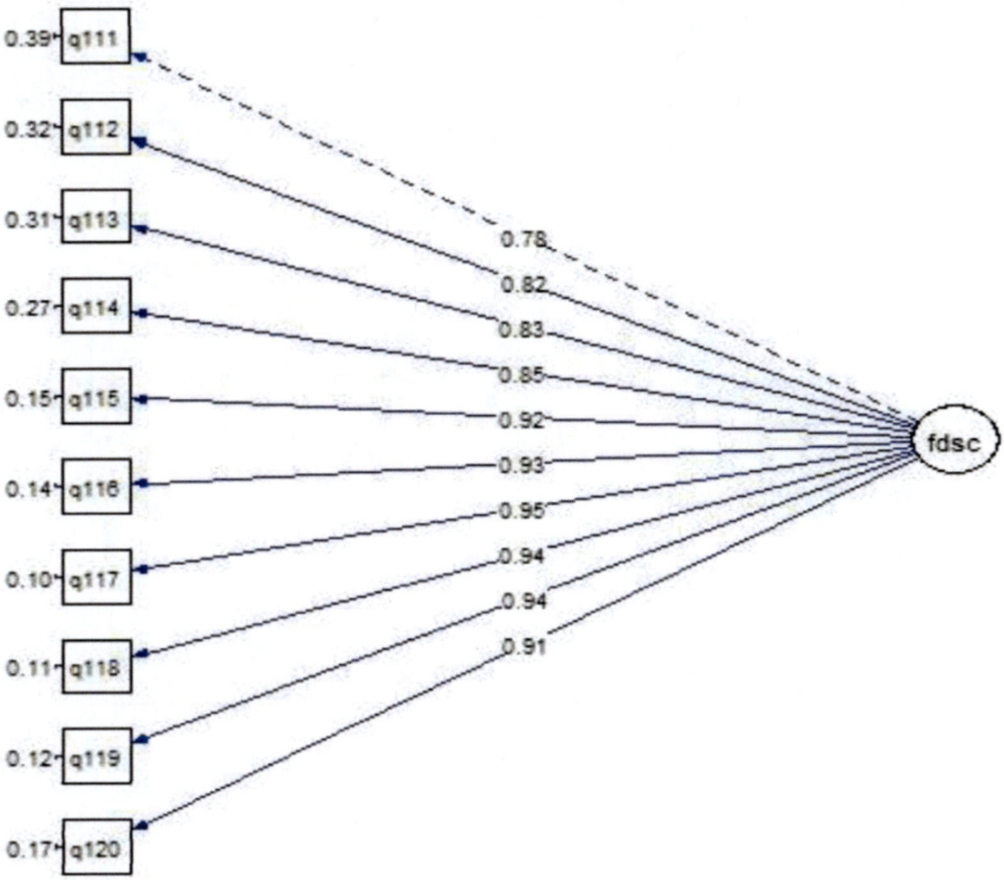

Food insecurity (fdsc), factor loadings (middle) and errors (end).

Figure 3. Food insecurity factor model structure.

The consistency in the scale used in the ten questions of HFIAS was assessed using a reliability analysis test, yielding Cronbach alpha of $\alpha = 0.96$, indicating a high level of internal validity. Construction of latent variables for food insecurity and validation was done using EFA and CFA respectively. The results of EFA shows that one factor accounted for much of the variation in food insecurity in Windhoek informal settlements (Figure 1), implying that only one factor was sufficient to describe food insecurity. All variables contained on HFIAS indicators had sufficient factor loadings greater than 0.30 (Figure 2). According to Hair et al. (2009) factor loadings of 0.30 are sufficient for a sample size of 350 or larger. The result of the hypothesised latent model in Figure 2 was tested using CFA to confirm if it fitted the data well. Table 1 shows that the overall goodness of fit of the one-factor model was adequate with $\chi^2(35) = 679.42$, $p < 0.001$, greater than the required $\chi^2 = 200$. Additionally, the results of absolute fit suggested that the model had a good fit with a SRMR = 0.05 and a RMSEA = 0.22 (90% CI = 0.20; 0.23). Further, the results yielded a TLI = 0.85 and a CFI = 0.89, that are below the suggested cut-off point of 0.95 or greater, with a value of one indicating a perfect fit. Since these fit indexes measures comparability between the proposed model and the null model, their resulting low values might be due to low correlation in the data. Prudon (2014) argued that the values of TLI and CFI are likely to approach one, when there is high correlations among variables. The result of the final model is summarised in Figure 3.

Quality of Life Using WHOQOL-BREF

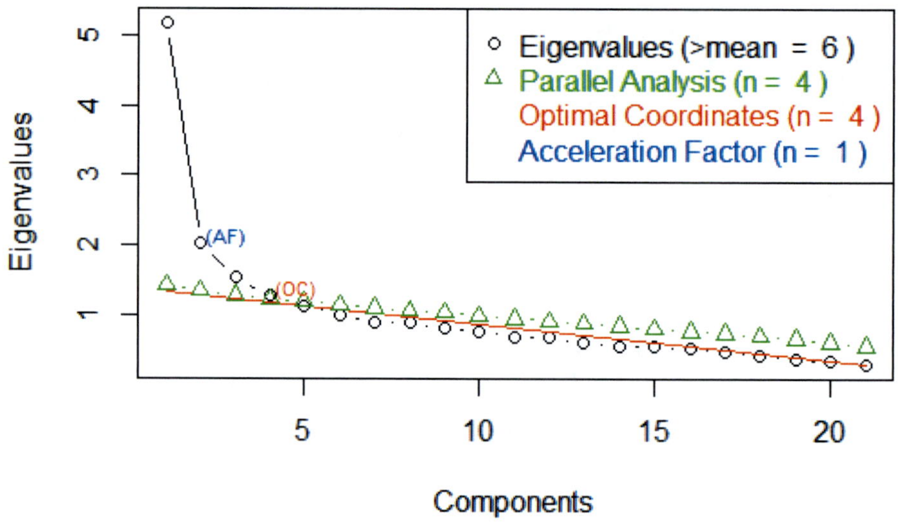

Figure 4. Scree plot of WHOQOL-BREF.

The data reduction result of WHOQOL-BREF in Figure 4 shows that four factors accounted for much of the variation in quality of life in the informal settlements of Windhoek. All 21 variables included on WHOQOL-BREF had sufficient factor loadings greater than 0.30, and the results are shown in Figure 5. The scale used in 21 items of WHOQOL-BREF yielded α= 0.80, indicating a high level of internal consistency in the ranking scale. The validity of the assumed latent model of WHOQOL-BREF domains was assessed using CFA to test if the underlying model described the data well. Table 2 displays that the overall goodness of fit of the four-factor model was sufficient with

$\chi^2(183) = 488.85$, $p < 0.001$, larger than the required $\chi^2 = 200$. The results of the absolute fit also suggested that we have a good fit with a SRMR = 0.07 and a RMSEA = 0.07 (90% CI 0.06; 0.08). However, the model yields a TLI = 0.79 and a CFI = 0.81 that were below the suggested cut-of points. This indicates that there was low correlation in the data, since the two indexes measures comparability between the proposed and the null model (Prudon, 2014). The results of the final four-factor model are summarised in Figure 6. We conclude that this is the best model for qualify of life in Windhoek informal settlements. However, there was a need to modify the model, due nonsignificant parameters in the model identified by inspecting Figure 6. The model was modified by removing psychological factor from the model.

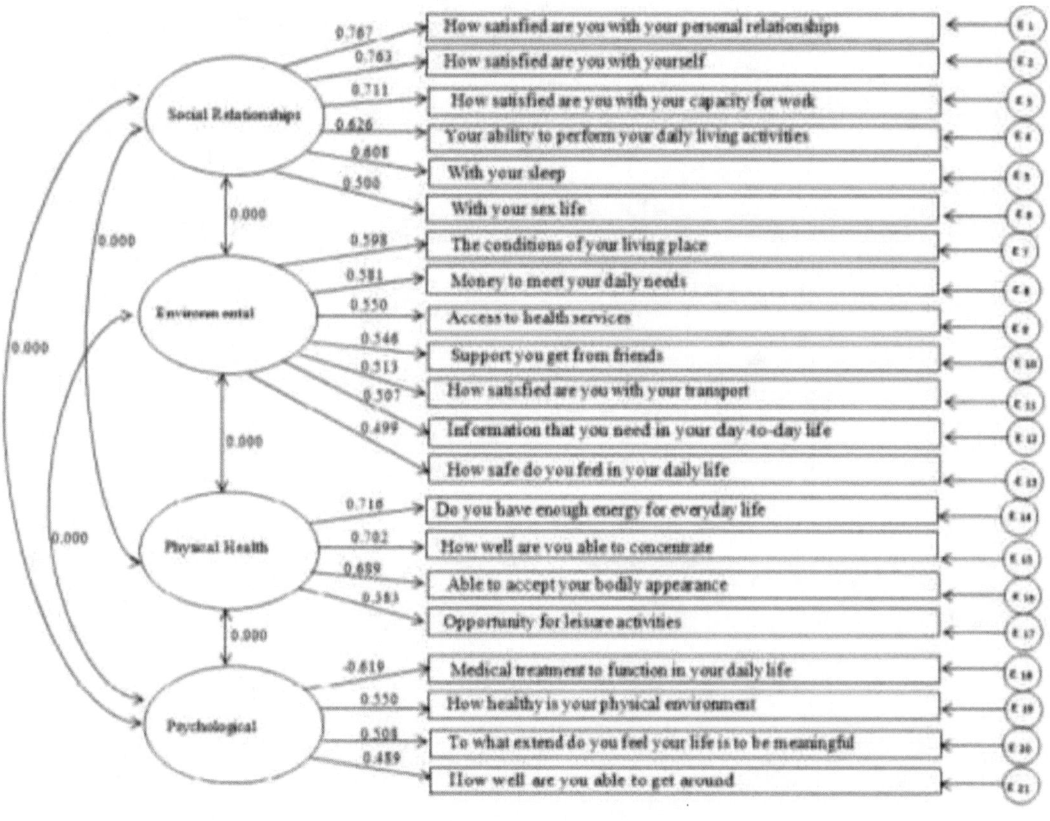

Figure 5. Summary for WHOQOL-BREF domains.

The results in Table 2 shows that the overall goodness of fit of the modified model three-factor model was adequate with $\chi^2(116) = 282.60$, $p < 0.001$, greater than the required $\chi^2 = 200$. The results of the absolute fit further suggested that the study has a best fit with a SRMR = 0.07 and a RMSEA = 0.07 (90% CI=0.06; 0.08) indicating that we have an acceptable fit. Moreover, the results of the baseline model fit yielded a TLI = 0.86 and a CFI = 0.88, that were not far from the suggested cut-off point of 0.95 or greater. Since these two fit indexes measures comparability between the proposed and the null model, their resulting low values could be due to low correlation in the data (Prudon, 2014).

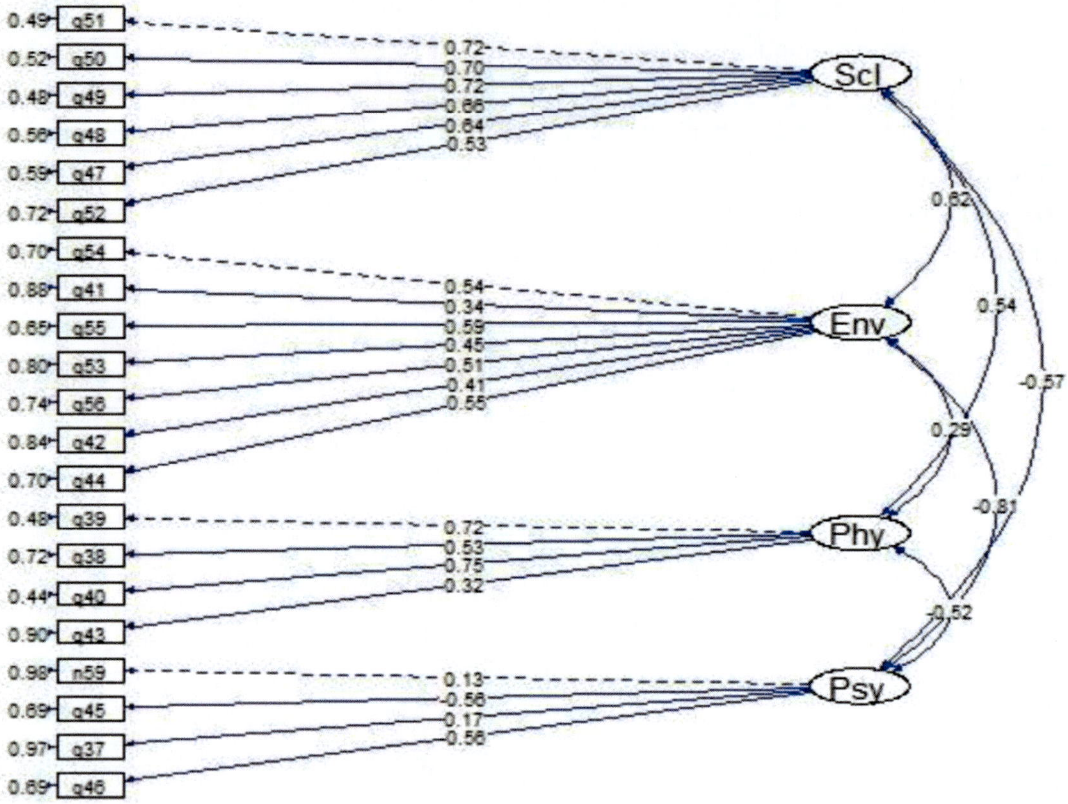

Scl (social health), **Envr** (environmental health), **Phys** (physical health), **Psyc** (physiological health), factor loadings (middle) and errors (end).

Figure 6. WHOQOL-BREF model structure (unmodified).

Table 2. Model fit analysis for WHOQOL-BREF

Indicator	Primary model	Modified model	Ideal Magnitude
Chi.sq.	488.85	282.60	> 200
RMSEA	0.07	0.07	≤ 0.05
SRMR	0.07	0.07	< 0.1
TLI	0.79	0.88	> 0.9
CFI	0.81	0.86	> 0.9
GFI	0.9	0.91	> 0.9

Table 3 presents the comparison findings of the two models. The results supported that the three-factor model appeared to be the best model compared to the four-factor model. Based on the study results, the three-factor model had lower absolute fit index compared to the four-factor model. According to the rule of thumb, we select the model that has the lowest values of the expected cross-validation index (ECVI) and Akaike Information Criterion (AIC) (Byrne, 2013). The results of the three-factor model are summarised in Figure 7.

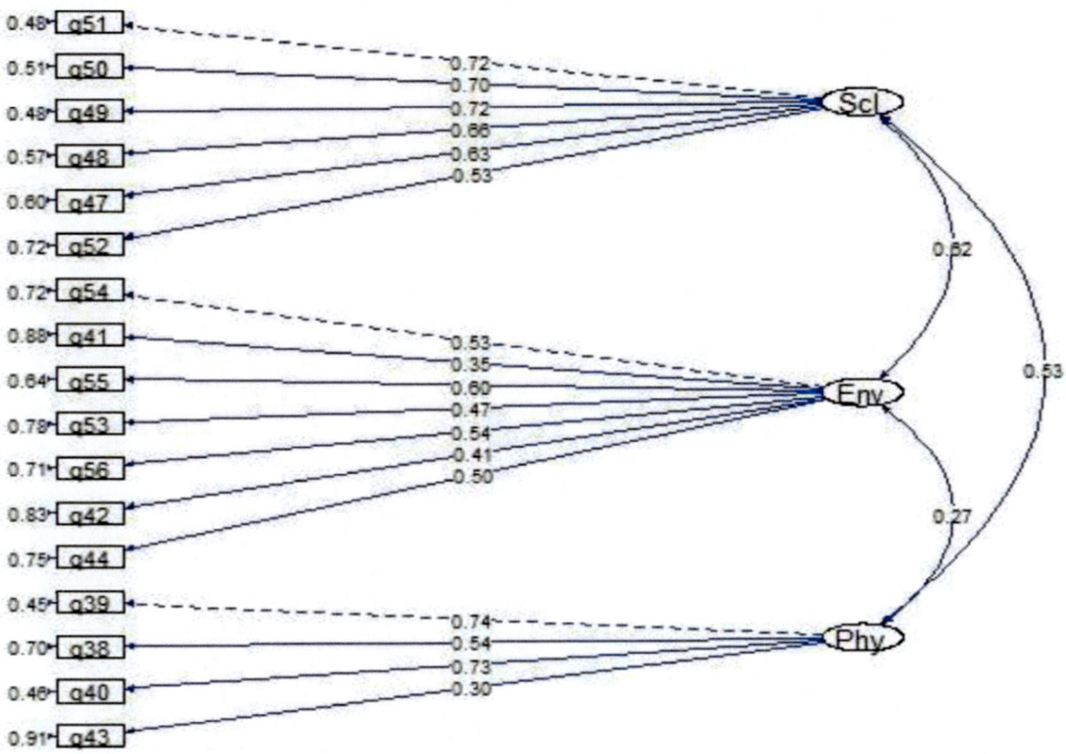

Scl (social health), Envr (environmental health), Phys (physical health), factor loadings (middle) and errors (end).

Figure 7. WHOQOL-BREF model structure (modified).

Table 3. Four-factor model vs. three-factor (modified) model (WHOQOL-BREF)

Indicator	Four-factor model	Three-factor model	Fitness evaluation
ECVI	1.77	1.07	Not ideal
AIC	20852.68	16680.48	Not ideal

Quality of Life Using EQ-5D

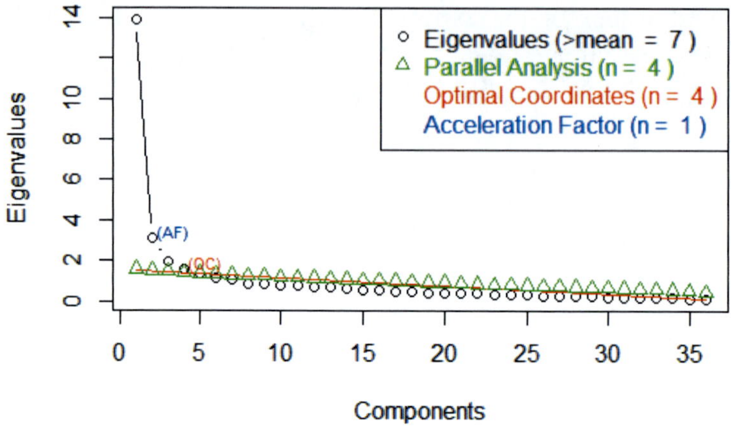

Figure 8. Scree plot of EQ-5D components.

Figure 9. Summary for EQ-5D domains.

The EFA results of EQ-5D in Figure 8 shows that four factors accounted for much of the variation in quality of life in Windhoek informal settlements. All 36 variables contained with the EQ-5D module had adequate factor loadings greater than 0.30, the results are displayed in Figure9. The scale used in 36 questions of EQ-5D had $\alpha = 0.92$, indicating a high level of internal validity. The result of the theorised model in Figure 9 was examined using CFA to verify if the measured model fit the data well.

Table 4 shows that the overall goodness of fit of the model was sufficient with $\chi^2(588) = 2693.74$, $p < 0.001$, that is greater than the required χ^2 value = 200. The results of the absolute fit also suggested that the study has a good fit with a SRMR = 0.07 and a RMSEA = 0.09 (90% CI 0.09; 0.10). However, study results found that the model had comparative fit indexes that were below the cut-off point, with a TLI = 0.44 and a CFI = 0.75. Indicating a low correlation in the data (Prudon, 2014). Thus, we conclude that the study has identified the best model for quality of life in Windhoek informal settlements. However, after inspecting Figure 10 we identified a need to modify the model due to nonsignificant parameters in the model. The results suggested that there were two manifest variables that had negative factor loadings (*"how much of a problem did you have with feeling sad, low or depressed (q78)"* and *"how much difficulty did you have with work or household activities (q60)"*). Hence, the two nonsignificant manifest variables were removed to improve the model performance.

Table 4. Model fit of EQ-5D

Indicator	Primary model	Modified model	Ideal Magnitude	
Chi.sq.	2693.74	2283.54	> 200	Yes
RMSEA	0.09	0.09	≤ 0.05	No
SRMR	0.07	0.08	< 0.1	Yes
TLI	0.44	0.76	> 0.9	No
CFI	0.75	0.78	> 0.9	No
GFI	0.69	0.711	> 0.9	Yes

Results in Table 4 shows that the overall goodness of fit of the modified model was sufficient with $\chi^2(521) = 2283.54$, $p < 0.001$, that is greater than the required $\chi^2 = 200$. The results further showed that the study had a good fit with a SRMR = 0.06 and a RMSEA = 0.09 (90% CI = 0.09: 0.10), indicating that the study has an acceptable fit. Furthermore, the results shows a TLI = 0.78 and the CFI = 0.76. Based on Table 5 the study concluded that this is the best-fit model for quality of life in Windhoek informal settlements.

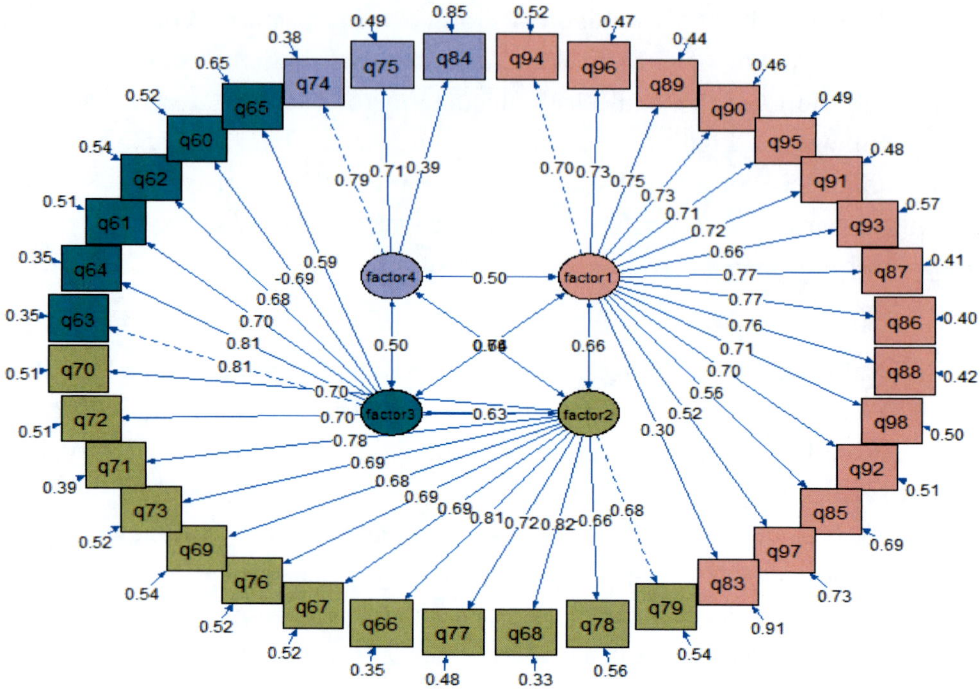

Factor 1 (pain/discomfort), factor2 (anxiety/depression), factor3 (self-care), factor 4 (usual activities), factor loadings (middle) and errors (end).

Figure 10. EQ-5D model structure (unmodified).

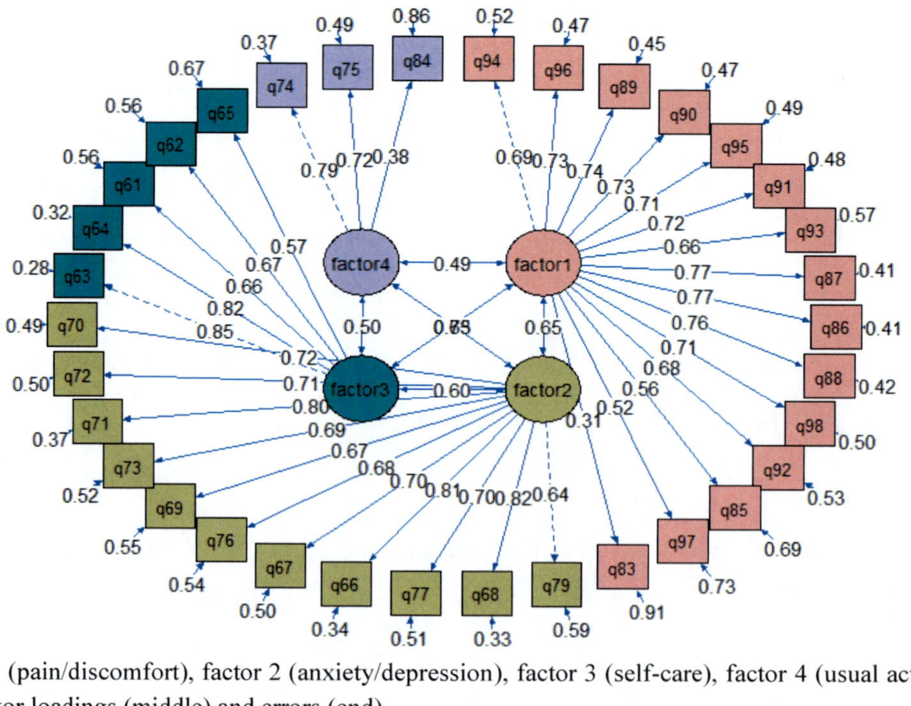

Factor 1 (pain/discomfort), factor 2 (anxiety/depression), factor 3 (self-care), factor 4 (usual activities), factor loadings (middle) and errors (end).

Figure 11. EQ-5D four-factor model structure (modified).

Table 5. Unmodified model vs. modified model EQ-5D

Indicator	Unmodified	Modified	Fitness evaluation
ECVI	7.83	6.55	Not ideal
AIC	31895.94	30613.24	Not ideal

Unmodified (full model), Modified (q78 and q60 deleted).

Structural Equation Modelling

Quality of life was linked to food insecurity using SEM based on the results achieved using CFA and we captured their level of relation by quantifying the relationship between food insecurity (access) and quality of life in Windhoek informal settlements.

Table 6. Structural equation model summary

Indicator	Primary model	Ideal Magnitude	
Chi.sq.	4527.90	> 200	Yes
RMSEA	0.07	≤ 0.05	No
SRMR	0.07	< 0.1	Yes
TLI	0.76	> 0.9	No
CFI	0.78	> 0.9	No
GFI	0.65	> 0.9	No

The results in Table 6 shows the goodness of fit of the model, the results indicated that the model was significant with $\chi^2(1741) = 4527.90$, $p < 0.001$, that is greater than the required $\chi^2 = 200$. The results of the absolute fit also suggested that we have a best fit with a SRMR = 0.07 and a RMSEA = 0.07 (90% CI=0.07; 0.08), indicating that the study had identified an acceptable model for food insecurity and quality of life in Windhoek informal settlements. Additionally, the results of the base line model found a TLI = 0.76 and the CFI = 0.78, that were below the suggested cut-off point of 0.95 or greater, with a value of one indicating a perfect fit.

Table 7 shows the regression coefficient of the latent variables. The results supported that the model has a good internal validity with α=0.86. Considering the linkages in the structural model, the results indicated that all seven measures were statistically significant. The results show that the first six (social, environmental, physical, pain/discomfort, anxiety/depression, and self-care) factors were statistically significant at $p < 0.001$, and for these factors, we tested the null hypothesis that the estimated coefficient values are not zero at 1% level of significance. On the other hand, the factor usual-activities was statistically significant at $p < 0.05$, and we tested the null hypothesis that the estimated coefficient value is not zero at 5% level of significance.

Table 7. Structural equation model

Parameter Estimates						
Information Standard Errors		Expected Standard				
Latent variables	**Estimate**	**Std.Err**	**z-value**	**P(>|z|)**	**Standardised factor loadings**	**R-Squared**
Social ~ food insecurity	-0.28	0.05	-5.37	<0.001***	-0.35	0.13
Environmental ~food insecurity	-0.37	0.06	-6.20	<0.001***	-0.51	0.26
Physical ~ food insecurity	-0.25	0.06	-4.05	<0.001***	-0.28	0.08
Pain/Discomfort ~ food insecurity	0.17	0.03	5.17	<0.001***	0.33	0.11
Anxiety/Depression ~ food insecurity	0.30	0.05	5.97	<0.001***	0.39	0.16
Self-care ~ food insecurity	0.27	0.05	5.42	<0.001***	0.35	0.12
Usual Activities ~ food insecurity	0.18	0.05	3.42	0.001**	0.24	0.06
Cronbach's alpha = 0.86						

'~'Regress on, *** $p < 0.001$, ***p* <0.05.

The results further showed that the factors social, environmental, and physical health had estimated coefficients with negative signs, implying a negative relationship with food insecurity. Conversely, the factors pain/discomfort, anxiety/depression, self-care, and usual activities had estimated coefficients with positive signs, denoting a positive relationship with food insecurity. Based on the regression-estimated coefficients the study concluded that, a unit increase in a households' food insecurity decreases social health by 0.28 units. The same can be said about environmental health, with a unit increase in food insecurity, decreasing environmental health by 0.37 units. The results further disclosed that, when food insecurity increases by one unit, the physical health of an individual in an informal settlement in Windhoek decreases by 0.25 units.

On the other hand, the results further indicated that as the household food insecurity increases by one unit, it increases pain/discomfort by 0.17 units. Similarly, a unit increase in food insecurity at household level increases the level of anxiety/depression among individuals living in Windhoek informal settlements by 0.30 units. In the same way, the results revealed that as the household food insecurity increases by one unit, individuals' level of self-care increases by 0.27 units. Furthermore, the results supported that as the household food insecurity rises by one unit, individuals in these particular households in Windhoek informal settlements increase their level of engagement in their usual activities by 0.18 units.

With regard to relative contribution of food insecurity to quality of life constructs or measures, the results showed that food insecurity had a strong negative influence on environmental health (-0.51), followed by a moderate influence on anxiety/depression (0.39) and negative moderate influence on social health (-0.35). Additionally, food insecurity also had a moderate influence on self-care (0.35), followed by moderate influence on pain/discomfort (0.33), a negative week influence on physical health (-0.28), and a weak influence on usual activities (0.24). The results of food insecurity relative contribution to quality of life are summarised in Figure 12.

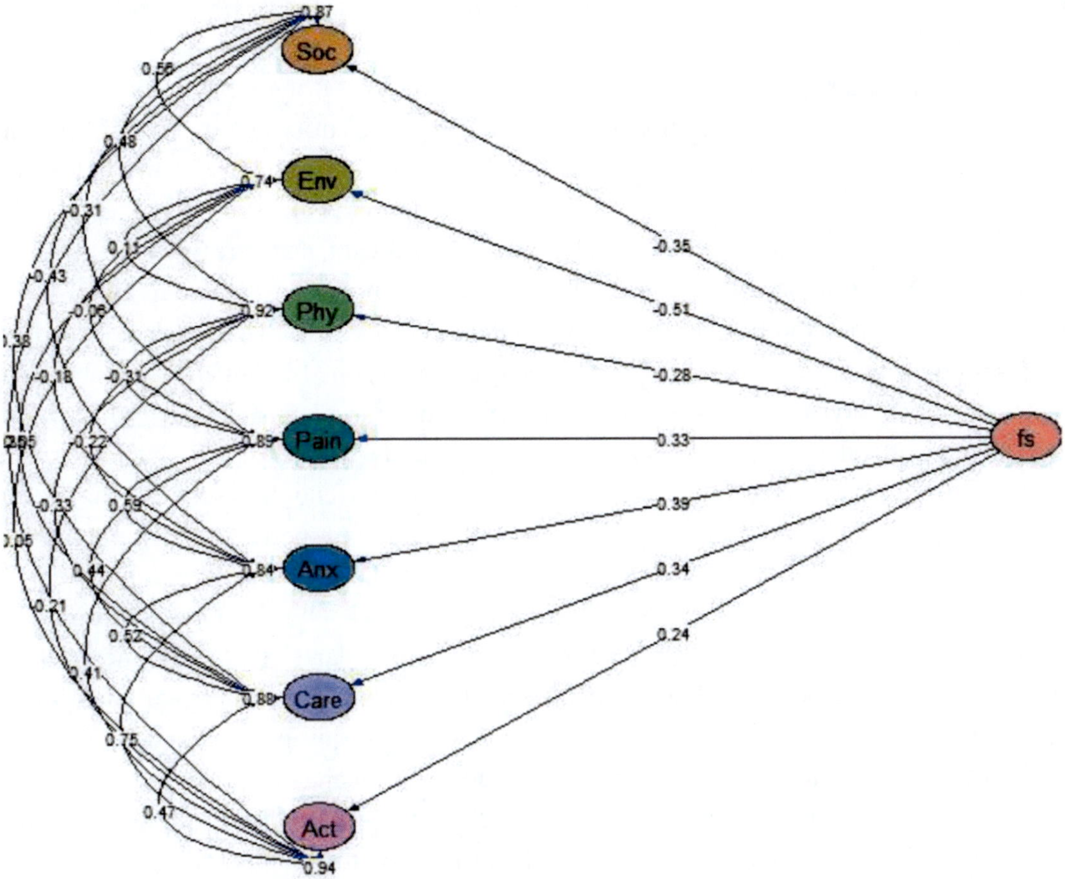

fs (food insecurity), Soc (social health), Env (environmental health), Phy (physical health), Pain (pain/discomfort), Anx (anxiety/depression), Care (self-care), Act (usual activities), factor loadings (middle), correlation between latent variables (end)

Figure 12. Structural model for food insecurity and quality of life.

Considering coefficient of determination, which is simply the standardised loading squared. The results revealed that 26% variation in environmental health was explained by food insecurity, followed by 16% variation in anxiety/depression. Furthermore, Table 7 also showed that 6% variation in usual activities was explained by food insecurity.

DISCUSSION

The study's main objective was to relate quality of life constructs to food insecurity in Windhoek informal settlements. We used SEM approach to link two means of quality of life constructs to food insecurity construct. The study revealed that all quality of life constructs were statistically significant in the model. More importantly, the study results demonstrated that food insecurity was a significant predictor of quality of life in Windhoek informal settlements. The quality measures of the study constructs were assessed by inspecting their validity.

Based on the structural relationship identified using SEM, the study examined seven hypotheses to appreciate the relationship between food insecurity and quality of life. The hypotheses were tested using standardised coefficients between quality of life constructs and the construct of food insecurity. The study fully supported all seven hypotheses.

Findings showed that food insecurity had a significant negative impact on social health. This implies that as households' food insecurity increases, their social well-being also worsens. This could be because most people do not like disclosing their households' food shortage, because they fear to be judged and questioned about their capability to supply for their households. As a result, people tend to isolate themselves from their neighbours and other close relatives. However, a study by Pendleton, Nickanor, and Pomuti (2012) in Windhoek pointed out food insecure households resort to using social protection methods in times of food shortage such as, sharing meals with neighbours and borrowing food. Even though the study by Pendleton et al. (2012) found social protection to be practiced by food insecure households in Windhoek, the findings of the current study dictate otherwise. The study findings were also similar to those of Jones (2017); Nanama and Frongillo (2012) who found that borrowing and sharing of meals with neighbours is regarded as a socially unacceptable way of acquiring food and it often lead to the feeling of alienation, shame, powerless, feeling guilty and social deprivation.

Furthermore, the study results also supported the hypothesis that, food insecurity had a significant impact on environmental health. From the study findings, it is concluded that environmental health was negatively related to food insecurity. Implying that, as households become more food insecure, their environmental or living conditions are likely to deteriorate, given the high level of poverty in informal settlements. Hence, as a households tend to spend more on food, their ability to spend on other items reduces due to limited resources that they have. Not having enough money to meet their daily need affects the living conditions of the poor, this findings were similar to those of Tarasuk, Mitchell, McLaren, and McIntyre (2013), who found that resources constrained households put more effort in reducing their food insecurity by using labour intensive strategies to acquire food at low cost and carefully portion it. They further identified that due to limited resources, these households at times tend to delay payments of other bills and seek assistance from friends or relative for them to source food.

Our analysis further supported the claim that food insecurity had a significant negative impact on physical health. This suggests that as the household food insecurity increases, individual's physical health is also likely to decrease. This is an indication that insufficient food intake can results in poor health outcomes (Nanama & Frongillo, 2012).

A study by Weigel et al. (2016) found that food insecurity is related to chronic diseases that affect physical well-being. Similarly, Black et al. (2008) also found that undernutrition pose a significant threat on well-being. They also argued that over-time lack of nutrition can cause the body to weaken, and thus distressing the functioning of the body organs. This findings were also similar to those of Friel and Ford (2015) who found that food insecurity was associated with poor mental development among most children in the US and it was also linked to lower participations in jobs requiring physical strength among adults.

On the other hand, the study tested a hypothesis if whether; food insecurity had a significant impact on individual's pain/discomfort. The study findings suggest that, food insecurity had a significant positive impact on pain/discomfort. This implies that as households' food insecurity increases, individuals in these households are likely to experience pain/discomfort that is triggered by stress caused by food shortage. Likewise, Weigel et al. (2016) also found that individuals from food insecure households are reported to have had experienced pain/discomfort/tiredness and chest pain.

Additionally, the study found that food insecurity had a significant impact on anxiety/depression. This suggests that as households' food insecurity increases, individuals in these households are likely to be anxious or depressed. This findings agreed with those of Leung, Epel, Willett, Rimm, and Laraia (2014), who found that adult from food insecure households were likely to experience troubles in sleeping, feeling, and hopeless. Other studies by Atuoye and Luginaah (2017); Weigel et al. (2016) also made similar discovery, when they found that food insecurity was the main cause of depression and mental distress.

Furthermore, the study findings also supported the claim that, food insecurity had a significant positive impact on self-care. Heerman et al. (2016) made a similar discovery when they reported that food insecurity was significantly associated with self-care. However, Weigel et al. (2016) claimed that food insecurity impede individuals' ability to take care of themselves, and they found it to be very common in women. They pointed out that in times of food shortage; women spend a lot of their time searching for good quality food at cheaper prices from one place to another. On top of that, they also spend much of their time preparing food and caring for other affected household members, other than taking care of themselves.

The study also supported that food insecurity had a significant impact on usual activities. This implies that as households become food insecure in Windhoek informal settlements, individuals living in these households are more likely to continue engaging in their usual activities. However, this findings are in contrast with findings of To, Frongillo, Gallegos, and Moore (2014) who found that food insecure individuals were less likely to engage in usual physical activities, than their counter parts from food secure households.

Our study has several limitations. Firstly, SEMs are only valid when certain assumptions are met, and model identification solely depends on model fit indices. Nonetheless, there are no standard grounds to infer that the fitted model is in fact the true model. Even though, the study has identified models that fitted the data well, there might have been insignificant factors that were included in the models. Secondly, the study encountered the issues of specification errors. Some latent variables in the path models contributed to other constructs other than on their intended constructs, resulting in some manifest and latent variables to be excluded from the final model. Thirdly, the study used exploratory data analysis method to identify the models other than the theoretical approach, and model modifications were not theoretically motivated. Lastly, the issue of missing values and omitted variables in the analysis resulted in the loss of statistical power due to reduced sample size. Hence, the study only included some variables; however, there might have been other multilevel community factors and household or individual factors that may influence the relationship between food insecurity and quality of life, and such variables were excluded from the models.

CONCLUSION AND RECOMMENDATIONS

In conclusion, the study found that food insecurity was a significant predictor of quality of life outcomes in Windhoek informal settlements. The study further supported that food insecurity plays a predominant role in determining individuals' well-being. Equally important, the study discovered that the high level of urban poverty in Windhoek's informal settlements has affected households' food security and this has resulted in poor quality life outcomes among individuals living in these areas.

The study recommends that, there is a need to re-evaluate food insecurity prevalence in these informal settlements to assess if the situation has changed after the government had introduced the food bank programme in these areas in the second half of 2016. The programme was introduced with the aim to provide food for the needy and reduce youth unemployment in the informal settlements. Additionally, the study recommends the government to give priority to policies addressing poverty and inequality, in order to end poverty. The study further recommends the government to explore other new ways to alleviate poverty and it need to create partnerships with non-governmental organisations to formulate and implement projects that are aimed to improve the livelihoods of people living in informal settlements around Windhoek to eradicate poverty, end hunger, and maintain good health for all by 2030.

APPENDIX A

R Sample Codes

Confirmatory Factor Analysis

```
install.packages(c("lavaan", "semPlot", "semTools", "nonnest2", "MVN", "GPArotation",
    "gridExtra", "knitr", "dplyr", "tidyr", dependencies =T ))
library (lavaan);library (semPlot); library(semTools); library (nonnest2); library(MVN);
    library (lattice)
library (GPArotation); library(gridExtra); library(qgraph); library(knitr); library(dplyr);
    library(tidyr)
thesisdata<-read.csv("C:\\Users\\tshinyemba\\Documents\\Academics\\Thesis\\
Thesis_QoL_FS_Tobias_27_Sept_17_With_Factors.csv", header=T, sep=",")
```

Fitting Food Insecurity Model

```
#fitting a one-factor CFA model
#Specifying the one-factor model
FS.model1<-'fdsc=~q111+q112+q113+q114+q115+q116+q117+q118+q119+q120'
#Performing the regression and storing the regression model
fit1f<-cfa(FS.model1, data=thesisdata)

#obtaining the regression coefficients
summary(fit1f, fit.measures=TRUE, standardized=TRUE)

#Obtaining the goodness of fit indicators of the model
fitMeasures(fit1f, c("chisq", "rmsea", "srmr", "gfi", "ecvi"))
    reliability(fit1f)
inspect(fit1f)

# Obtaining the factor loadings of the one-factor model
parameterEstimates(fit1f, standardized = TRUE, ci = TRUE, level = 0.95) %>% filter(op
= = " = ~") %>% select('Latent Factor' = lhs, Indicator = rhs, B = est, SE = se, Z = z,
'p-value' = pvalue, Beta = std.all) %>% kable(digits = 3, format = "pandoc", caption =
"Factor Loadings") # Summarises the table
#plotting the one-factor model
semPaths(fit1f, what = "paths", whatLabels = "std", style = "lisrel", layout = "tree",
intercepts = TRUE, residuals = TRUE, edge.label.cex = 0.8, thresholds = TRUE,
intStyle = "multi", rotation = 4, curve = 2, curvature = 1, nCharNodes = 4, nCharEdges =
3, edge.color = "blue", reorder = TRUE, structural = FALSE, equalizeManifests =
```

FALSE, optimizeLatRes = TRUE, node.width = 1.1, edge.width = 0.5, shapeMan = "rectangle", shapeLat = "ellipse", shapeInt = "triangle", sizeMan = 4, sizeInt = 4, sizeLat = 6, node.label.cex = .5, edge.label.color = "black")

Food insecurity END/

Acknowledgments

Lawrence Kazembe and Ndeyapo Nickanor were supported by National Commission for Research, Science and Technology grant number: NCRST/1nc/0814/010.

References

Altman, M., Hart, T., & Jacobs, P. (2009). Household Food Security Status in South Africa. *Agrekon, 48*(4), 345-361.

Atuoye, K. N., & Luginaah, I. (2017). Food as a social determinant of mental health among household heads in the Upper West Region of Ghana. *Social science & medicine, 180*, 170-180. doi:https://doi.org/10.1016/j.socscimed.2017.03.016

Black, R. E., Allen, L. H., Bhutta, Z. A., Caulfield, L. E., De Onis, M., Ezzati, M., . . . Rivera, J. (2008). Maternal and child undernutrition: global and regional exposures and health consequences. *The lancet, 371*(9608), 243-260. doi:https://doi.org/10.1016/S0140-6736(07)61690-0

Buhaug, H., & Urdal, H. (2013). An urbanization bomb? Population growth and social disorder in cities. *Global Environmental Change, 23*(1), 1-10. doi:https://doi.org/10.1016/j.gloenvcha.2012.10.016

Byrne, B. M. (2013). *Structural Equation Modeling with Mplus: Basic Concepts, Applications, and Programming*. New York: Routledge: Taylor & Francis Group.

Cella, D. F. (1994). Quality of Life: Concepts and definition. *Journal of pain and symptom management, 9*(3), 186-192. doi:https://doi.org/10.1016/0885-3924(94)90129-5

Chung, H.-K., Kim, O. Y., Kwak, S. Y., Cho, Y., Lee, K. W., & Shin, M.-J. (2016). Household Food Insecurity Is Associated with Adverse Mental Health Indicators and Lower Quality of Life among Koreans: Results from the Korea National Health and Nutrition Examination Survey 2012–2013. *Nutrients, 8*(12), 819. doi:https://doi.org/10.3390/nu8120819

Coates, J., Swindale, A., & Bilinsky, P. (2007). Household Food Insecurity Access Scale (HFIAS) for Measurement of Food Access: Indicator Guide (v.3). Retrieved 12 June 2017, from USAID http://www.fao.org/fileadmin/user_upload/eufao-fsi4dm/doc-training/hfias.pdf

Crush, J., & Frayne, B. (2010) The Invisible Crisis: Urban Food Security in Southern Africa. *Urban Food Security Series No. 1*. Kingston and Cape Town: Queen's University and AFSUN.

Crush, J., Frayne, B., & Pendleton, W. (2012). The Crisis of Food Insecurity in African Cities. *Journal of Hunger & Environmental Nutrition, 7*(2-3), 271-292. doi:https://doi.org/10.1080/19320248.2012.702448

Dunteman, G. H. (1989). Principal component analysis. Quantitative applications in the social sciences series (vol. 69). Thousand Oaks: SAGE Publications.

Everitt, B., & Hothorn, T. (2011). *An Introduction to Applied Multivariate Analysis with R*. New York: Springer Science + Business Media.

Everitt , B. S., & Dunn , G. (2001). *Applied multivariate data analysis* (Vol. 2): Wiley Online Library.

FAO. (2003). Trade reforms and Food Security: Concepualizing the linkages. Retrieved 5 May 2017, from Food and Agriculture Organisation of the United Nations http://www.fao.org/3/a-y4671e.pdf

FAO, IFAD, UNICEF, WFP, & WHO. (2017). The State of Food Security and Nutrition in the World 2017. Retrieved 7 March 2018, from UNICEF http://www.fao.org/3/a-I7695e.pdf

Felce, D., & Perry, J. (1995). Quality of life: Its definition and measurement. *Research in developmental disabilities, 16*(1), 51-74. doi:https://doi.org/10.1016/0891-4222(94)00028-8

Fox, S. (2014). The political economy of slums: Theory and evidence from Sub-Saharan Africa. *World development, 54*, 191-203. doi:https://doi.org/10.1016/j.worlddev.2013.08.005

Franklin, S. B., Gibson, D. J., Robertson, P. A., Pohlmann, J. T., & Fralish, J. S. (1995). Parallel analysis: a method for determining significant principal components. *Journal of Vegetation Science, 6*(1), 99-106. doi:https://doi.org/10.2307/3236261

Friel, S., & Ford, L. (2015). Systems, Food Security and Human Health. *Food Security, 7*(2), 437-451. doi:https://doi.org/10.1007/s12571-015-0433-1

Garrett, J. L. (2000) Achieving Urban Food and Nutrition security in the Developing World. In J. L. Garrett & M. T. Ruel (Series Ed.), *A 2020 vision for Food, Agriculture, and the Environment*. Wandshington D.C: International Food Policy Research Institute.

Garvelink, W. J., & Wedding, K. (2013). *Nutrition and Food Security in the City*. Retrieved from Washington, DC: https://www.csis.org/analysis/nutrition-and-food-security-city

Gregory, P. J., Ingram, J. S. I., & Brklacich, M. (2005). Climate change and food security. *Philosophical Transactions of the Royal Society B: Biological Sciences, 360*(1463), 2139-2148. doi:https://doi.org/10.1098/rstb.2005.1745

Gundersen, C. (2013). Food Insecurity is an Ongoing National Concern. *Advances in Nutrition: An International Review Journal, 4*(1), 36-41. doi:https://doi.org/10.3945/an.112.003244

Hair, J. F., Anderson, R. E., Babin, B. J., & Black, W. C. (2009). *Multivariate data analysis* (7 ed.). New Jersey: Pearson.

Hayton, J. C., Allen, D. G., & Scarpello, V. (2004). Factor Retention Decisions in Exploratory Factor Analysis: a Tutorial on Parallel Analysis. *Organizational Research Methods, 7*(2), 191-205. doi:https://doi.org/10.1177/1094428104263675

Heerman, W., Wallston, K., Osborn, C., Bian, A., Schlundt, D., Barto, S., & Rothman, R. (2016). Food insecurity is associated with diabetes self-care behaviours and glycaemic control. *Diabetic Medicine, 33*(6), 844-850. doi:https://doi.org/10.1111/dme.12896

Hu, L. t., & Bentler, P. M. (1999). Cutoff criteria for fit indexes in covariance structure analysis: Conventional criteria versus new alternatives. *Structural Equation Modeling: A Multidisciplinary Journal, 6*(1), 1-55. doi:https://doi.org/10.1080/10705519909540118

Indongo, N., Angombe, S., & Nickanor, N. (2013). *Urbanisation in Namibia: Views from semi-informal and Informal settlement*. Retrieved from Windhoek: www.unam.edu.na/wp-content/uploads/2014/09/Urbanisation-in-Namibia.pdf

Ivers, L. C., & Cullen, K. A. (2011). Food Insecurity: Special Considerations for Women. *The American journal of clinical nutrition, 94*(6), 1740S-1744S. doi: https://doi.org/10.3945/ajcn.111.012617

Jha, D. K., & Tripathi, V. (2014). Quality of Life in Slums of Varanasi City: A Comparative study. *Transactions, 36*(2). doi:https://ssrn.com/abstract=2597839

Johnson, R. A., & Wichern, D. W. (2007). *Applied multivariate statistical analysis* (6 ed.). New Jersey: Prentice-Hall.

Jones, A. D. (2017). Food Insecurity and Mental Health Status: A global Analysis of 149 Countries. *American Journal of Preventive Medicine*. doi:https://doi.org/10.1016/j.amepre.2017.04.008

Kimani-Murage, E. W., Schofield, L., Wekesah, F., Mohamed, S., Mberu, B., Ettarh, R., . . . Ezeh, A. (2014). Vulnerability to Food Insecurity in Urban Slums: Experiences from Nairobi, Kenya. *Journal of Urban Health : Bulletin of the New York Academy of Medicine, 91*(6), 1098-1113. doi:https://doi.org/10.1007/s11524-014-9894-3

Leung, C. W., Epel, E. S., Willett, W. C., Rimm, E. B., & Laraia, B. A. (2014). Household Food Insecurity Is Positively Associated with Depression among Low-Income Supplemental Nutrition Assistance Program Participants and Income-Eligible Nonparticipants. *The Journal of nutrition, 145*(3), 622-627. doi:https://doi.org/10.3945/jn.114.199414

Naicker, N., Mathee, A., & Teare, J. (2015). Food insecurity in households in informal settlements in urban South Africa. *SAMJ: South African Medical Journal, 105*(4), 268-270. doi:http://dx.doi.org/10.7196/SAMJ.8927

Nanama, S., & Frongillo, E. A. (2012). Altered social cohesion and adverse psychological experiences with chronic food insecurity in the non-market economy and complex households of Burkina Faso. *Social science & medicine, 74*(3), 444-451. doi:https://doi.org/10.1016/j.socscimed.2011.11.009

NSA. (2013). *Namibia 2011 Population and Housing Census Main Report*. Retrieved from http://cms.my.na/assets/documents/p19dmn58guram30ttun89rdrp1.pdf

Pendleton, W., Crush, J., & Nickanor, N. (2014). Migrant Windhoek: Rural–Urban Migration and Food Security in Namibia. *Urban Forum, 25*(2), 191-205. doi:http://dx.doi.org/10.1007/s12132-014-9220-x

Pendleton, W., Nickanor, N., & Pomuti, A. (2012) The State of Food Insecuritity in Windhoek, Namibia. *Urban Food Security Series (14)*. Capet Town: African Food Security Urban Network (AFSUN).

Prudon, P. (2014). Confirmatory factor analysis: a brief introduction and critique. *Qualtrics, P. UT, USA*.

Rencher, A. C. (2002). *Methods of Multivariate Analysis* (2 ed.). New York: John Wiley & Sons Inc.

Sanchez, B. N., Budtz-Jorgensen, E., Ryan, L. M., & Howard, H. (2005). Structural Equation Models: A Review with Applications to Environmental Epidemiology. *Journal of the American Statistical Association, 100*(472), 1443-1455. doi:https://doi.org/10.1198/016214505000001005

Tacoli, C., McGranahan, G., & Satterthwaite, D. (2015). *World Migration Report 2015: Urbanisation, Rural-urban Migration and Urban Poverty* (Background paper). Retrieved from London: http://pubs.iied.org/pdfs/10725IIED.pdf

Tarasuk, V., Mitchell, A., McLaren, L., & McIntyre, L. (2013). Chronic Physical and Mental Health Conditions among Adults May Increase Vulnerability to Household Food Insecurity. *The Journal of nutrition, 143*(11), 1785-1793. doi:https://doi.org/10.3945/jn.113.178483

Teo, T., Tsai, L. T., & Yang, C.-C. (2013) Application of Structural Equation Modeling in Educational Research and Practice. *Vol. 7. Applying Structural Equation Modeling (SEM) in Educational Research: An Introduction*. Rotterdam: Sense Publisher.

To, Q. G., Frongillo, E. A., Gallegos, D., & Moore, J. B. (2014). Household Food Insecurity Is Associated with Less Physical Activity among Children and Adults in the US Population. *The Journal of nutrition, 144*(11), 1797-1802. doi:https://doi.org/10.3945/jn.114.198184

UN-DESA. (2015). *World Urbanization Prospects: The 2014 Revision, Highlights (ST/ESA/SER.A/352)*. Retrieved from New York: https://esa.un.org/unpd/wup/Publications/Files/WUP2014-Highlights.pdf

UN-Habitat. (2015). *22 - Informal Settlements*. Retrieved from https://unhabitat.org/habitat-iii-issue-papers-22-informal-settlements/

UN-Habitat. (2016). *Urbanization and Development: Emerging Futures; World Cities Report 2016*. Retrieved from https://unhabitat.org/wp-content/uploads/2014/03/WCR-%20Full-Report-2016.pdf

van Reenen, M., & Janssen, B. (2015). *EQ-5D-5L User guide: Basic information on how to use the EQ-5D-5L instrument (V. 2.1)*. Retrieved from Rotterdam: https://euroqol.org/wp-content/uploads/2016/09/EQ-5D-5L_UserGuide_2015.pdf

Verma, K. (2017). Quality of Life in Informal Settlement: Jaipur *International Journal on Emerging Technologies, 8*(1), 586-591.

Weigel, M. M., Armijos, R. X., Racines, M., Cevallos, W., & Castro, N. P. (2016). Association of Household Food Insecurity with the Mental and Physical Health of Low-Income Urban Ecuadorian Women with Children. *Journal of Environmental and Public Health, 2016*, 1-14. doi:http://dx.doi.org/10.1155/2016/5256084

Wekesa, B. W., Steyn, G. S., & Fred-Otieno, F. A. O. (2011). A review of physical and socio-economic characteristics and intervention approaches of informal settlements. *Habitat international, 35*(2), 238-245. doi:https://doi.org/10.1016/j.habitatint.2010.09.006

WHO. (1996). WHOQOL-BREF: Introduction, Administration, Ccoring and Generic version of the Assessment: Field Trial version, December 1996. Geneva: World Health Organization.

Wokekoro, E., & Owei, O. B. (2014). An assessment of residential quality of life in informal settlements in Port Harcourt municipality. *Journal of Humanities and Social Science, 9*(3), 11-26.

Zelterman, D. (2015). *Applied Multivariate Statistics with R*. Cham: Springer.

In: Statistical Modelling …
Editors: N-B. Kandala and L. Kazembe
ISBN: 978-1-53615-981-3

Chapter 7

A Zero-Truncated Negative Binomial Regression Model for Dietary Diversity in Namibian under-5 Children

Lillian Pazvakawambwa, PhD and Ndeyapo M. Nickanor, PhD*
Department of Statistics and Population Studies,
University of Namibia, Windhoek, Khomas, Namibia

Abstract

A nutritious diet in appropriate quantity and quality is important for children to meet their nutrient needs, support growth and reduce morbidity and mortality. This study aimed at exploring appropriate count models to establish determinants of dietary diversity (a measure of food security) in Namibian under-5 children based on secondary data from the NDHS 2013, a national cross-sectional survey of representative woman in the reproductive age group (15-49) from all 13 provinces. Based on a dietary diversity score (DDS) calculated from consumption of 12 food groups. Two zero-truncated count regression models (Poisson and Negative Binomial) were explored and adjudicated based on AIC. The best model was then deployed to model dietary diversity with independent variables such as socio-economic status, region, mother's level of education, religion, ethnicity, marital status and place of residence. The Zero Truncated Negative Binomial regression results indicated significant differentials in dietary diversity due to maternal age, level of education, religion, socio-economic status and cultural factors. Generally, the majority of Namibian children consumed a diet with little diversity. Policy efforts could focus on further regulating healthy food prices to increase household access to diverse diets, boost agricultural production, and step-up maternal education to influence positive dietary decision making.

* Corresponding Author's Email: lpazvakawambwa@unam.na.

Keywords: dietary diversity, food security, children, Namibia

INTRODUCTION

The three major components underlying food security are food availability, accessibility and utilization. Critical to ensuring adequate intake of essential nutrients that promotes good health and well-being is the intake of a variety of foods and food groups (Ruel, 2002). Dietary diversity is not only viewed as a proxy measure of food security (Hoddinott &Yohannes, 2002), but has become an important measure of the diverse diet and by implication an indicator of nutritional adequacy and quality (Foote, Murphy, Wilkens, Basiotics and Carlos, 2004), thus nutritional deficiencies are not only as a result of low food quantity consumed but also of poor quality and diversity. Consumption of a higher number of food items and food groups is associated with positive health outcomes such as birth weight, child anthropometric status (Hoddinott &Yohannes, 2002); improved nutritional adequacy it also highly correlated with caloric and protein adequacy. Among adults, diversified diet is associated with reduced complications of diabetes (Wahlqvist & Myers, 1989) reduced incidence of cancers (Lucenteforte, Garavello, Bosetto, Talamini, Zambon, Fransceshi, Negri, and La Vecchia, 2008). A number of studies have shown, using Dietary diversity score (DDS) which counts the number of food groups consumed over a specified reference period to assess dietary diversity, that DDS is positively associated with overall dietary quality (Steyn, Nel, Nantel, Kennedy, and Labadarios, 2006; Steyn, Nel, Labadarios, Maunder, and Kruger, 2014; Kennedy, Pedro, Seghieri, Nantel, and Brouwer, 2007).

In developing countries, deficiencies of selected micronutrients prevalent among young children and women of childbearing age were found to be associated with birth defects, growth restriction, impaired cognition and increased morbidity and mortality (Black, Allen, Bhutta, Caufield, de Onis, Ezzati, Mathers, and Rivera, 2008). A limited or lack of dietary diversity is primarily responsible these deficiencies particularly among population groups of concerns in developing countries.

Malnutrition in Namibia is widespread. The NDHS 2013 indicated that 24% of children under 5 years were stunted (short for age); 6% were wasted (thin for height) and 13% underweight (thin for age), an indication of both acute and chronic under nutrition, possibly due to lack of dietary diversity. Literature on breastfeeding in Namibia is limited, with the only existing one (Indongo & Mutorwa, 2017), indicating a much lower practice of exclusive breastfeeding practise 19% than that shown by the NDHS 2013 of 49%. Understanding dietary intakes among children is important, as breast milk from the age of six months is no longer able to meet the energy and micronutrient needs of children (Rakotonirainy, Razafindratovo, Remonja, Rasoloarjaona, Piola, Raharintsoa, and Randremanana, 2018), dietary diversity is shown to improve energy and nutrient intake, and by implication dietary adequacy as well as child growth, development and nutrition (Faber, Laubscher, and Berti, 2016; Nti, 2011); Temesgen, Yeneabat, and Teshome, 2018; Amugsi, Dimbuene, Kimani-Murage, Mberu, and Ezech, 2017). Furthermore, literature on the relationship between dietary diversity and child anthropometry has shown that a diet that lacks sufficient diversity is associated with poor anthropometric outcome (Gausman, Perkins, Lee H-Y, Merjia-Guevara, Nam, Lee J-K, Oh, and Subramanian, 2018). The purpose of the present study was to assess dietary diversity among children under five in Namibia as a proxy of food security. In addition, we examined the relationships between dietary diversity and socio-demographic variables.

METHODS

Regression Models for Count Data

In a number of regression applications, the response variable may be given in the form of event counts, for example, the number of traffic accidents per week at a certain road intersection, or even the number of food groups consumed (dietary diversity score) by a child to assess dietary diversity.

The response variable Y can be considered as a non-negative integer-valued random variable $\{Y \in 0,1,2,3,\dots\}$. At times an upper bound exists for the response because the number of occurrences may be finite e.g., potential medical aid claims but usually the number is often very large and therefore considered as irrelevant in modelling.

The basic model for count data is the Poisson distribution, which can be treated within the framework of a generalized linear model. The Poisson distribution was derived as a limiting case of the Binomial distribution. The discrete random variable Y is distributed with intensity parameter $\lambda, \lambda > 0$ if the probability density function is given by

$$P(Y = y) = \begin{cases} \frac{\lambda^y e^{-\lambda}}{y!} \; y \in \{0,1,2,3,\dots\} \\ 0 \text{ otherwise} \end{cases} \text{abbreviated } Y \sim P(\lambda)$$

The mean and variance of the Poisson distribution are equal $E[Y] = Var[Y] = \lambda$ and this property is referred to as the equi-dispersion property of the Poisson distribution. Unlike the normal distribution, whereby the mean and the variance are not linked, the Poisson distribution implicitly models stronger variability for larger means, a property that is often found in real life data. However, with real life data the variance often exceeds the mean, an effect referred to as over-dispersion which calls for separate modelling.

Poisson Regression Model

Let (y_i, x_i) denote n independent observations and assume the mean $\mu_i = E(y_i|x_i)$, where y_i is a response and x_i is a predictor, then $y_i|x_i \sim P(\mu_i)$ so that $\mu_i = h(\boldsymbol{x}_i^T\boldsymbol{\beta})$or $g(\mu_i) = \boldsymbol{x}_i^T\boldsymbol{\beta}$ where g is a known link function of $\mu = g^{-1}$denotes the response function. Since the Poisson distribution is from the exponential family, the model is a generalized linear model. The most widely used model uses the canonical link function by specifying

$$\mu_i = \exp(\boldsymbol{x}^T\boldsymbol{\beta})$$

called a log linear model since the logarithm of the conditional mean is linear in the parameters. This log-linear form makes interpretation of the parameters easy. The model implies that the conditional mean given $x_t = (x_1, x_2, \cdots, x_p)$ has a multiplicative form given by $h(\boldsymbol{x}) = \exp(\boldsymbol{x}^T\boldsymbol{\beta}) = e^{x_1\beta_1 \cdots x_p\beta_p}$.

Thus e^{β_j}represents the multiplicative effect on $\mu(\boldsymbol{x})$ if the variable x_j changes by one unit to $x_j + 1$, (given that the rest of the variables are fixed) one obtains $\frac{\mu(x_1,\cdots x_{j+1},\cdots x_p)}{\mu(x_1,\cdots x_j,\cdots x_p)} = e^{\beta_j}$ or equivalently $\log \mu(x_1, \cdots x_{j+1}, \cdots x_p) - log\mu(x_1, \cdots x_j, \cdots x_p) = \beta_j$ where β_j is the change in the log means if x_j increases by one unit, e^{β_j} is the multiplicative effect, which is easier to interpret because it directly effects upon the mean (Lawless, 1987).

Negative Binomial Model

A more flexible model than the Poisson model is the Negative Binomial distribution which is a two parameter distribution and can model over-dispersed counts. The negative binomial model can be derived as a mixture of Poisson distributions. If we consider the Gamma distribution $b_i = \Gamma(v, \alpha)$ with probability density

$$f(b_i) = \begin{cases} 0 & b_i \le 0 \\ \frac{\alpha^v}{\Gamma(v)} b_i^{v-1} e^{-\alpha b_i} & b_i > 0 \end{cases}$$

The mean and variance are $E(b_i) = {}^{v}/_{\alpha}, and\ Var(b_i) = {}^{v}/_{\alpha^2}$

If one assumes for the random parameter b_i the Gamma distribution $\Gamma(v, v)$, the mean fulfils $E(b_i) = 1$ and obtains for the marginal probability

$$P(y_i) = \int f(y_i|b_i)f(b_i)db_i$$

$$= \int \left(e^{-b_i\mu_i}\frac{(b_i\mu_i)^{y_i}}{y_i!}\right)\left(\frac{v^v}{\Gamma(v)}b_i^{v-1}e^{-vb_i}\right)db_i$$

$$= \frac{\Gamma(y_i+v)}{\Gamma(v)\Gamma(y_i+v)}\left(\frac{\mu_i}{\mu_i+v}\right)^{y_i}\left(\frac{v}{\mu_i+v}\right)^{v}$$

which represents the Negative Binomial distribution $NB(v, \mu_i)$ with mean and variance given by

$$E(y_i) = \mu_i = \exp(\boldsymbol{x}_i^T\boldsymbol{\beta}) \text{ and } Var(y_i) = \mu_i + \frac{{\mu_i}^2}{v}$$

The mean is the same as for the simple Poisson distribution but the variance is greater than the Poisson variance by $\frac{{\mu_i}^2}{v}$. The Poisson case may be seen as a limiting case $(v \to \infty)$. The scaling of v is such that small values signal strong over-dispersion when compared to the Poisson model while for large values of v the model is similar to the Poisson Model and $\frac{1}{v}$ is considered the dispersion parameter. The additional parameter makes the model more flexible than the simple Poisson model (Lawless, 1987).

Zero Truncated Poisson and Negative Binomial Regression Models

The Negative Binomial model may not be appropriate if the over-dispersion is due to excess zeros because it underestimates the probability of zeros and consequently underestimates the variability present in the outcome. In such situations, alternative models such as zero inflated/hurdle models that account for over-dispersion due to excess zeros are useful. Often, because of the way data is collected, or in the case where zero counts are meaningless e.g., the number of food groups consumed by an individual can only be greater than or equal to 1, otherwise the person cannot survive. That is, our data is truncated at 0. This fact is incorporated by using a zero-truncated Poisson or zero-truncated Negative Binomial regression model. Zero-truncated models calculate the probability of response variable based on positive count data using Bayes's Theorem. The probability mass function, mean, variance, likelihood function and response surface of a zero-truncated negative binomial (ZTNB) model are shown in equation (1) to (5) respectively (Liu, Saat, Qin, and Barkan, 2013; Shonkwiler, 2016).

$$Pr(y_i|y_i > 0) = \frac{(\Gamma(y_i+\alpha^{-1})/y_i!\Gamma(\alpha^{-1}))(\alpha^{-1}/(\alpha^{-1}+\mu_i))^{\alpha^{-1}}(\mu_i/(\alpha^{-1}+\mu_i))^{y_i}}{1-(1+\alpha\mu_i)^{-\alpha^{-1}}} \tag{1}$$

$$E(y_i|y_i > 0) = \frac{\mu_i}{Pr(y_i>0)} = \frac{\mu_i}{1-(1+\alpha\mu_i)^{-\alpha^{-1}}} \tag{2}$$

$$Var(y_i|y_i > 0) = \frac{E(y_i|y_i > 0)}{Pr(y_i>0)^{\alpha}}[1 - \Pr(y_i = 0)^{1+\alpha} E(y_i|y_i > 0)] \quad (3)$$

$$L = \prod_{i=1}^{N} \frac{(\Gamma(y_i+\alpha^{-1})/y_i!\Gamma(\alpha^{-1}))(\alpha^{-1}/(\alpha^{-1}+\mu_i))^{\alpha^{-1}}(\mu_i/(\alpha^{-1}+\mu_i))^{y_i}}{1-(1+\alpha\mu_i)^{-\alpha^{-1}}} \quad (4)$$

$$\log(\mu_i) = \beta_0 + \beta_1 X_{1i} + \cdots + \beta_k X_{ki} \quad (5)$$

where $P_r(y_i|y_i > 0)$ is the probability mass function of zero-truncated negative binomial distribution, $E(y_i|y_i > 0)$ is the expectation of zero-truncated negative binomial distribution, $\text{Var}(y_i|y_i > 0)$ is the variance of zero-truncated negative binomial distribution, α is the over-dispersion parameter, L is the likelihood function, μ_i is the estimated number of food groups consumed for the ith observation, y_i is the observed food groups for the ith observation, β_k is the parameter coefficient of the kth predictor variable ($k = 0$ for intercept), X_{ki} is the value of the kth predictor variable for the ith observation.

Statistical Analysis

R packages have been developed to fit zero truncated Poisson and Negative Binomial models using the VGAM packages (Yee, 2008). Secondary data from the Namibia Demographic and Health Survey NDHS (2013), from a cross-sectional sample of representative mothers from all the regions was used. The woman responded to dietary questions regarding the feeding of their children. The responses were used to compute a dietary diversity score (DDS) given by the total number of food groups consumed by the children out of the recommended 12 (Bread, noodles, other made from grains; Potatoes, cassava, or other tubers; Eggs; Meat (beef, pork, lamb, chicken, etc); Pumpkins, carrots, squash (yellow or orange inside); Dark green leafy vegetables; Mangoes, papayas, other vitamin A fruits; Any other Fruits; Liver and heart organs; Fish or shell fish; Food made from beans, peas, and lentils; and Cheese, yorghut, and other milk products). A DDS of $<$ 5 was regarded as reflecting poor dietary diversity and poor food security.

The Zero Truncated Poisson (ZTP) and Zero Truncated Negative Binomial (ZTNB) regression models were compared based on the Akaike Information Criterion (AIC) with independent variables such as wealth index, region, mother's highest educational level, religion, ethnicity (Main Language spoken at home), marital status, Sex of household head and place of residence, to establish predictors of dietary diversity in children.

RESULTS

Background Characteristics of the Sample

The proportion of children consuming foods from the various groups are presented in Figure 7.1. The most consumed food group was "Bread, Noodles and other made from grain" group (41.2%) and the least consumed food group was "beans, peas, lentils, and others" (2.1%).

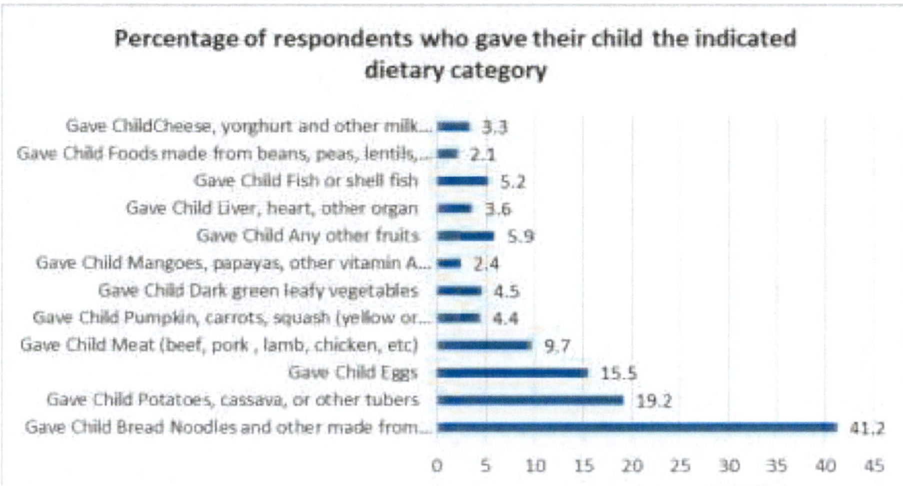

Figure 7.1. Percentage of respondents who gave their child food from the indicated dietary food group.

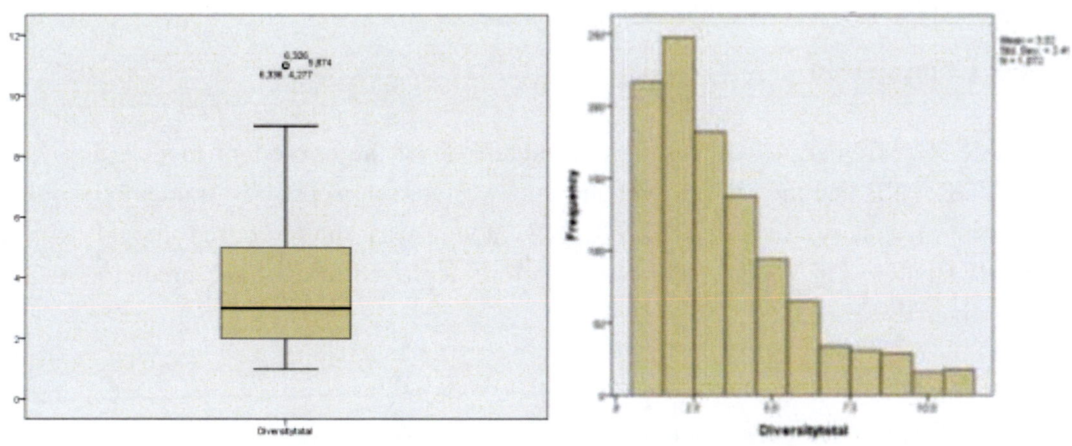

Figure 7.2. Box plot and histogram of dietary diversity scores for under 5 children.

Only 27% of the children had a good dietary diversity (DDS score below 5). From the food groups proposed by FANTA (Swindale & Paula, 2006), the dietary diversity score, which is the sum of the different food groups consumed by each child, ranged from 1 to 11 with a mean of 4 food groups out of the 12 and a standard deviation of 2. The boxplot and histogram of the dietary diversity scores are presented in Figure 7.2. The dietary diversity scores were skewed towards very few food groups with only a few children consuming a variety of more than 5 food categories.

Association between the Dietary Diversity Score and Potential Predictor Variables

Chi-square tests of association between the dietary diversity score and potential predictor variables are given in Table 7.1. The results of Chi-square tests of association at 5% level of significance indicated that dietary diversity in children was significantly associated with region (Chi-square = 95.077, $p < 0.001$); place of residence (Chi-square = 49.716, $p < 0.001$); highest educational level (Chi-square = 53.322, $p < 0.001$); wealth index (Chi-square = 72.753, $p < 0.001$); marital status (Chi-square = 19.083, $p = 0.002$); and main language spoken at home (Chi-square = 30.326, $p < 0.001$). There was no significant association between dietary diversity in children with the mother's age-group (Chi-square = 7.972, $p = 0.240$); religion (Chi-square = 4.315, $p = 0.505$) and the sex of the head of that household (Chi-square = 2.848, p = 0.0.091). Those potential indicator variables with significant associations were input into the binary logistic regression model to establish the extent of their effects on dietary diversity in children.

Model Comparison

Table 7.2 summarizes the results of comparison of the two count models that were fitted. The ZTNB had the lowest AIC and thus was chosen as the best model to establish predictors of dietary diversity. The ZTNB model was the preferred model with a relatively smaller AIC. This model was therefore deployed to establish predictors of the dietary diversity score.

Determinants of Dietary Diversity

Results of the Zero Truncated Negative Binomial Regression model are presented Table 7.3. The log count of the dietary diversity score was more for children of mothers aged 35-39 than for children of mothers aged 15 to 19 ($z = 2.215$, $p = 0.0268$). There were no significant regional disparities in dietary diversity. Even though the log of dietary diversity score was 0.1139 times lower for children from rural areas compared to their urban counterpart, the differentials were not significant at 5% level. Dietary diversity notably increased as the maternal level of education increased. Children from mothers with secondary ($z = 2.465$, $p = 0.014$) and higher ($z = 2.691$, $p = 0.0071$) education had significantly higher dietary diversity compared to those with mothers who had no formal education. With regard to religion, children of mothers belonging to the Seventh Day Adventist Church ($z = -2.323$, $p = 0.0202$) had significantly lower dietary diversity compared to their counterparts from the Roman Catholic Church.

Table 7.1. Results of Chi-Square Tests of Association between the dietary diversity score and potential predictor variables (*p < 0.05, **p < 0.01, ***p < 0.001)

Variable	Dietary Diversity		Chi-square	p-value
	Poor	Good		
Maternal Agegroup 15-19	80.8	19.2	8.503	0.204
20-24	76.8	23.2		
25-29	75.2	24.8		
30-34	71.9	28.1		
35-39	67.2	32.8		
40-44	75.5	24.5		
45-49	85.7	14.3		
Region Zambezi (Ref)	75.9	24.1	83.923	P < 0.001***
Erongo	51.9	48.1		
Hardap	84.6	15.4		
!Karas	62.4	37.6		
Kavango	76.2	23.8		
Khomas	49.4	50.6		
Kunene	78.0	22.0		
Ohangwena	86.5	13.5		
Omaheke	84.6	15.4		
Omusati	90.4	9.6		
Oshana	77.6	22.4		
Oshikoto	79.1	20.9		
Otjozonjupa	75.3	24.7		
Place of Residence Urban (Ref)	65.2	34.8	42.423	p < 0.001***
Rural	82.6	17.4		
Highest Educational Level No Formal Education (Ref)	93.1	6.9	50.970	p < 0.001***
Primary	82.8	17.2		
Secondary	72.5	27.5		
Higher	38.1	61.9		
Religion Roman Catholic (Ref)	73.9	26.1	3.490	0.625
Protestant/ Anglican	76.1	23.9		

Table 7.1. (Continued)

Variable	Dietary Diversity		Chi-square	p-value
	Poor	Good		
ELCIN (Evangelical Lutheran Church in Namibia)	75.1	24.9		
Seventh Day Adventist Church	75.4	24.6		
No religion	91.7	8.3		
Other Religion	69.6	30.4		
Sex of Household Head Male (Ref)	73.4	26.6	0.899	0.343
Female	76.1	23.9		
Wealth Index Poorest (Ref)	86.6	13.4	64.427	p < 0.001***
Poorer	82.2	17.8		
Middle	77.2	22.8		
Richest	54.2	45.8		
Marital Status Never in Union (Ref)	77.0	23.0	15.467	0.009**
Married	66.7	33.3		
Living with Partner	77.1	22.9		
Living with Partner	77.1	22.9		
Widowed	33.3	66.7		
Divorced	76.1	23.9		
No longer living together/ separated				
Currently working No (Ref)	79.2	20.8	15.018	p < 0.001***
Yes	68.7	31.3		
Main Language Spoken at home Afrikaans (Ref)	62.4	37.6	28.301	p < 0.001***
Damara / Nama	77.3	22.7		
English	50.0	50.0		
Herero	72.2	27.8		
Kwangali	77.9	22.1		
Lozi	75.9	24.1		
Oshiwambo	78.0	22.0		
San	86.7	13.3		
Other Languages	50.0	50.0		

Table 7.2. Model comparison

Model	Log-Likelihood	AIC
Zero Truncated Poisson (ZTP)	-2111.948	4319.895
Zero Truncated Negative Binomial (ZTNB)	-2063.144	4224.288*

*best model based on AIC and.

There were no significant disparities in dietary diversity scores of children from male and female headed households ($z = 0.486$, $p = 0.6271$). With regard to socio-economic status, dietary diversity scores initially increased for children from the poorer to the richer wealth quintile and then slightly declined from the richer to the richest wealth quintile.

Table 7.3. Results of zero truncated negative binomial regression to establish the determinants of dietary diversity (*p < 0.05, **p < 0.01, ***p < 0.001)

Variable	Estimate	Standard Error	Z- Value	p-value
Intercept 1	0.6013	0.3076	1.955	0.0506
Intercept 2 (Over dispersion parameter)	1.6232	0.1547	10.492	p < 0.001***
Agegroup 15-19 Ref				
20-24	0.0449	0.0948	0.473	0.6359
25-29	0.0639	0.1021	0.626	0.5315
30-34	0.1293	0.1156	1.119	0.2632
35-39	0.2887	0.1303	2.215	0.0268*
40-44	0.1933	0.1639	1.180	0.2382
45-49	0.1536	0.2844	0.540	0.5891
Region Zambezi (Ref)				
Erongo	0.2867	0.2439	1.176	0.2398
Hardap	-0.1390	0.2486	-0.559	0.5759
!Karas	0.2298	0.2379	0.966	0.3342
Kavango	0.1138	0.2437	0.467	0.6405
Khomas	0.1473	0.2390	0.616	0.5377
Kunene	-0.0459	0.2565	-0.179	0.8581
Ohangwena	-0.1851	0.2475	-0.748	0.4545
Omaheke	-0.0299	0.2516	-0.083	0.9335
Omusati	-0.2743	0.2488	-1.103	0.2702
Oshana	-0.0407	0.2516	-0.162	0.8715
Oshikoto	-0.0135	0.2418	-0.056	0.9553
Otjozonjupa	-0.0834	0.2452	-0.340	0.7339
Place of Residence Urban (Ref)				
Rural	-0.1139	0.0669	-1.703	0.0886
Highest Educational Level No Formal Education (Ref)				
Primary	0.2127	0.1395	1.524	0.1275
Secondary	0.3414	0.1385	2.465	0.014*
Higher	0.4952	0.1840	2.691	0.0071**
Religion Roman Catholic (Ref)				
Protestant/ Anglican	-0.0111	0.0782	-0.143	0.8866
ELCIN (Evangelical Lutheran Church in Namibia)	-0.0131	0.0694	-0.189	0.8498
Seventh Day Adventist Church	-0.3372	0.1451	-2.323	0.0202*
No religion	-0.3601	0.3046	-1.182	0.2371
Other Religion	-0.0088	0.0970	-0.089	0.9287
Sex of Household Head Male (Ref)				
Female	0.0261	0.05369	0.486	0.6271
Wealth Index Poorest (Ref)				
Poorer	0.0123	0.0881	0.140	0.8886
Middle	0.1113	0.0902	1.233	0.2177
Richer	0.2151	0.1012	2.126	0.0335*
Richest	0.1725	0.1222	1.412	0.1581
Total Children ever Born	-0.0408	0.0220	-1.852	0.0640
Marital Status Never in Union (Ref)				

Table 7.3. (Continued)

Variable	Estimate	Standard Error	Z- Value	p-value
Married	0.1044	0.0772	1.353	0.1761
Living with Partner	0.0081	0.0700	0.116	0.9076
Widowed	-0.3847	0.3695	-1.041	0.2979
Divorced	0.3463	0.2978	1.163	0.2449
No longer living together/ separated	0.1248	0.1217	1.026	0.305
Currently Working No (Ref)				
Yes	0.0844	0.0531	1.592	0.1115
Main Language Spoken at home Afrikaans (Ref)				
Damara / Nama	0.0326	0.1068	0.305	0.7603
English	0.0203	0.2204	0.092	0.9265
Herero	0.0599	0.1357	0.441	0.6590
Kwangali	0.0933	0.1450	0.643	0.5120
Lozi	0.2064	0.2438	0.846	0.3973
Oshiwambo	0.1031	0.1060	-1.128	0.2594
San	-0.3421	0.3033	-1.128	0.2594
Other Languages	0.4559	0.1564	2.915	0.0036**

Children of richer mothers ($z = 2.216$, $p = 0.0335$) had higher dietary diversity scores compared to their counterparts from the poorest mothers. Even though dietary diversity scores decreased with increasing number of children in the family, the effect was not significant ($z = -1.852$, $p = 0.0640$). The marital status and working status of the mother did not significantly influence the child's dietary diversity score. There were significant disparities in dietary diversity due to cultural factors (measured by the "main language spoken at home"). Children of mothers who spoke other languages had significantly higher dietary diversity scores compared to those whose mothers mainly spoke Afrikaans ($z = 2.915$, $p = 0.0036$). Although children from San speaking mothers had relatively lower dietary diversity scores compared to those of their Afrikaans speaking counterparts, the disparities were not that significant ($z = -1.128$, $p = 0.2594$).

Discussion

In this cross-sectional analysis, we examined the dietary diversity of children under-five regardless of breast feeding status. This study identified that only 27% of the children received five or more food groups within the 24 hours preceding the survey. The results are consistent with findings of other studies where maternal age, education, religion, cultural background, and socio-economic status were found to be significant predictors of functional food consumption (Mullie, Guelinckx, Clarys, Degrave, Hulens and Vansant, 2009; Senarath, Godakandage, Jayawickrama, Siriwardena, and Dibley, 2012; Aemro, Musele, Birhanu, and Atenafu, 2013). The quality of the diet of children during their early years depends mainly on the behavior and decisions of the mothers or those who usually care for the child

In children, food neophobia and "picky/fussy" eating have been associated with rejection/ acceptance of fruits and vegetables. Other possible reasons for low dietary might be related to the practise of complimentary feeding after 6 months. Eating, drinking and food choices are complex behaviours determined by a multiplicity of factors and their interactions thereby demanding interdisciplinary approaches (Koster, 2009). There is a need to re-look into certain food types which might be in the form of liquids.

Our results shows that there is no significant difference in dietary diversity between rural and urban children and the level is low. Strategies such as the need for investments to boost agricultural production, to keep food prices low; increase incomes; and formulation of targeted programs to compliment these investments by supporting livelihoods; enhancing access to diverse diets in poor populations; and fostering woman's empowerment have been well documented (Ruel &Alderman, 2013).

Overall, our results of the regression tend to confirm the findings of previous work in neighbouring South Africa, as measured in the NFCS in 1999, where DDS had a mean of 3.6. Comparing it to Namibia, although different dietary methods were used, it should be recognised that the majority of children in Namibia have a very limited dietary variety. The adequacy of intake of some specific food items (especially lentils, Vitamin A rich fruits) implied means that micronutrients was particularly low. Folate intakes were also very low due to the limited quantities consumed by the children in lentils and green leafy vegetables (2.1% and 4.5% respectively). Only 3.3% of children consumed milk or milk products which are essential for providing Calcium. The children's diet consst primarily of starchy foods and by implication certain micronutrients were deficient. These children should be regarded as being at risk of undernutrition.

CONCLUSION

Our study shows dietary diversity among young children in Namibia is alarmingly low and primarily starch based. This limits the consumption of essential micronutrients leading to lack sufficient nutrient- dense foods to achieve micronutrient adequacy as shown by the limited dietary diversity. Although a small number of children did consume from a diverse food group this study was not able to determine the sufficiency and frequency of the quantities in order to assess the adequacy, as a limitation of the study. Another limitation was that the foods consumed by the children was based on the recall from their mothers who may not recall some food items consumed in between meals.

Recommendations

From the results of this study, the following recommendations are presented. Policy efforts could:

1. Subsidize healthy food prices to enhance access to diverse diets in dry regions.
2. Formulate Targeted agricultural intervention programs in certain regions to boost agricultural production
3. Efforts to improve education of the girl child for them make informed dietary choices for their children
4. Initiate income generating projects to raise the standard of living of households

Acknowledgment

The authors acknowledged support from the University of Namibia.

References

Aemro, M. Mesele, M. Birhanu, Z. and Atenafu, A. (2013). Dietary diversity and meal frequency practices among infant and young children aged 6–23 months in Ethiopia: a secondary analysis of Ethiopian Demographic and Health Survey 2011. *Journal of nutrition and metabolism, 2013*.

Amugsi, D. A. Dimbuene, Z. Kimani-Murage, E. W. Mberu, B. and Ezech, A. (2017). Differential effects of dietary diversity and maternal characteristics on linear growth of children aged 6 -59 months in sub-Saharan Africa: a multi-country analysis. *Public Health Nutrition*: doi:10.1017/S1368980016003426.

Black, R. E. Allen, L. H. Bhutta, ZA. Caulfield, L. E. de Onis, M. Ezzati, M. Mathers, C. and Rivera, J. (2008). Maternal and Child Undernutrition Study Group. Maternal and Child undernutrition: global and regional exposures and health consequences. *The Lancet*. 371:243-260.

Faber, M. Laubscher, R. and Berti, C. (2016). Poor dietary diversity and low nutrient density of the complementary diet for 6 –to 24-month-old children in urban and rural KwaZulu-Natal, South Africa. *Maternal and Child Nutrition*, 12:528 – 545.

Foote, J. A. Murphy, S. P. Wilkens, L. R. Basiotics, P. P. and Carlos, A. (2004). Dietary Variety increases the probability of nutrient adequacy among adults. *Journal of Nutrition*, 134:1779-1785.

Gausman, J. Perkins J. Lee H-Y. Mejia-Guevara I. Nam Y-S. Lee J-K. Oh J. m and Subramanian S. (2018). Ecological and social patterns of child dietary diversity in India: a population- based study. *Nutrition*, 53: 77 – 84.

Hoddinott, J. and Yohannes, Y. (2002). Dietary Diversity as a Household Food Security Indicator. Food and Nutrition Technical Assistance Project, Academy for educational Development. Washington DC.

Indongo, N. and Mutorwa, K. (2017). The Practice of Exclusive Breastfeeding in Namibia. *International Journal of Sciences: Basic and Applied Research*, 1(36): 159 – 169.

Kennedy, G. L. Pedro, M. R. Seghieri, C. Nantel, G. and Brouwer, I. (2007). Dietary diversity score is a useful indicator of micronutrient intake in non-breast-feeding Filipino children. *The Journal of Nutrition* 137: 472–477.

Köster, E. P. (2009). Diversity in the determinants of food choice: A psychological perspective. *Food quality and preference*, *20*(2), 70-82.

Lawless, J. F. (1987). Negative Binomial and Poisson Regression, *The Canadian Journal of Statistics*, 15 (3) 209-225.

Liu, X., Saat, M. R., Qin, X., & Barkan, C. P. (2013). Analysis of US freight-train derailment severity using zero-truncated negative binomial regression and quantile regression. *Accident Analysis & Prevention*, *59*, 87-93.

Lucenteforte, E. Garavello, W. Bosetto, C. Talamini, R. Zambon, P. Franceschi, S. Negri, E. and La Vecchia, C. (2008). Diet diversity and the risk of squamous cell esophageal cancer. *Int. J. Cancer*, 123:2397-2400.

Mullie, P. Guelinckx, I. Clarys, P. Degrave, E. Hulens, M. and Vansant, G. (2009). Cultural, socioeconomic and nutritional determinants of functional food consumption patterns. *European Journal of Clinical Nutrition*, *63*(11), 1290-1296.

Nti, C. (2011). Dietary Diversity is Associated with Nutrient Intakers and Nutritional Status of Children in Ghana. *Asian Journal of Medical Sciences*, 2:105 – 109.

Rakotonirainy, N. H. Razafindratovo, V. Remonja, C. R. Rasoloarjaona, R. Piola, P. Raharintsoa, C. Randremanana, V. (2018). Dietary Diversity of 6- to 59-month-old children in rural areas of Moramanga and Morondava districts, Madagascar. *PLoS ONE* 13(7):e0200235. https://doi.org/10.1371/journal.pone.0200235.

Ruel, M. T. Alderman, H. Maternal and Child Nutrition Study Group. (2013). Nutrition-sensitive interventions and programmes: how can they help to accelerate progress in improving maternal and child nutrition? *The Lancet*, *382*(9891), 536-551.

Ruel, M. (2002). Is Dietary Diversity and Indicator of Food Security or Dietary Quality? A Review of measurement issues and research needs. Food Consumption and Nutrition Division (FNCD) Discussion paper 140. International Food Policy Research Institute (IFPRI), Washington DC.

Senarath, U. Godakandage, S. S. Jayawickrama, H. Siriwardena, I. and Dibley, M. J. (2012). Determinants of inappropriate complementary feeding practices in young children in Sri Lanka: secondary data analysis of demographic and health survey 2006–2007. *Maternal & child nutrition*, *8*(s1), 60-77.

Shonkwiler, J. S. (2016). Variance of the truncated negative binomial distribution. *Journal of Econometrics*. 195(2), 209-10.

Steyn, N. Nel, J. Labadarios, D. Maunder, E. and Kruger, H. (2014). Which Dietary diversity indicator is best to assess micronutrient adequacy in children 1-9y, *Nutrition*, 30:55-60.

Steyn, N. P. Nel, J. H. Nantel, G. Kennedy, G. L. Labadarios, D. (2006). Food variety and dietary diversity scores in children: are they good indicators of dietary adequacy? *Public Health Nutr* 9: 644–650.

Swindale, A., and Paula B. (2006) "Household dietary diversity score (HDDS) for measurement of household food access: indicator guide." *Washington, DC: Food and Nutrition Technical Assistance Project, Academy for Educational Development.*

Temesgen, H. Yeneabat, T. and Teshome, M. (2018). Dietary Diversity and Associated factors among children aged 6 -23 in Sinan Woreda, Northwest Ethiopia: a cross-sectional study. *BMC Nutrition*, 4:5 https://doi.org/10.1186/s40795-018-0214-2.

Wahlqvist, M. L. Lo, C. S. and Myers, K. (1989). Food variety associated with less macrovascular disease in those with type II diabetes and their healthy controls. *J Am Coll Nutr*, 8:515-523.

Yee, T. W. (2008). The VGAM package. *R News* 8 (2), 28 39. URL http/CRAN.R-project.org/doc/RNews.

In: Statistical Modelling …
Editors: N-B. Kandala and L. Kazembe
ISBN: 978-1-53615-981-3

Chapter 8

A Copula Approach to Sample Selection Modelling of Treatment Adherence and Viral Suppression among HIV Patients on Antiretroviral Therapy (ART) in Namibia

Jason Nakaluudhe and Lawrence N. Kazembe*
Department of Statistics and Population Studies,
University of Namibia, Windhoek, Namibia

Abstract

Namibia has a generalized human immunodeficiency virus (HIV) epidemic, with HIV mainly transmitted through heterosexual transmission. Although the number of people receiving antiretroviral therapy (ART) has increased, the achievement of the 90-90-90 strategy on testing, treatment and suppression has not been evaluated. In this study, we focused on treatment adherence which is a critical element in achieving viral suppression. Particularly, we examined factors associated with treatment adherence and viral suppression, which may assist in designing appropriate interventions for accelerated achievement of the 90-90-90 prevention and control strategy. However, modelling treatment adherence and viral suppression may result in biased estimates if sample selection is ignored. We therefore fitted a joint distributional model of ART treatment adherence and viral suppression, to adjust for sample selection bias among HIV patients on ART in Namibia, to examine the presence of tail dependence in sample selection bias, and to investigate factors associated with viral suppression and ART adherence. The study used data of HIV+ patients, aged above 16 years, on ART in selected health facilities in Windhoek, Namibia. A Heckman-type selection analysis was carried out using a variety of copula families to capture dependence in the outcomes. The best

* Corresponding Author's Email: lkazembe@yahoo.com.

models were selected based on the lowest AIC and BIC. Results showed a strong negative correlation between adherence to ART treatment and viral load suppression, with high adherence of >95% associated with viral suppression at both 6 and 12 months. Findings further showed that current body weight was statistically significant with viral suppression. However, demographic and clinical characteristics were not associated with ART adherence nor viral suppression. We can conclude that patients should be retained-in-care to maintain high adherence to treatment, hence achieve viral suppression within a short space of time, and for accelerated and sustainable HIV prevention and control in Namibia. A more comprehensive study covering more health facilities is proposed for better evidence-based decision making. Similar studies in the future should consider socio-economic factors in addition to those considered in this study.

Keywords: Heckman-type sample selection, joint bivariate copula modeling, HIV/AIDS, antiretroviral therapy (ART) adherence, viral suppression, Namibia

INTRODUCTION

Human immunodeficiency virus (HIV) epidemic has become one of the major challenges to public health care sector worldwide. According to 2017 World Health Organization (WHO) report, more than 70 million people have been infected with the HIV virus and about 35 million people have died of HIV attributable illness. Globally, 36.7 million people were living with HIV at the end of 2016, with an estimated 0.8% of adults aged 15 to 49 years worldwide affected. Although the burden of the epidemic continues to vary considerably between countries and regions, the sub-Saharan Africa (SSA) remains the most severely affected, with nearly 1 in every 25 adults (4.2%) living with HIV and accounting for nearly two-thirds of people living with HIV worldwide (WHO, 2017). In Namibia, the proportion of adults presenting with HIV infection is relatively high compared to other countries in SSA, but decreasing (Namibia Population-Based HIV Impact Assessment [NAMPHIA], 2018). According to NAMPHIA, HIV prevalence, in Namibia, was 12.6% among adults aged 15-64 years in 2017, a drop of 10.2% since 2002 when it was the highest.

In 2015, the Joint United Nations Program on HIV and AIDS (UNAIDS) developed the 90-90-90 targets to be achieved by 2020 with the aim of eliminating AIDS epidemic globally by 2030 (UNAIDS, 2016). In order for a country to eliminate the HIV epidemic, health practitioners need to test 90% of all the people who are suspected to be HIV positive and let them know their status, provide antiretroviral therapy (ART) to 90% of all those who tested HIV positive and retain them on ART so that 90% of all those on treatment have viral load suppressed. The achievement of the 90-90-90 target remains a big challenge. Current estimates show that there is a huge percentage of those living with HIV that do not know their status, nor have initiated treatment. Globally in 2015, for instance, around 11.9 million people living with HIV did not know their HIV status, while 12.7 million people were in need of ART, whereas 13.0 million people on treatment had not suppressed the virus (UNAIDS, 2016).

Confronted with such challenges, at country level, the Namibia National Policy on HIV/AIDS, implemented together with National Strategic Framework for HIV and AIDS (2017/18 to 2021/22), highlights the need of giving HIV care and support services to people living with HIV and their families (MoHSS, 2016), and accelerate the provision of free antiretroviral therapy (ART), which started in 2003, to the public. According to MoHSS (2016), since the roll-up of ART, 93% of HIV patients have been enrolled into the programme. Although the number of people receiving ART has increased, very little is known on the achievement of optimal levels of 90-90-90 (UNAIDS, 2017).

To achieve such optimal level, requires continuous adherence to treatment. However, a number of studies have shown that non-adherence is a major issue and a rising problem (Heestermans et al., 2016; Kheswa, 2017). The World Health Organization (WHO) defined treatment adherence as the extent to which a person's behavior corresponds with the agreed recommendations of a healthcare provider (WHO, 2017). Although, adherence behavior is influenced by the clinician and the healthcare system, the disease stage and its treatment, and socio-economic factors (Noens et al., 2009). Adherence to antiretroviral therapy (ART) is essential to ensure viral load suppression, decrease the risk of disease progression and reduce drug resistance (Rougemont, Stoll, Elia & Ngang, 2009). However, sustaining ART adherence over the long term requires accurate and consistent monitoring. In particular, according to Nachega, Mills and Schechter (2010), and Reda and Biadgilign (2012), sustaining adherence to ART is a challenge for many countries in the sub-Saharan Africa region. Moreover, despite many studies elsewhere, there are no published results on outcome and determinants of treatment adherence in Namibia.

The impact of ART adherence in suppressing the HIV virus and improving survival rates for those living with HIV has been well documented (Peltzer et al, 2010; Rougemont, 2009). However, successful antiretroviral treatment is dependent on sustaining high rates of adherence. Although, more strong ART regimens can allow for effective viral load suppression at moderate level of adherence, many studies have shown that a very higher adherence level of >95% is required for ART regimens to work effectively (Castro, 2005; Peltzer et al, 2010; Reda & Biadgilign, 2012). Thus designing appropriate and informed interventions that promote high adherence is critical. Consequently, there is need to determine factors that are influential on optimizing adherence and viral suppression.

In identfying factors associated with both adherence and viral suppression, the issue of sample selection arises, because adherence is an important factor of viral suppression. Sample selection, as introduced by Heckman (1979), accounts for problems arising when the outcome of the endogenous variable is observed only for a selected sample. In our case study, viral suppression can only be observed among those (a non-random sample) who are highly adherent to ART, thus inducing a bias on those excluded for not fully adhering. The sample selection model is not only very useful when the outcome variable is a nonrandom part of the population, but it can easily be extended to the switching regression model (Van der Klaauw & Koning, 2003). There are several ways in which sample selecton bias can be corrected. In this study, we propose to use a copula approach to capture dependence and endogeneity inherently introduced by sample selection.

The copula approach is a modelling strategy whereby a joint distribution is induced by specifying marginal distributions on both the sample selection and outcome equations, and specify a function that link them together (Smith, 2003; Kuethe et al., 2009). Generally, copula models provide a generic flexible framework for specifying the joint distribution of any two random variables (Nelsen, 1999; Cunanan & Koopmeiners, 2014). Moreover, this framework allows the copula dependence parameter to be specified as a function of an additive predictor (Marra et al, 2017).

The link between treatment adherence and viral load suppression based on sample selection model has not been explored in the context of HIV/AIDS, except in estimating HIV prevalence (McGovern et al, 2015). This study used a joint bivariate copula-based sample selection model to accommodate dependence structure between ART treatment adherence and viral load suppression equations.

SAMPLE SELECTION MODEL

In the sample selection context, the aim is to fit a regression model when some observations for the response of interest, W_{2i}, are missing not at random (Wojtyś, Marra & Radice, 2016). The process of selection can be typically assumed that there is a variable W_{1i} which governs whether or not an observation on the variable of primary interest is generated. Based on Marra, Papageorgiou and Radice, (2014); Wojtyś, Marra and Radice, (2016), the latent variable representation can be written as

$$W^*_{1i} = \eta_{1i} + \varepsilon_{1i} = x_{1i}\alpha^T + \sum_{k_1=1}^{K_1} s_{1k_1}(u_{1k_{1i}}) + \varepsilon_{1i},$$

$$\text{for } i = 1, 2, \dots, n \tag{1}$$

such that

$$W^*_{1i} \sim N(\eta_{1i}, 1), \tag{2}$$

$$W_{2i} = \begin{cases} 1 \; if \; W^*_{1i} > 0 \\ 0 \;\; \text{otherwise.} \end{cases} \tag{3}$$

Here n is a sample size, and W^*_{1i} is independent and identically distributed, η_{1i} is a linear predictor with the obvious definition, α is a P_1 dimensional coefficient vector of all parametric components, x_{1i} is a vector of (binary and categorical) covariates, $s_{1k_1}(u_{1k_{1i}})$ are unknown smooth functions of the K_1 continuous covariates u_1k_{1i} and ε_{1i} is an error term with zero mean and constant variance (Marra, Radice, Bärnighausen, Wood & McGovern, 2017). The outcome equation can be written as (Marra et al., 2017)

$$W_{2i} \sim Ƒ(\mu_i, \sigma, v) \tag{4}$$

where $\mu_i = E(W_{2i}) = exp(\eta_{2i}) = exp\left(\alpha^T x_i + \sum_{k_2=1}^{K_2} s_{2k_2}(u_{2k_{2i}})\right)$, and Ƒ is a discrete distribution, η_{2i} is a linear predictor, and σ and v are scale and shape distribution specific parameters. The number of parameters that characterize Ƒ depends on the chosen distribution. Vector α has length P_2 which represents the parameters of all parametric components, x_{2i} is a vector of factor variables and $s_{2k_2}(u_{2k_{2i}})$ are unknown smooth functions of the K_2 continuous covariates $u_{2k_{2i}}$ (Marra et al., 2017).

The smooth functions are approximated using regression splines as (Marra and Wyszynski, 2016)

$$\sum_{j=1}^{J_{vkv}} \alpha_{vkvj} b_{vkvj}(u_{vki}) = \boldsymbol{B}(u_{vkvi})^T \boldsymbol{\alpha}_{vkv}\,, \tag{5}$$

where the $b_{vkvj}(u_{vki})$ are known spline basis function, the α_{vkvj} are regression parameters, J_{vkv} is the number of bases used to represent the smooth term, $\boldsymbol{B}(u_{vki}) = [b_1(u_{vkvi}), b_2(u_{vkvi}), \ldots, b_J\ (u_{vkvi})]^T$ is a vector containing J_{vkv} basis functions and α_{vkv} is the corresponding parameter vector. The subscript indicating which equation each smooth component belongs to has been suppressed for simplicity. The smoothing function is subject to the centering constraint (Wojtyś et al., 2016; Marra & Radice, 2017).

Linking Model Equations

In a sample selection model concept, it is assumed that the selection and outcome equations are linked through unobservable (Marra & Wyszynski, 2016). This link is formulated, in a standard case, using a bivariate normal distribution. The relationship between a genetic pair of variables (W_1, W_2) can be represented using a copula function (Sklar, 1959). Let $F_1(w_1)$ and $F_2(w_2)$ be the cumulative density functions (cdfs) of w_1 and w_2, respectively, then F_1 relates to the selection equation and F_2 to the outcome equation. Their joint cdfs is represented as follows:

$$F(w_1, w_2) = C(F_1(w_1), F_2(w_2); \theta), \quad (6)$$

for some two-place copula function C which is unique only on closure $\text{Ran}(F_1) \times \text{Ran}(F_2)$, where $\text{Ran}(\cdot)$ denotes the range of argument and θ is an association parameter measuring the dependence between the two marginal cdfs. For more details see Marra and Wyszynski (2016), Wojtyś, Marra and Radice (2016) and Smith (2003).

Copula Functions

There are a number of families of copula functions currently implemented in literature. These are summarized in Table 8.1. They are normal, Farlie-Gumbel-Morgenstern (FGM), and some belong to Archimedean class of copula. Archimedean copulas form a large family of copulas with a number of convenient properties and they allow for a large number of dependence structures (Manner, 2007). Since in some cases θ cannot be interpreted directly, it can be transformed into Kendall's τ ranging on the interval $[-1, 1]$ yielding a general interpretation for all copulas.

Table 8.1. Example of the families of bivariate copulas

Name	Copula $C(u, v; \theta)$	Parameter space of θ	Parameter space of Kendall's τ	Kendall's τ in terms of θ
FGM	$uv\big(1 + \theta(1-u)(1-v)\big)$	$-1 \le \theta \le 1$	$-2/9 \le \tau \le 2/9$	$\frac{2}{9}\theta$
Normal	$\Phi_2(\Phi^{-1}(u), \Phi^{-1}(v); \theta)$	$-1 \le \theta \le 1$	$-1 \le \tau \le 1$	$\frac{2}{\pi}\ arcsin(\theta)$
AMH	$uv/\big(1 - \theta(1-u)(1-v)\big)$	$-1 \le \theta \le 1$	$-0.1817 \le \tau < \frac{1}{3}$	$1 - \frac{2}{3\theta^2}(\theta + (1-\theta)^2 \log(1-\theta))$
Clayton	$\left(u^{-\theta} + v^{-\theta} - 1\right)^{-1/\theta}$	$1 \le \theta \le \infty$	$0 < \tau < 1$	$\frac{\theta}{\theta + 2}$
Frank	$-\theta^{-1} \log\left(1 + \left(e^{-\theta u} - 1\right)\left(e^{-\theta v} - 1\right)/ \left(e^{-\theta} - 1\right)\right)$	$\theta \in \mathbb{R}\{0\}$	$-1 < \tau < 1$	$1 - \frac{4}{\theta}[1 - D_1(\theta)]$
Gumbel	$exp\left(-\left((-\log u)^\theta + (-\log v)^\theta\right)^{1/\theta}\right)$	$1 \le \theta \le \infty$	$0 \le \tau < 1$	$1 - \frac{1}{\theta}$
Joe	$1 - \left((1-u)^\theta + (1-v)^\theta - (1-u)^\theta(1-v)^\theta\right)^{1/\theta}$	$1 \le \theta \le \infty$	$0 < \tau < 1$	$1 + \frac{4}{\theta^2} D_2(\theta)$

Note: this table present the example of families of bivariate copulas. u and v represent marginal cdfs. $\Phi_2(\cdot,\cdot;\theta)$ denotes bivariate standard normal cdf with correlation coefficient θ, and $\Phi^{-1}(\cdot)$ is the inverse cdf of a standard normal. $D_1(\theta) = \frac{1}{\theta}\int_0^\theta \frac{t}{exp(t)-1} dt$ is the Debye function and $D_2(\theta) = \int_0^1 t \log(t)(1-t)^{\frac{2(1-\theta)}{\theta}}$.

Kendall's τ is not a margin-free measurement of dependence and should therefore be used with caution especially when the choice of margins is based on dependence (Marra and Wyszynski, 2016). The copulas currently available in literature and the respective conversions from θ to Kendall's τ are also summarized in Table 8.1.

A copula function may help to explain the relevance of rotation but can only make sense for ones with an asymmetric dependence structure (Manner, 2007). The Clayton, Joe and Gumbel copulas can be rotated by 90, 180 and 270 degree. These rotations are clearly shown as

$$\begin{aligned} C_{90} &= v - C(1-u, v; \theta), \\ C_{180} &= u + v - 1 + C(1-u, 1-v; \theta), \\ C_{270} &= u - C(u, 1-v; \theta), \end{aligned} \tag{7}$$

where u and v are the margins (Brechmann and Schepsmeier, 2013 as cited by Marra and Wyszynski, 2016).

Tail Dependence

In this concept, tail dependence measures the strength of dependence in lower and upper tail of a multivariate distribution. According to Joe (1997), tail dependence describes the limits of a proportion that one margin exceeds a certain threshold given that the other margin has already exceeded that threshold.

Let $W_1 \sim F_1$ and $W_2 \sim F_2$ be two random variables with copula *C*, define the coefficients of tail dependency

$$\begin{aligned} \lambda_L &= \lim_{u \downarrow 0} \mathbb{P}[F_1(W_1) < u | F_2(W_2) < u] = \lim_{u \downarrow 0} \frac{C(u,u)}{u} \\ \lambda_U &= \lim_{u \uparrow 1} \mathbb{P}[F_1(W_1) > u | F_2(W_2) > u] = \lim_{u \uparrow 1} \frac{1-2u+(u,u)}{1-u} \end{aligned} \tag{8}$$

where λ_L is the lower tail and λ_U is the upper tail, C is said to have lower or upper tail dependence if and only $\lambda_L \neq 0$ ($\lambda_U \neq 0$).

Likelihood

The sample selection model occurs when observations on a pair of random variables (*R*, *W*) are generated according to the following relationships (Smith, 2003; Marra & Wyszynski, 2016; Wojtyś et al, 2016; Rüschendorf, 2013):

$$R = 1\{W_1{}^* > 0\} \quad and \quad W = 1\{W_1 > 0\}W_2, \tag{9}$$

where $1\{\omega\}$ denotes the indicator function, taking value 1 if event ω holds, and 0 otherwise. In effect, W_2 can be observed only when $W_1 > 0$. The participation mechanism is represented by the Bernoulli variable R, and it derives its properties from those of $W_1{}^*$. Note that when $R = 0, W_2$ cannot be observed, and W is assigned a dummy value of 0.

Let $r_1, \dots, r_m$ denote m observations generated on $R\ (r_j \in \{0,1\}, j = 1, \dots, m)$, and $w_1, \dots, w_m$ the corresponding m observations generated on $W\ (w_j \in \mathbb{R}, j = 1, \dots m)$. For a random sample of m observations, the likelihood function for the sample selection model is formulated as follows (Smith, 2003 as cited by Marra & Wyszynski, 2016)

$$L = \prod_0 Pr(W_1 = 0) \prod_1 Pr(W_2 = w_2, W_1 = 1)$$

$$= \textstyle\prod_0 Pr(W_{1j}{}^* \leq 0) \prod_1 f_{2|1}(w_2 | w_{1j}{}^* > 0)\, Pr(W_{1j}{}^* > 0), \tag{10}$$

where, $\prod_0$ indicate the product over those observations for which $r_j = 0$, and $\prod_1$ the product over those observations for which $r_j = 1$. For convenience, j is the index which can be dropped from now and the function $f_{2|1}$ denotes the probability density function (pdf) of W_2, given event $W_1{}^*$ and it can be derived as follows (Smith, 2003):

$$f_{2|1}(w_2 | w_1{}^* > 0) = \frac{\partial}{\partial w_2} \frac{F_2(w_2) - F(0, w_2)}{F_1(1)}$$

$$= \frac{1}{1 - F_1(0)} \frac{\partial}{\partial w_2} \left(F_2(w_2) - F(0, w_2)\right)$$

$$= \frac{1}{1 - F_1(0)} \left(f_2(w_2) - \frac{\partial}{\partial w_2} F(0, w_2) \right) \tag{11}$$

The general likelihood is the form of the sample selection model. Particularly, likelihood functions arise from specification assumptions. Therefore, the likelihood function is given by

$$L = \textstyle\prod_0 Pr(W_{1j}{}^* \leq 0) \prod_1 f_{2|1}(w_2 | w_{1j}{}^* > 0)\, Pr(W_{1j}{}^* > 0). \tag{12}$$

Model Specification and Estimation

In copula approach, there exist a number of methods for estimating these models. The first method is a one-step approach referred to as the Exact Maximum Likelihood (EML), which estimates all parameters of the model at the same time (Manner, 2007; Payandeh, Farid-Rohani, & Qazvini, 2013). The EML estimator can be specified by first considering the following representation of the joint density known as the copula decomposition of a joint distribution and the resulting log-likelihood function,

$$h(w_{1i}, w_{2i}, \theta) = f(w_{1i}; \varphi) \cdot g(w_{2i}; \gamma) \cdot c(F(w_{1i}; \varphi), G(w_{2i}, \gamma); \delta)$$
$$L_{W_1W_2} = \sum_{i=1}^{T} ln\big(f(w_{1i}, \varphi)\big) + \sum_{i=1}^{T} ln\big(g(w_{2i}, \gamma)\big) + \sum_{i=1}^{T} ln\big(c(F(w_{1i}; \varphi), G(w_{2i}, \gamma); \delta)\big) \tag{13}$$

where $\theta \in \Theta$ is the parameter vector to be estimated. This parameter vector can be split up into the parameters for the marginal and the copula function as follows: $\theta = [\varphi', \gamma', \delta']'$, such that $\varphi \in \phi$ denotes the parameter(s) of $f(w_{1i})$, $\gamma \in \Gamma$ indicates the parameter(s) of $g(w_{2i})$, and $\delta \in \Delta$ is the parameter(s) of copula $c\big(F(w_{1i}), G(w_{2i})\big)$. Based on Manner (2007), the EML estimator is then given as

$$\hat{\theta} = arg\,\max L_{W_1W_2}, \tag{14}$$

The second method is a two-step approach called Inference Functions for Margins (IFM), which first estimates the parameters φ and γ of the marginal densities through MLE, that is

$$\hat{\varphi} = arg\,maxL_{W_1} \tag{15}$$

$$\hat{\gamma} = arg\,maxL_{W_2} \tag{16}$$

Using these estimates one would proceed to transform the marginals into uniform (0, 1) variables, then estimate the copula parameter(s) δ by maximizing the copula density, that is

$$\hat{\delta} = arg\,maxLc(\hat{\varphi}, \hat{\gamma}, \delta). \tag{17}$$

Heckman (1979) proposed an estimation in parametric sample selection models which is typically based on the two-step estimation procedure.

The third method is the semiparametric estimation, known as canonical maximum likelihood method (CML), which leaves the marginal densities unspecified and uses the empirical probability integral transform in order to obtain the uniform marginal needed to estimate the copula parameters (Manner, 2007). The empirical distribution function is defined as Payandeh et al. (2013).

$$\hat{F}(\cdot) = \frac{1}{T}\sum_{t=1}^{T} 1_{(X_t \leq \cdot)}, \tag{18}$$

where $1_{(X_n \leq \cdot)}$ is the indicator function. The copula parameter δ can be estimated by maximizing the log-likelihood function of the copula density using the transformed variables given by Araichi, Peretti and Belkacem (2014)

$$L_C(\delta) = \sum_{t=1}^{T} ln\left(c\left(\hat{F}(w_{1t}), \hat{G}(w_{2t})\right)\right)$$
$$= \sum_{t=1}^{T} ln\left(c(\hat{u}_{1t}, \hat{u}_{2t})\right). \tag{19}$$

The semi parametric estimator is then given as

$$\hat{\delta} = arg\, maxL_C(\delta). \tag{20}$$

Chen, Fan and Tsyrennikov (2006) proposed sieve maximum likelihood estimation procedure for semiparametric multivariate copula models. The proposed estimation achieved efficiency gains in finite samples, especially when prior information of the marginal distribution is incorporated.

The last two methods are nonparametric ways of estimating the copula. The first one is estimating the empirical copula directly from the data, leaving the whole specification nonparametric. The empirical copula is the function C_n given as (Manner, 2007)

$$C_n\left(\frac{i}{T}, \frac{j}{T}\right) = \frac{h_i}{T}, \quad \text{for } i, j = 1, 2, \ldots, T \tag{21}$$

where h_i denotes the number of pairs (w_{1t}, w_{2t}) in the sample such that $w_{1t} \leq w_{1i}$ and $w_{2t} \leq w_{2i}$. The other is obtaining a nonparametric estimate for Kendall's tau. The sample of Kendall's tau is then given as

$$\hat{\tau} = \frac{c-d}{c+d}, \tag{22}$$

where c denote the number of concordant pairs in the sample and d the number of discordant pairs in the sample. Using this estimate and the relationship between Kendall's tau and the copula parameter to obtain nonparametric estimate if a closed form expression for this relationship exist. On the other hand, if a closed form expression does not exists one can still estimate the copula parameter by using the general form of Kendall's tau for Archimedean copulas and the form is given as

$$\tau_C = 1 + 4\int_0^1 \frac{\varphi(t)}{\varphi'(t)} dt. \tag{23}$$

APPLICATION TO HIV TREATMENT

Data

Study Sites

Two health facilities, Katutura Health Center and Okuryangava Clinic, in Windhoek were selected for evaluation. The two clinics form part of the public health care services in Windhoek, the capital city of Namibia. In total, Namibia has 343 health care centres, of these only seven public clinics are in Windhoek (i.e the Khomasdal, Robert Mugabe, Donkerhoek, Blackchain, Wanaheda, Okuryangava and Hakahana clinics). As in all public health facilities, these two provide free HIV care, including counselling and giving ART medication for free.

Study Design

A quantitative retrospective cohort study design was applied. Data were collected among patients who initiated ART from January 2010 to December 2015 at Katutura Health Centre (KHC) and Okuryangava Clinic (OC). The target age group was all HIV positive patients aged 16 years and above.

Sample Size and Sampling

The sample size was estimated at 944 under the following assumption: (1) to detect a relative risk of 2 between the exposed and non-exposed; (2) two-sided significance of 95%; a power of 80%, ratio of sample equal to 1 between the exposed and un-exposed and (3) percent of unexposed with outcome equal to 5% and percent of exposed with outcome equal to 10% (EpiTools, 2018). The exposed were defined as those patients with less than 95% adherence to ART, while the unexposed were considered those who achieved at least 95% adherence to ART. The sample was selected from two health facilities using the following approach: First the simple random sampling of the patients' files was done, followed by a systematic sampling whereby every third file was selected until the desired sample was achieved. An equal sample was planned for each clinic, however, Okuryangava was oversampled as it had more patients.

Eligibility and Ineligibility Criteria

Included all HIV positive patients, aged 16 years and above, whoever started on ART between January 1, 2010 to December 31, 2015 and aged 16 to 50 years, and excluded all those that did not meet this criteria.

Variables

In this context, two health outcomes were considered. The primary outcome was the viral load suppression, and the secondary outcome was body weight. Both were evaluated at 6 months and 12 months after the ART initiation. Treatment adherence was the main effect modifier. Treatment adherence was defined as those who achieved at least 95% adherence to ART, and non-adherence if there was less than 95% inconsistent uptake of ART. The covariates included were age, sex, marital status, facility, WHO stage, CD4 at ART initiation, ART regimens and hepatitis B. Table 8.2 shows a summary of variables in the data.

Table 8.2. Description of the outcome and selection variables and observed confounders among patients sampled in the study

Variable name	Values of the variable	Type
Selection		
Treatment adherence	1 = yes if >95% adherence, 0 = no if <95% adherence	Categorical
Outcome		
Viral load suppression	1 = yes, 2 = no	Categorical
Viral load	None	Continuous
Demographic		
Age	1 = < 25 years, 2 = 25-49 years, 3 = ≥50 years	Categorical
Sex	1 = male, 2 = female	Categorical
Marital status	1 = single, 2 = married, 3 = others	Categorical
Health-related		
Facilities	1 = KHC, 2 = OC	Categorical
Year of initiation		Categorical
WHO stage	1 = stage 1, 2 = stage 2, 3 = stage3, and 4 = stage 4	Categorical
CD4 at ART initiation	1 = <200 CD4 counts, 2 = ≥200 CD4 counts	Categorical
Weight at ART initiation	1 = <50 weights, 2 = 50-70 weights, 3 = >70 weights	Categorical
Weight at 6 months	1 = <50 weights, 2 = 50-70 weights, 3 = >70 weights	Categorical
Weight at 12 months	1 = <50 weights, 2 = 50-70 weights, 3 = >70 weights	Categorical

Analysis

Binary Logistic Regression Analysis

Separate univariate and multivariate logistic regression analyses were used to assess the factors associated with ART adherence and viral load suppression. In multivariate analysis, adjustments were made for sex, age, facility, WHO stages, CD4 cell count at ART initiation, weight at ART initiation, weight at 6 months, weight at 12 months and ART regimen. The equations for ART treatment adherence and viral suppression were then given as

$$log \frac{P(adherence \geq 95\%)}{P(adherence < 95\%)} = \alpha_{10} + \sum_{i=1}^{n} \alpha_{1i} x_{1i}$$

and

$$log \frac{P(viral\ suppression)}{P(no\ viral\ suppression)} = \alpha_{20} + \sum_{1=i}^{n} \alpha_{2i} x_{2i} \ ,$$

where α_{10}, α_{20}, α_{1i} and α_{2i} are unknown parameters, x_{1i} and x_{2i} are the explanatory variables mentioned above and Σ indicates the summation of the associated parameters (excluding intercepts) and explanatory variables.

Heckman-Type Model Using a Joint Bivariate Copula Model

Heckman-type selection analysis using a joint bivariate copulas was used to investigate the dependence structure between ART treatment adherence and viral suppression. This further aimed at adjusting model estimates for endogeneity in the two outcome equations (McGovern, Bärnighausen, Marra and Radice, 2015). The semi-parametric predictors for the selection (Adherence to ART) and outcome (viral load suppression) equations and for the copula parameter (θ) were then specified as:

$$log \frac{P(adherence \geq 95\%)}{P(adherence < 95\%)} = \alpha_{10} + \alpha_{11}(sex) + \alpha_{12}(age_i) + \alpha_{13}(marital\ status_i) + \alpha_{14}(WHOstage_i) + \alpha_{15}(CD4counts_i) + \alpha_{16}(facilities_i)$$

$$log \frac{P(viral\ suppression)}{P(no\ viral\ suppression)} = \alpha_{20} + \alpha_{21}(sex) + \alpha_{22}(age_i) + \alpha_{23}(marital\ status_i) + \alpha_{24}(WHOstage_i) + \alpha_{25}(CD4counts_i) + \alpha_{16}(facilities_i)$$

$$Copula\ parameter\ (\theta) = \alpha_{30} + \alpha_{31}(facilities_i), \qquad \text{for } i = 1, \ldots, n$$

The copula function is then specified as:

$$F\left(log \frac{P(adherence \geq 95\%)}{P(adherence < 95\%)}, log \frac{P(viral\ suppression)}{P(no\ viral\ suppression)}\right) = C\left(F_1\left(log \frac{P(adherence \geq 95\%)}{P(adherence < 95\%)}\right), F_2\left(log \frac{P(viral\ suppression)}{P(no\ viral\ suppression)}\right); \theta\right),$$

where θ is an association parameter measuring the dependency between two marginals (Marra & Radice, 2016). We applied the Akaike Information criterion (AIC) and Bayesian information criteria (BIC) for model selection, with smaller AIC and BIC suggesting a better model.

Implementation in R: GJRM Package

The models were analyzed in the R package GJRM (Marra & Radice, 2017; Wyszynski & Marra, 2017). It implements maximum likelihood inference for copulas, and allows the specification of a wide range of non-random selection models, with varied dependence structures. The non-random sample selection model is specified as the combination of a distributional assumption and structural assumptions. The function gjrm( ) provides an interface to fit sample selection models for binary distributional margins. The function is given as

$$\text{gjrm}(formula, data = list(), model =, margins =, BivD = \text{"}N\text{"}, \ldots) \quad (24)$$

where *formula* is a list of two compulsory equations, which in our case are generalized linear models, and *data* is a data frame with a list containing the variables in the models. The component *model* permits specification of a possible sampling scheme- in our case a sample selection, while *margins* indicate the distribution for the two margins. The model also specifies the mean, variance and shape of the selection, and outcome equations and copula parameter (Marra & Radice, 2017). The families of copula linking the selection and outcome equations are found in gjrm and can be specified through *BivD* function, with Normal ("N") considered as default copula. Other possible copula families currently implementable in the package are Frank, FGM, AMH, Student-t, Joe, Gumbel and Clayton and rotations by 0, 90, 180 and 270 degrees.

RESULTS

Descriptive Summaries

A total of 943 patients (45.6% from KHC and 54.4% from OC) were included in the study. Baseline demographic characteristics of all infected patients disaggregated by adherence to ART and viral load suppression at 12 months are presented in Table 8.3. Adherence to ART was 66.7% and 71.2%, respectively, between the two health facilities. About 72.2% male patients adhered to ART treatment compared to 66.6% females. The proportions that achieved viral supporession were similar between the two study health facilities (86.3% and 85.9% at KHC and OC respectively).

Table 8.3. Demographic and clinical characteristics by adherence to ART treatment and viral load suppression at 12 months

	Adherence to ART treatment						Viral load suppression 12 months					
Variables	Yes		No		Total		Yes		No		Total	
	N	%	N	%	N	%	N	%	N	%	N	%
Facility												
KHC	286	66.7	143	33.3	429	100	201	86.3	32	13.7	233	100
OC	365	71.2	148	28.8	513	100	244	85.9	40	14.1	284	100
Sex												
Male	282	72.7	106	27.3	388	100	193	84.3	36	15.7	229	100
Female	369	66.6	185	33.4	554	100	252	87.5	36	12.5	288	100
Age												
< 25 years	79	61.2	50	38.8	129	100	59	92.2	5	7.8	64	100
25-49 years	545	70.6	227	29.4	772	100	366	85.1	64	14.9	430	100
≥ 50 years	26	65.0	14	35.0	40	100	19	86.4	3	13.6	22	100
Marital Status												
Single	536	68.5	246	31.5	782	100	364	86.1	59	13.9	423	100
Married	82	71.3	33	28.7	115	100	58	85.3	10	14.7	68	100
Others	4	66.7	2	33.3	6	100	2	100	0	0.0	2	100
WHO stage												
Stage-I	490	71.1	199	28.9	689	100	354	87.8	49	12.2	403	100
Stage-II	65	69.9	28	30.1	93	100	36	80.0	9	20.0	45	100
Stage-III	65	64.4	36	35.6	101	100	40	80.0	10	20.0	50	100
Stage-IV	15	55.6	12	44.4	27	100	6	66.7	3	33.3	9	100
CD4 at ART initiation												
< 200	236	66.7	118	33.3	354	100	152	79.2	40	20.8	192	100
≥ 200	383	73.8	136	26.2	519	100	271	90.3	29	9.7	300	100
Weight at ART initiation												
< 50	85	60.7	55	39.3	140	100	51	79.7	13	20.3	64	100
50-70	431	72.1	167	27.9	598	100	293	85.4	50	14.6	343	100
> 70	115	71.4	46	28.6	161	100	87	92.6	7	7.4	94	100
Weight at ART 6 months												
< 50	60	65.9	31	34.1	91	100	37	80.4	9	19.6	46	100
50-70	418	77.4	122	22.6	540	100	286	86.7	44	13.3	330	100
> 70	135	81.3	31	18.7	166	100	101	91.7	10	8.3	111	100
Weight at ART 12 months												
< 50	57	76.0	18	24.0	75	100	41	87.2	6	12.8	47	100
50-70	389	84.7	70	15.3	459	100	262	85.9	43	14.1	305	100
> 70	141	84.9	25	15.1	166	100	107	88.4	14	11.6	121	100
Hepatitis B												
Positive	68	73.1	25	26.9	93	100	41	87.2	6	12.8	47	100
Negative	532	69.6	232	30.4	764	100	374	85.6	63	85.6	437	100
ART Regimen												
TDF/FTC/EFV	189	67.5	91	32.5	280	100	146	91.3	14	8.7	160	100
TDF/3TC/EFV	255	70.1	109	29.9	364	100	171	85.9	28	14.1	199	100
TDF/3TC/NVP	186	71.0	76	29.0	262	100	114	81.4	26	18.6	140	100
Others	18	69.2	8	30.8	26	100	12	75.0	4	25	16	100

Slightly more female patients achieved viral load suppression (87.3%) compared to males (84.3%). Of those that achieved viral suppression, 366 (85.1%) were aged between 25 to 49 years, 364 (86.1%) were single, 271 (90.3%) patients had CD4 load of more than 200 at ART initiation, and 41 (9.9%) patients had hepatitis B. Table 3 further shows that patients with body weights between 50 to 70 kg achieved viral load suppression at six months (n=286, 86.7%), and twelve months (n=262, 85.9%). Both ART adherence and viral suppression decreased with advanced clinical stages of the disease. Adherence for WHO clinical stage I was 71.1%, while those at Stage IV were at 55.6%. Similarly viral suppression was 87.8% for stage I patients, and 66.7% for patients at stage IV. Further, adherence did not vary much with type of treatment regimen, but the majority of patients were treated with TDF/FTC/EFV achieved viral suppression (91.3%).

Table 8.4 presents the relationship between ART adherence and health related outcomes: patients' viral load suppression and body weights both at 6 and 12 months. Results revealed that viral load suppression at both 6 and 12 months was associated with ART treatment adherence (p-value<0.000), but only with body weight at 6 months (p-value = 0.009).

Table 8.4. Relationship between adherence level, viral load suppression, and body weight at 6 months and 12 months

Variable	**Adherence**		**P-value**
	at least 95%	**less than 95%**	
Viral load suppression - 6 months	%	%	p<0.000
Yes	77.8	34.8	
No	22.2	65.2	
Viral load suppression-12 months			p<0.000
Yes	89.4	66.7	
No	10.6	33.2	
Weight at ART 6 months			p = 0.009
less than 50	9.8	16.8	
50-70	68.2	66.3	
more than 70	22	16.8	
Weight at ART 12 months			p = 0.168
less than 50	9.7	15.9	
50-70	66.3	61.9	
more than 70	24	22.1	

Binary Regression Analysis of ART Adherence

Unadjusted univariate and adjusted multivariate odd ratios (OR) of factors associated with adherence to ART treatment are given in Table 8.5. In the adjusted multivariate analyses, the table revealed that patients with age between 25 to 49 years of age (OR

1.201, 95% CI 1.032-4.510), female (OR 1.676, 95% CI 1.029-2.730) and those who initiated ART in Okuryangava clinic (OR 1.405, 95% CI 1.087-2.267) were more likely to adhere to ART. Patients with CD4 count greater than or equal to 200 at ART initiation (OR 1.402, 95% CI 1.084-2.321) were more likely to be associated with adherence to ART treatment than those individual with less than 200 CD4 count at ART initiation. Adherence among patients on WHO stage-II (OR= 0.771, 95% CI 0.191-3.119), stage-III (OR 0.580, 95% CI 0.121-2.782) and stage-IV (OR= 0.739, 95% CI 0.167-3.266) were not different from individual on stage-I.

Table 8.5. Univariate and multivariate odd ratios (OR) of the factors associated with adherence

Factors	Unadjusted univariate OR (95% CI)	P-value	Adjusted multivariate OR (95% CI)	P-value
Sex (ref. male)				
Female	0.75 (0.56, 0.99)	0.047	1.676 (1.029, 2.730)	0.038
Age (ref. < 25 years)				
25-49 years	1.17 (0.56, 2.46)	0.669	1.201 (1.032, 4.510)	0.786
≥50 years	0.77 (0.39, 1.50)	0.451	0.85 (0.26, 2.72)	0.787
Facility (ref. KHC)				
OC	1.23 (0.93, 1.67)	0.138	1.45 (1.087, 2.267)	0.164
WHO stage (ref. stage-I)				
stage-II	0.58 (0.23, 1.14)	0.087	0.77 (0.19, 3.11)	0.715
stage-III	0.53 (0.22, 1.29)	0.167	0.58 (0.12, 2.78)	0.496
stage-IV	0.69 (0.29, 1.68)	0.403	0.73 (0.17, 3.266)	0.690
CD4 at ART Initiation (ref. < 200)				
≥ 200	1.41 (1.04 1.89)	0.023	1.40 (1.085, 2.321)	0.190
Weight at ART Initiation (ref. < 50)				
50-70	1.62 (1.02,2.61)	0.050	0.96 (0.25, 3.55)	0.951
> 70	0.96 (0.65, 1.43)	0.872	0.72 (0.26, 1.96)	0.523
Weight at 6 months (ref. < 50)				
50-70	2.25 (1.25, 4.03)	0.006	1.23 (0.25, 6.08)	0.793
> 70	1.27 (0.81, 1.97)	0.285	1.59 (0.49, 5.12)	0.432
Weight at 12 months (ref. < 50)				
50-70	1.78 (0.90, 3.51)	0.096	1.65 (0.45, 5.97)	0.44
> 70	1.02 (0.62, 1.66)	0.953	0.99 (0.38, 2.46)	0.947
Regimen (ref. TDF/FTC/EFV)				
TDF/3TC/EFV	1.08 (0.45,2.58)	0.857	0.41 (0.14, 1.22)	0.109
TDF/3TC/NVP	0.96 (0.46, 2.28)	0.929	0.51 (0.18, 1.43)	0.201
Other	0.92 (0.38, 2.20)	0.851	0.48 (0.16, 1.42)	0.187

Table 8.5 further revealed that patients of weight 50 to 70 kg (OR 0.960, 95% CI 0.258-3.565) and those with greater than 70 kg weight (OR 0.721, 95% CI 0.265-1.962) were less likely to adhere to ART treatment than those individual with less than 50 kg weight.

Finally, patients on baseline ART regimens having tenofovir (TDF), emtricitabine (FTC) and nevirapine (NVP) (with OR 0.412, 95% CI 0.139-1.218) and tenofovir (TDF), lamivudine (3TC) and efavirenz (EFV) (with OR 0.509, 95% CI 0.181-1.434) were less likely to adhere to ART treatment than those whose first regimen contained tenofovir (TDF), emtricitabine (FTC) and efavirenz (EFV).

Binary Regression Analysis of Viral Suppression

Table 8.6 shows the unadjusted univariate and adjusted multivariate odd ratios (OR) of factors associated with viral load suppression. In the adjusted multivariate analyses, patients greater than 50 years (OR 4.798, 95% CI 0.540-42.618) and female (OR 1.003, 95% CI 0.523-1.922) were more likely to achieve viral load suppression. Higher CD4 cell counts (greater than or equal to 200) at ART initiation was associated with viral suppression (OR 2.734, 95% CI 1.397-5.352, p-value = 0.003). Patients receiving treatment in Okuryangava clinic (OR 0.790 95% CI 0.408-1.527) were less likely to achieve suppression than those individual in Katutura hospital center. Patients on baseline clinical stage-II (OR 0.424, 95% CI 0.070-2.554), stage-III (OR 0.468, 95% CI 0.065-3.366) and stage-IV (OR 0.683, 95% CI 0.100-4.678) were also less likely to achieve viral suppression than individual on baseline clinical stage-I (Table 8.6).

Patients with body weight at ART initiation between 50 and 70 kg (OR 3.641, 95% CI 0.520-25.510), or greater than 70 kg (OR 3.650, 95% CI 0.739-18.025), and body weight at six months of between 50 and 70 (OR 6.105, 95% CI 0.813-45.864), and those with greater than 70 kg (OR 1.804, 95% CI 0.392-8.303) were more likely to achieve viral load suppression than those individual with less than 50 kg body weight. However, body weight at twelve months, regardless of category, was not associated viral load suppression. Similarly, viral suppression was not a factor of type of treatment regimens (Table 8.6).

Heckman-Type Model of ART Adherence and Viral Suppression

Table 8.7 summarizes the estimation of the association parameters together with 95% confidence interval bounds from the selection model based on the Normal, Frank, FGM, Ali-Mikhail-Haq (AMH), Student-t and the rotated versions of Joe, Clayton and Gumbel models. The copula models are summarized in Table 8.7. There is support for the hypothesis of selection bias. The rotated copula models indicated a strong negative association between adherence to ART treatment and viral load suppression. In fact, copula by Joe, Gumbel and Clayton exhibited a negative skewed distribution. Further, the presence of tail dependence in sample selection bias, determined by examining the results from Joe, Gumbel and Clayton rotated 90 and 270 degrees, indicated asymmetric dependence between ART adherence and viral load suppression margins.

Table 8.6. Univariate and multivariate odd ratios of the factors associated with viral suppression

Factors	Unadjusted univariate OR (95% CI)	P-value	Adjusted multivariate OR (95% CI)	P-value
Sex (ref. male)				
Female	1.31(0.79, 2.15)	0.294	1.03 (0.53, 1.92)	0.994
Age (ref. < 25 years)				
25-49 years	0.53(0.17, 2.49)	0.423	1.48 (0.18,5.77)	0.763
> 50 years	1.17(0.318,3.85)	0.872	4.79 (0.54, 42.62)	0.159
Facility (ref. KHC)				
OC	0.97 (0.58,1.63)	0.909	0.79 (0.48, 1.57)	0.483
WHO stage (ref. stage-I)				
stage-II	0.27 (0.067, 1.14)	0.076	0.44 (0.07, 2.54)	0.349
stage-III	0.50 (0.14, 2.39)	0.386	0.48 (0.06, 3.36)	0.450
stage-IV	0.50 (0.16, 2.35)	0.381	0.63 (0.10, 4.68)	0.698
CD4 at ART Initiation (ref. <200)				
≥ 200	2.49 (1.45, 4.17)	0.001	2.74 (1.37, 5.35)	0.003
Weight at ART Initiation (ref. <50)				
50-70	3.18 (1.18, 8.46)	0.021	3.64 (0.52, 25.51)	0.193
> 70	2.12 (0.928, 4.86)	0.075	3.65 (0.79, 18.05)	0.112
Weight at 6 months (ref. <50				
50-70	2.47 (0.96, 6.51)	0.071	6.15 (0.83, 45.86)	0.079
>70	1.55 (0.74, 3.22)	0.232	1.84 (0.32, 8.33)	0.449
Weight at 12 months (ref. <50kg)				
50-70	1.18 (0.43,3.18)	0.830	0.19 (0.03, 1.21)	0.078
>70	1.25 (0.69, 2.38)	0.490	0.51 (0.173, 1.88)	0.357
Regimen (ref. TDF/FTC/EFV)				
TDF/3TC/EFV	0.28 (0.08, 1.02)	0.052	0.44 (0.08, 2.12)	0.320
TDF/3TC/NVP	0.49 (0.18, 1.63)	0.246	0.90 (0.17, 4.13)	0.892
Other	0.64 (0.24,2.22)	0.538	1.15 (0.20, 5.32)	0.851

Table 8.8 shows the AIC and BIC for the sample selection models with different copulas. Frank copula followed by Joe copula rotated by 270 degrees had the lowest AIC and BIC values. On the other hand, Vuong test and Clarke test did not indicate any preferred model (results not shown).

Table 8.9 presents the results of the Frank copula model which was a better model. It was observed that ART adherence was marginally associated with being of age between 25 and 49 years, and being a patient who received ART treatment at Okuryangava clinic. On the other hand, patients who had CD4 counts of more than 200 were more likely to adhere to ART treatment (p-value = 0.008). With regards to viral suppression, results show that patients who adhered at least 95% were not associated with viral suppression (p-value = 0.609), while CD4 count of more that 200 was important at achieving viral suppression (p=0.022). Baseline clinical stages, and ART regimens were not associated with both margins. Be as it may, overall there was a viral suppression achieved (based on the intercept, p=0.004).

Table 8.7. Measure of the association between ART adherence and viral load suppression in Heckman-type model

Copula Model	Theta ($\hat{\theta}$) (95% CI)		Kendall's Tau ($\hat{\tau}$) (95% CI)	
Normal	0.049	(-0.857, 0.851)	0.034	(-0.671, 0.669)
Frank	0.645	(-7.570, 7.560)	0.060	(-0.583, 0.544)
FGM	-0.076	(-0.985, 0.955)	-0.017	(-0.219, 0.212)
AMH	0.217	(-0.872, 0.952)	0.082	(-0.161, 0.307)
Student-t	0.074	(-0.833, 0.863)	0.052	(-0.631, 0.711)
Joe 90	-1.460	(-24.20, -1.090)	-0.173	(-0.734, -0.048)
Joe 270	-2.690	(-39.10, -1.230)	-0.356	(-0.943, -0.100)
Gumbel 90	-1.370	(-42.70, -1.100)	-0.222	(-0.860, -0.078)
Gumbel 270	-1.640	(-39.50, -1.080)	-0.308	(-0.924, -0.068)
Clayton 90	-1.740	(-58.10, -0.240)	-0.953	(-0.953, -0.096)
Clayton 270	-0.532	(-46.90, -0.130)	-0.179	(-0.809, -0.057)

Table 8.8. Akaike and Bayesian information criteria (AIC and BIC) for copula functions in Heckman-type model

Model	AIC	BIC	Model	AIC	BIC
Normal	778.69	895.67	Joe 270	779.07	896.05
Frank	778.56	895.54	Gumbel 90	779.20	896.18
FGM	778.86	895.84	Gumbel 270	779.16	896.14
AMH	779.24	896.22	Clayton 90	779.07	896.05
Student-t	778.89	895.88	Clayton 270	779.22	896.21
Joe 90	779.21	896.20			

Table 8.9. Frank copula parameter estimates of the Heckman-type model

Factors	ART treatment adherence			Viral suppression		
	Estimate	Z-value	P-value	Estimate	Z-value	P-value
Intercept	-0.43	-1.520	0.129	-0.557	-2.863	0.004
Sex (ref. Male)						
Female	-0.019	-0.125	0.901	-0.035	-0.211	0.833
Age (ref. <25 years)						
25-49 years	-0.360	-1.677	0.094	0.330	1.099	0.272
≥50 years	0.062	0.168	0.866	0.231	0.539	0.597
Facility (ref. KHC)						
OC	-0.273	-1.810	0.070	0.177	0.701	0.484
WHO Stage(ref. Stage-I)						
Stage-II	-0.097	-0.377	0.706	-0.032	-0.121	0.903
Stage-III	0.041	0.175	0.861	0.093	0.391	0.696
Stage-IV	-0.250	-0.413	0.679	0.626	1.267	0.205
Regimen (ref. TDF/FTC/EFV)						
TDF/3TC/EFV	0.047	0.265	0.791	0.189	0.942	0.346
TDF/3TC/NVP	0.055	0.281	0.778	0.298	1.460	0.144
Other	0.265	0.693	0.488	0.625	1.468	0.142
CD4 at ART Initiation(ref. <200)						
≥200	-0.406	-2.640	0.008	-0.384	-2.299	0.022
ART adherence (ref. <95%)						
≥95%				0.473	0.511	0.609

DISCUSSION

In this retrospective cohort study, the main objective was to fit a joint distributional model to adjust for sample selection bias of treatment adherence and viral load suppression among HIV patients on ART in Windhoek, Namibia. It trying to achieve this we also examined if there was any tail dependence in sample selection bias and to investigate the factors associated with viral load suppression and treatment adherence.

With regard to ART adherence, we found that 71.2% patients treated in Okuryangava clinic, 70.6% patients' age between 25 to 49 years, 66.6% female and 68.5% never married were adherence to ART treatment. This is consistent with the study conducted by Peltzer, Friend-du Preez, Ramlagan and Anderson (2010) where 427 patients (82.9%) were 95% adherent in the month prior to the survey in KwaZulu-Natal, South Africa, nearly three-quarters (73.5%) of the 519 patients who had initiated ARVs were female. However, 62.2% patients' age between 30 and 49 years old and nearly three-quarters (73.3%) were never married. Our finding is inconsistent with the study conducted by Noens, et al. (2009) where 202 patients were recruited from 34 centers in Belgium, of whom 169 were evaluable and the majority of patients were male (55.0%) considered to be non-adherent. We further found that ART adherence was less likely with currently preferred ART regimens having tenofovir (TDF) either emtricitabine (FTC) or nevirapine (NVP) and tenofovir (TDF) either lamivudine (3TC) or efavirenz (EFV) than those first regimen contained tenofovir (TDF) either emtricitabine (FTC) or efavirenz (EFV). These findings in line with those found by Sethi et al. (2003) and Reis et al. (2013).

With regard to viral load suppression, we found that patients' age between 25 to 49 years old, female sex, never married and having CD4 more than 200 at ART initiation, patients with available hepatitis B and treated in Okuryangava clinic rather than in Katutura hospital center achieved viral load suppression. We also found that viral suppression was less likely with baseline clinical stage-II, stage-III and stage-IV than stage-I.

In the adjusted multivariate analyses (ART adherence), we found that female patients were less likely to be associated with ART adherence than male. We also found that ART adherence has not been associated with age group, facility, WHO stage, CD4 load more than 200 at ART initiation, body weight at ART initiation between 50-70 weights and body weight at 6 months, and ART regimen. These results in line with those found by Adiga, Adiga and Malawadi (2016) where adherence differ from CD4 counts but did not find significant between four treatment regimens and CD4 counts.

In the adjusted multivariate analysis (viral load suppression), we found that Higher CD4 cell counts (greater than or equal to 200) at ART initiation was associated with viral suppression. We also found that patients treated with a combination ART regimens containing tenofovir (TDF) either emtricitabine (FTC) or nevirapine (NVP) and tenofovir (TDF) either lamivudine (3TC) or efavirenz (EFV) were less likely to achieve viral load suppression than those whose first regimen contained tenofovir (TDF) either emtricitabine (FTC) or efavirenz (EFV). Our findings were consistent with those found by other studies in demonstrating the importance of the initial ART regimen in the rate of achieving viral load suppression at 12 months (Noens et al, 2009; Heestermans et al., 2016; Olowookere et al., 2016).

In the Heckman-type analysis, the framework allows the user to specify the dependence structure between the margins as documented by Mara and Radice (2017). We found that Joe, Gumbel and Clayton models rotated by 90 and 270 showed a strong negative association between adherence to ART treatment and viral load suppression margins. These agreed with the results found by McGovern et al. (2015) where nine of copula models (normal, Frank, Student-t and Clayton, Joe and Gumbel copulas rotated by 90, 270 degrees) employed and show negative association. A study done by Cunanan (2014) disagreed with the present results of which the copula models show a very strong degree of positive association, but this depends with the nature of the study outcomes.

CONCLUSION

Some demographic and clinical characteristic factors have not been associated with ART adherence and viral load suppression. ART adherence and viral suppression were dependent, even though some demographic and clinical characteristics were not associated with ART adherence and viral suppression. The rotated models indicate a strong negative association between adherence to ART treatment and viral load suppression margins, suggesting that high adherence is required for a reduced viral load.

REFERENCES

Araichi, S., Peretti, C. & Belkacem, L. (n.d.). Time Varying Copula Model for claims reserving in Non-life insurance. Retrieved from: https://www.emlyon.com/minisiteen/content/download/166676/6553128/version/1/file/Christian+de+Perretti.pdf.

Castro, A. (2005). Adherence to antiretroviral therapy: Merging the clinical and social course of AIDS. *PLoS Med* 2(12): e338.

Chen, X., Fan, Y. & Tsyrennikov, V. (2006). Efficient Estimation of Semiparametric Multivariate Copula Models. *Journal of the American Statistical Association.* 101:1228–1240.

Cunanan, K. & Koopmeiners, S. J. (2014). Evaluating the performance of copula models in phase I-II clinical trials under model misspecification. Retrieved from: http://www.biomedcentral.com/1471-2288/14/51.

EpiTools (2018). Simple risk-based surveillance - calculation of sample size. Accessible at http://epitools.eu/content.php?page=RiskBasedSSSimple.

Heckman, J.J. (1979). Sample selection bias as a specification error. *Econometric a* 47, 153-162.

Heestermans, T., Browne, L. J., Aitken, C. S., Vervoort, C. S. & Klipstein-Grobusch, K. (2016) Determinants of adherence to antiretroviral therapy among HIV-positive adults in sub-Saharan Africa: a systematic review. *BMJ Glob Health* 2016; 1:e000125. doi: 10.1136/bmjgh-2016-000125.

Joe, H. (1997). Multivariate Models and Dependence Concepts, Chapman & Hall/CRC. New York. Retrieved from: https://doi.org/10.1201/b13150.

Kheswa, G. J. (2017). Exploring the Factors and Effects of Non-Adherence to Antiretroviral Treatment by People Living with HIV/AIDS. *Indo-Pacific Journal of Phenomenology*. 17:1, 1-11, doi: 10.1080/20797222.2017.1280923.

Kuethe, T. H., Hubbs, T., & Waldof, B. (2009). Copula models for spatial point patterns and processes. Department of Agricultural Economics, Purdue University, West

Lafayette, Indiana, USA. Retrieved from: http://www.ub.edu/sea2009.com/Paper/162.pdf.

Manner, H. (2007). Estimation and Model Selection of Copulas with an Application to Exchange Rates. Department of Quantitative Economics, Maastricht University, Netherlands.

Marra, G. & Radice, R (2017). A joint regression modeling framework for analyzing bivariate binary data in R. De Gruyter Open. Retrieved from https://doi.org/10.1515/demo-2017-0016.

Marra, G., Radice, R., Bärnighausen, T., Wood, S. N., & McGovern, M. E. (2017). A Simultaneous Equation Approach to Estimating HIV Prevalence with Non-Ignorable Missing Responses. *Journal of the American Statistical Association,* 112(518), 484-496. doi: 10.1080/01621459.2016.1224713.

Marra, G. & Wyszynski, K. (2016). Semi-parametric copula sample selection models for count responses. *Computational Statistics & Data Analysis*, 104:110-129. doi: 10.1016/j.csda.2016.06.003.

Marra, G., Papageorgiou, G. & Radice, R (2014). Estimation of a Semiparametric Recursive Bivariate Probit Model with Nonparametric Mixing. Australian & New Zealand Journal of Statistics, https://doi.org/10.1111/anzs.12043.

McGovern, M. E., Bärnighausen, T., Marra, G., & Radice, R. (2015). On the Assumption of Bivariate Normality in Selection Models: A Copula Approach Applied to Estimating HIV Prevalence. *Epidemiology,* 26(2), 229-237. doi:10.1097/EDE.0000000000000218.

Ministry of Health and Social Services (2016). National Guidelines for Antiretroviral Therapy. Fifth edition 2016. Windhoek. Namibia.

Ministry of Health and Social Services (2017). National Strategic Framework for HIV and AIDS Response in Namibia 2017/18 to 2021/22, Directorate of Special Programmes, Namibia.

Nachega, B. J., Mills, E. J. & Schechter, M. (2010). Antiretroviral therapy adherence and retention in care in middle-income and low-income countries: current status of knowledge and research priorities, *Current Opinion in HIV and AIDS*. vol. 5, no. 1, pp. 70–77.

Namibia Population-based HIV Impact Assessment (NAMPHIA) (2018). Preliminary results. Retrieved from: https://www.cdc.gov/globalhealth/countries/namibia/pdf/Namibia_FactSheet.pdf.

Nelsen, R. B. (1999). An Introduction to Copulas. Lecture Notes in Statistics. New York: Springer.

Noens, L., Van Lierde, M., De Bock, R., Verhoef, G., Zache´e, P., Berneman, Z., Martiat, P., Mineur, P., Van Eygen, K., MacDonald, K., De Geest, S., Albrecht, T. & Abraham, I. (2009). Prevalence, determinants, and outcomes of nonadherence to imatinib therapy in patients with chronic myeloid leukemia: the ADAGIO study.

Prepublished online as Blood First Edition paper; doi: 10.1182/blood-2008-12-196543.

Olowookere, S. A., Fatiregun, A. A., Ladipo, M. M. A., Abioye-Kuteyi, E. A. & Adewole, I. F. (2016). Effects of adherence to antiretroviral therapy on body mass index, immunological and virological status of Nigerians living with HIV/AIDS. Alexandria University Faculty of Medicine; *Alexandria Journal of Medicine* (2016) 52 51-54.

Payandeh, T. A., Farid-Rohani, R. M., & Qazvini, M. (2013). A GLM-Based Method to Estimate a Copula's Parameter(s). *JIRSS* (2013). Vol. 12, No. 2, pp 321-334.

Peltzer, K., Friend-du Preez, N., Ramlagan, S. & Anderson, J. (2010). Antiretroviral treatment adherence among HIV patients in KwaZulu-Natal, South Africa. Retrieved from: http://www.biomedcentral.com/1471-2458/10/111; *BMC Public Health* 2010, 10:111.

Reda, A. A. & Biadgilign, S. (2012). Determinants of Adherence to Antiretroviral Therapy among HIV-Infected Patients in Africa. Hindawi Publishing Corporation AIDS Research and Treatment Volume 2012, Article ID 574656, doi:10.1155/2012/574656.

Reis, A., Guerra, M. P. & Lencastre, L. (2013). Treatment Adherence Quality of Life and Clinical Variables in HIV/AIDS Infection. Retrieved from http://dx.doi,org/10.4236/wja.2013.33032 Published Online September 2013 (http://www.scirp.org/journal/wja).

Rougemont, M., Stoll, E. B., Elia, N & Ngang, P. (2009). Antiretroviral treatment adherence and its determinants in Sub-Saharan Africa: a prospective study at Yaounde Central Hospital, Cameroon.

Rüschendorf, L. (2013). Copulas, Sklar's Theorem, and Distributional Transform. Mathematical Risk Analysis, Springer Series in Operations Research and Financial Engineering, doi: 10.1007/978-3-642-33590-7 1, © Springer-Verlag Berlin Heidelberg.

Sethi, K. A., Celentano, D. D., Gange, J. S., Moore, D. R. & Gallant, E. J. (2003). Association between Adherence to Antiretroviral Therapy and Human Immunodeficiency Virus Drug Resistance. Retrieved from: https://academic.oup.com/cid/article-abstract/37/8/1112/435290; Clinical Infectious Diseases Society of America 2003; 32:1112-3.

Sklar, A. (1959). Fonctions de répartition àndimensions et leurs marges. Publications de l'Institut de Statistique de l'Université de Paris 8, pp. 229–231.

Smith, M. D. (2003). Modelling sample selection using Archimedean copulas. Wiley on behalf of the Royal Economic Society. *The Econometrics Journal,* Vol. 6, No. 1 (2003), pp. 99-123.

UNAIDS (2016). Global AIDS update. Retrieved from http://www.unaids.org/en/resources/documents/2016/Global-AIDS-update-2016. *Accessed on May 19, 2017.*

UNAIDS (2016). 90-90-90 On the right track towards the global target. Retrieved from https://reliefweb.int/sites/reliefweb.int/files/resources/90_90_90_Progress_ReportFINAL pdf.

UNAIDS (2017). UNAIDS data 2017. Retrieved from http://www.unaids.org/sites/default/files/media_asset/20170720_Data_book_2017_en.pdf.

Van der Klaauw, B. & Koning, R. H. (2003). Testing the normality assumption in the sample selection model with an application to travel demand. *Journal of Business and Economic Statistics,* 21, 31-42. doi: 10.1198/073500102288618739.

WHO (2017). Antiretroviral Therapy. Available at http://www.who.int/topics/antiretroviral_therapy/en. *Accessed on May 20, 2017.*

Wojtyś, M., Marra, G., Radice, R. (2016). Copula Regression Spline Sample Selection Models: The R Package Semi Par Sample Sel. *Journal of Statistical Software.*

Wyszynski, K. & Marra, G. (2017). Sample selection models for count data in R. *Computational Statistics,* 33(3):1385–1412. 10.1007/s00180-017-0762-y.

In: Statistical Modelling …
Editors: N-B. Kandala and L. Kazembe
ISBN: 978-1-53615-981-3

Chapter 9

COPULA-LINKED GENERALIZED JOINT REGRESSION MODEL FOR WATER, SANITATION AND HYGIENE (WASH) COVERAGE IN NAMIBIA

Anastasia Johannes[*] and Lawrence Kazembe
Department of Statistics and Population Studies,
University of Namibia, Windhoek, Namibia

ABSTRACT

Understanding the spatial distribution of indicators is of vital importance, as it help with national planning and decision-making. Different models such as conditional autoregressive (CAR) can be used in spatial modelling. However, these models pose analytic and statistical inferential challenges, as they assume a normal distribution in spatial processes thus inappropriate when non-continuous and multivariate response are involved. In this study, therefore, a flexible spatial approach using copula models is proposed. Copula provides a novel way to model multivariate outcomes by decomposing data structure from the joint distribution into marginal distributions to define the dependence structure. Copula-based spatial modelling approaches were applied to model the discrete outcome variables. Specifically, a generalized joint regression modelling (GJRM) was used to model the joint spatial dependency of WASH indicators, assuming data was aggregated at areal (region) level. The Markov random field (MRF) smoother was applied to region as a geographical variable, while for continuous variables such as age and household size, were modelled as smooth spline functions. Results indicated that residence and wealth index were likely to be associated with improved drinking water, sanitation and hygiene. In addition, the study showed that there was a little regional spatial variation in improved drinking water.

[*]Corresponding Author's Email: stacy.johannes@gmail.com/lkazembe@yahoo.com.

Keywords: Gaussian copula marginal regression, generalized joint regression modelling, hygiene, sanitation, water Namibia

INTRODUCTION

Improved drinking water, sanitation and hygiene (WASH) have been called one of the best remedial advances since 1840 and has been classified as the overlooked establishment of wellbeing (Magalhaes, Barnett, & Clements, 2011). According to Hutton and Chase (2016), WASH are basics for improved standard of living, which include the protection of health and the environment, improved educational outcomes, dignity and gender equality. Furthermore, poor populations have brought down progress to improved WASH services and poorly connected practices. Henceforth, drinking water and sanitation is part of the Sustainable Development Goals (SDGs), Goal 6, which focuses on accomplishing general access to WASH by 2030 (Hutton & Chase, 2016, p. 11). Additionally, the Human Right to Safe Drinking Water and Sanitation (HRTWS) was approved in 2010 under a United Nations (UN) Resolution to guarantee sheltered, moderate, adequate, available, and accessible drinking water and sanitation administration for all. Furthermore, one of the objectives of the Namibia vision 2030 is to have access to portable drinking water for all people (National Planning Commission [NPC], 2004).

Although progress has been made to achieve improved WASH, still many developing countries in the global South lag behind. Roughly, 80% of the general population utilising water from unimproved sources are amassed in three regions: sub-Saharan Africa, Eastern Asia and Southern Asia. In sub-Saharan Africa, progress was made from 49% coverage in 1990 to 56% in 2004. For sanitation, general levels of utilization of improved facilities are far lower than for drinking water. In 2004, 59% of the world population had access to any type of improved sanitation facility at home, from 49% in 1990 (WHO, UNICEF, 2006). Notwithstanding such enhancement, national and sub-national inequalities in WASH scope endure, including rural-urban and regional disparities.

In Namibia, the situation is similar. According to the Namibia Demographic and Health Survey (NDHS) (2014), 87% of households have access to improved drinking water, while only 34% have access to improved sanitation. The survey further indicated that inadequate sanitation is poor in rural areas with only 17% having access to improved sanitation, as compared to urban areas with 49%. Furthermore, 98% of households in urban areas have access to improved drinking water compared to 76% of households in rural areas. However, improved drinking water scarcity continued to be a concern for the government and the population at large.

According to the United Nations Children's Fund (UNICEF) (2016), 88% of deaths caused by diarrhoea are because of the absence of access to safe drinking water, poor sanitation and hygiene. Various studies have demonstrated that one of the most inequalities in access to improved drinking water and sanitation relates to location, ordinarily measured in terms of urban versus rural areas (Yu, Bain, Mansour, & Wright, 2014). An analysis that was done in 19 Latin American and Caribbean countries, showed that location adds to the gap in access to improved water for 17 countries (Yu et al. 2014) and this applies to sanitation.

Spatial modelling of WASH indicators is of importance as it can help with the investigation of distribution and patterns of WASH indicators for targeted interventions. Distinctive models, for example, the Conditional Autoregressive (CAR) models have been used in various spatial modelling studies utilizing the Markov Chain Monte Carlo (MCMC) algorithm. CAR models are for the most part used to provide results measured at areal level. When undertaking CAR modelling of areal data, the adjacency matric that portrays the neighbourhood of the data is characterized. However, CAR model proposed by Besag, York and Mollie (1991) exhibit a single global level of spatial smoothness determined by geographical adjacency and thus not flexible enough to capture complex structure. In addition, CAR model often assume normality, which may not be achieved in this case. According to Hughes and Haran (2013), CAR model poses collinearity problem that leads to poor estimation performance for the fixed effects.

Therefore, this study proposed a joint model for discrete outcomes using copulas as an alternative to CAR model. The significance of copulas lies in the fact that by a method of change, it can be used to generate joint distribution functions for random variables with arbitrary non-uniform marginal distributions (Winkelmann, 2011). Copula framework has attracted much consideration in spatial modelling in the course of recent years (Kazianka & Pilz, 2009). This is because of their adaptability in portraying conditionals and the likelihood of separating marginal and joint effects (Perrone & Muller, 2016). According to Perrone and Müller (2016), copula has been applied in many application areas of applied statistics such as insurances, econometrics, medicine, marketing, spatial extreme events, time series analysis and even sports.

Bhat and Sener (2009) used a copula-based approach for spatial dependence modelling based on a spatial logit structure rather than a spatial probit structure, to model teenagers' physical activity participation levels. They showed that copula method can be used to consider locational information of neighbouring observations as a way to control for spatial dependence.

Copula modelling has also been applied using a Bayesian approach (Smith, 2011). Previous Bayesian work on copula modelling includes that of Huard, Evin, and Favre (2006), who recommended a technique selected between various bivariate copulas and Silva and Lopes (2008) who use Markov Chain Monte Carlo (MCMC) methods to estimate low dimensional parametric copula functions. Pitt, Chan, and Kohn (2006), Hoff

(2007) and Danaher and Smith (2011) estimate Gaussian copula regression models using MCMC methods. However, adopting a Gaussian copula does not mean the data are normally distributed. Smith, Gan, and Kohn (2010) extended the work of Pitt et al. (2006) to copulas derived by inversion from skewed *t*-distributions constructed by hidden conditioning.

Regardless of their boundless use in other quantitative fields, copula methods have not yet been utilized widely in spatial or biomedical science. According to Kuethe, Hubbs, and Waldolf (2009), copula techniques in the spatial area have beforehand been recommended in a predetermined number of other geographic sciences, for example in geostatistics and geophysics. Kazianka and Pilz (2009) use copula-based models for spatial interpolation in a study of inadvertent arrivals of radioactive materials because of their flexible tool to perform spatial interpolation for non-Gaussian, multimodal and extreme values.

According to Zimmer and Trivedi (2006), many modelling approaches either use fully parametric model or use few data semiparametric framework that requires partial specification of the interdependent equation. However, these approaches poses challenges because their joint distribution of variables of interest maybe unknown, thus making it difficult to model. In addition, their joint distribution are less tractable, thus marginal distribution of joint modelling provide an alternative to model less difficulties. Hence, a joint copula approach then allows to effectively capture dependence structure of data, without giving up appealing properties of the marginal. Moreover, joint model have the prospect to decrease estimate bias (Sudell, Kolamunnage-Dona, & Tudur-Smith, 2016) and enable the inclusion of covariates that might have errors. However, this study will instead apply the Gaussian copula approach, to model spatial distribution of improved drinking Water, Sanitation and Hygiene (WASH) indicators in Namibia with respective to regional level.

Joint analysis of WASH indicators is of importance as improved drinking water and sanitation alone would not guarantee health, but rather adding hygiene is required to improve health (Garriga & Foguet, 2013), thus allowing to establish the association between these outcomes. In addition, WASH indicators are often handled jointly as without improved water or proper sanitation, hygiene can be an issue as well. According to Costello (2013), water is important for hygiene practice especially handwashing with soap. If a household lacks water, risky hygiene behavior such as not washing hands after critical time are likely to develop. Furthermore, Ntoziniet al. (2015) pointed out that access to improved drinking water and sanitation are important determinants of behavioral response to improved hygiene and intervention. The aim of this chapter, therefore, is to model spatial variability of WASH indicators in Namibia, through a copula-linked joint model that will permit simultaneously analyzed the three indicators.

MODELLING SPATIAL DATA

Consider data aggregated at areal levl, then $Y_1, \dots, Y_n$ consititute observed summaries in each area, $i = 1,2, \dots, n$. Conditional autoregressive (CAR) models are mostly used to analyse areal-aggregated spatial data. According to Paciorek (2013), CAR represents the spatial dependence structure such that areas that share boundaries are considered neighbours. This is called first Tobler's law of geography (1970), which states that nearer things are similar.

The spatial model structure for areal data proceeds as follows. Let the mean $\mu_i = E(Y|X_i, g)$ be related via a link function, $h(\cdot)$ to a linear predictor. The structure for areal data is defined as:

$$h(\mu_i) = X_i^T \beta + F_i g \tag{1}$$

where F_i is the ith row of a mapping matrix, X_i is the spatial predictor for the ith areal unit having regression coefficient β, while $g(\cdot)$ is the unknown latent spatial process represented as a piecewise constant surface on a fine rectangular grid, given as:

$$g \sim N(0, ({}_k H)^{-1}),$$

where H is a CAR precision matrix and k a precision parameter, recognizing the potential singularity of H by using the generalized inverse notation (Paciorek, 2013, p. 949). According to De Oliveira (2012), CAR models have been widely used for the analysis of spatial data in different areas, such as demography, economy, epidemiology and geography, as models for latent and observed variables. When undertaking CAR modelling of data at an areal level, it is important to describe a nearness grid that depicts the neighbourhood structures of the information being analysed. For a geographic region that is divided into sub regions indexed by integers $1,2, \dots, n$, nearness considered for the dependence structure for the CAR model follows a neighbourhood system defined by $\{N_i : i = 1, \dots, n\}$, where N_i denotes the sub regions that, in a well-defined sense, are neighbours of sub region i. The neighbourhood system must satisfy the conditions for any $(i,j) = 1, \dots, n$, where $j \in N_i$ if and only if $i \in N_j$ and $i \notin N_i$. A commonly used example in this application is the neighbourhood system defined in terms of geographic adjacency (De Oliveira, 2012),

$$N_i = \{j : sub\ regions\ i\ and\ j\ share\ a\ boundary\}, i = 1, \dots., n.$$

Dependency Modelling using Copula

Spatial modelling often relies on the Gaussian assumption, which is often not satisfied for environment process (Kazianka & Pilz, 2009; Kuetheet al. 2009; Blanchet and Davison, 2011) pointed out that spatial modelling are often difficult to estimate in applied work due to complex nonlinearities and non-typical aspect of data. Therefore, an approach using copula has recently emerged as an alternative to the parametric spatial modelling as it allows for flexible modelling spatial dependency.

Defining Copula

Bohdalova and Nanasiova (2006) stated that, an *n*-dimensional copula is a multivariate cumulative distribution function (CDF), ***C***, with uniformly distributed margins on [0,1] (i.e., ***U***(0,1)) and has the following properties:

1. ***C***: $[0, 1]^n \rightarrow [0,1]$;
2. ***C*** is grounded and *n*-increasing;
3. ***C*** has margins C_i which satisfy $C_i(u) = C(1, \dots, 1, u, 1, \dots, 1) = u$ for all $u \in [0,1]$.

If $F_1, \dots, F_n$ are univariate distribution functions, then $C\big(F_1(x_1), \dots, F_n(x_n)\big)$ is a multivariate CDF with margins $F_1, \dots, F_n$ since $U_i = F_i(X_i), i = 1, \dots, n$ is a uniform random variable. Copula functions are a useful tool to construct and simulate multivariate distributions (Bohdalova & Nanasiova, 2006). The Sklar's theorem (Sklar, 1959) is of importance with regard to copula and is used in all the application of copula (Baek, 2010). It is denoted as follows. Let F be an *n*-dimensional distribution function with margins $F_1, \dots, F_n$. Then there exists an *n*-copula C such that for all $x \in \mathrm{R}^n$,

$$F(x_1, \dots, x_n) = C\big(F_1(x_1), \dots, F_n(x_n)\big) \tag{2}$$

If $F_1, \dots, F_n$ are all continuous, then C is unique. Conversely, if C is an *n*-copula and $F_1, \dots, F_n$ are distribution functions, then the function F defined above is an *n*-dimensional distribution function with margins $F_1, \dots, F_n$ (Nelsen, 2006).

There are several copula functions that can be used in modelling and some are summarised in Table 9.1. The range of [-1, 1] for *t*-copula permit a typical correlation capturing both negative and positive correlation with range of -1 and 1. The Clayton and Gumble copulas capture only positive correlation and permit tail dependences, while the Frank copula allow both positive and negative dependence. As for the Gaussian, when the dependence parameter approaches −1 and 1, it attains the Frechet lower and upper bound.

Table 9.1. A summary of selected copula functions (Trivedi & Zimmer, 2005)

Copula type	Function $C(u_1, u_2)$	θ-domain
t-copula	$\int_{-\infty}^{t_{\theta 1}^{-1}(u_1)} \int_{-\infty}^{t_{\theta_2}^{-1}(u_2)} \frac{1}{2\pi(1-\theta_2^2)^{\frac{1}{2}}} * \left\{1+\frac{s^2-2\theta_2 st+t^2}{v(1-\theta_2^2)}\right\}^{(\theta_1+2)/2} dsdt,$	$\theta_1, \theta_2 \epsilon$ [-1,1]
Frank	$-\theta^{-1}\log\left\{1+\frac{(e^{-\theta u1})(e^{-\theta u2}-1)}{e^{\theta}-1}\right\}$	$\theta \epsilon (-\infty, \infty)$.
Clayton	$\left(u_1^{-\theta}+u_2^{-\theta}-1\right)^{-1/\theta}$	$\theta \epsilon (0, \infty)$.
Gumbel	$\exp\left(-(\bar{u}_1^{\theta}+\bar{u}_2^{\theta})^{\frac{1}{\theta}}\right)$	$\theta \epsilon [1, \infty)$
Gaussian	$\Phi_G(\Phi^{-1}(\mu_1), \Phi^{-1}(\mu_2); \theta)$	$\theta \epsilon$ [-1, 1]

The Gaussian Spatial Copula Model

In the spatial statistical framework, copula can be used to describe the joint multivariate distribution corresponding to the variables with the points of the studied area (Bardossy, 2006).

In the class of random fields of copula, the Gaussian random field is of importance where all multivariate distribution follows a Gaussian distribution, with the mean vector of 0 and covariance-variance matrix Σ. The dependency structure of the Gaussian copula is parameterized by assuming that its correlation function follows one of the classical geostatistical models with parameter θ, for example, the Matern model (Kazianka & Pilz, 2009). The vector of continuous, discrete or categorical dependent random variables can be expressed as $Y = (Y_1, \ldots, Y_n)^T$ and the corresponding realizations of $y = (y_1, \ldots y_n)^T$. Dependence can be of different forms, either in georeferenced, correlated or clustered data. In a situation where the primary scientific objective is to evaluate how the distribution of Y_i varies according to changes in a vector of p covariates $x_i = (x_{i1}, \ldots, x_{ip})^T$, dependence will be regarded as a secondary, but significant aspect (Masarotto & Varin, 2012).

According to Masarotto & Varin (2012, p.1519), let $p_i(y_i; \lambda) = p(y_i|x_i; \lambda)$ be the density function of Y_i given x_i so that covariates are allowed to affect not only the mean of Y_i but also the univariate marginal distribution. With the absent of other assumptions about the nature of the dependence among the responses, inference on the marginal parameters λ can be conducted with the pseudo-likelihood constructed under working independence assumptions. Consider,

$$\mathcal{L}_{ind}(\lambda, y) = \prod_{i=1}^{n} p_i(y_i; \lambda). \quad (3)$$

If the marginal $p_i(y_i;\ \lambda)$are correctly specified, then the maximum independence likelihood estimator $\lambda_{ind} = \arg_{max\lambda}\ \ \mathcal{L}_{ind}(\lambda; y)$consistently estimates λ with no specification of the joint distribution of Y given the model matrix X=$(x_1, \ldots, x_n)^T$. Despite this important robustness property, there are various causes of concern with the above estimator. Firstly, it may suffer from considerable loss of efficiency when dependence is appreciable. Secondly, standard likelihood theory does not apply and corrected standard errors should be based on sandwich-type formulas (Masarotto & Varin, 2012, p.1519). Finally, predictions that do not take into account dependence may be of poor quality.

It is for these reasons that complementing the regression model $p_i(y_i;\ \lambda)$with a dependence structure is important for attaining more precise inferential conclusions and predictions. The ideal model would contain all the possible joint distributions of Y with univariate marginals $p_i(y_i;\ \lambda)$, for $i = 1, \ldots, n$ (Masarotto &Varin, 2012).

GENERALIZED JOINT REGRESSION MODEL

Regression models are commonly fitted with one response variable and a set of covariates (Filippouet al. 2017). However, multiple response variables can also be fitted. Generalizee joint regression model (GJRM) can be applied to continuous, discrete and mixed correlated outcomes. The most common choice is the Gaussian copula given by

$$C(u_1, u_2, u_3|\rho) = \Phi_3(\Phi^{-1}(u_1), \Phi^{-1}(u_2), \Phi^{-1}(u_3)) \quad (4)$$

The dependency in eqn.5 has been captured using a trivariate Gaussian copula cumulative distribution function $C(u_1, u_2, u_3|\rho)$, where u_1, u_2, u_3 are the random variables, Φ is the standard normal distribution, and Φ_3 is the standard trivariate normal distribution, with dependence parameter ρ.

In general, a trivariate binary model can be expressed as (Filippouet al. 2017),

$$y_{xi} = V_{xi}^T\alpha_x + \sum_{v_x=1}^{\tilde{N}x} S_{xv_x}\left(Z_{xv_xi}\ \right) + \epsilon_{xi}, i = 1, \ldots, n, \forall_x = 1,2,3 \quad (5)$$

where n is the sample size, y_{xi} is a latent continuous variable corresponding to an observed outcome, w_{xi} such that if $y_{xi} > 0$ then $w_{xi} = 1$ or 0 otherwise, V_{xi} is the binary or categorical outcome, α_x is a vector that represent the effect of the variables in V_{xi} and $S_{xv_x}\left(Z_{xv_xi}\ \right)$ is a smooth function of continuous covariate$\left(Z_{xv_xi}\ \right)$, $\forall v_x =$ $1, \ldots, \tilde{N}_x$ with $\tilde{N}_x$ being the number of smooth terms. The error term, ϵ_{xi}, is assumed to follow a trivariate normal distribution with $(\epsilon_{1i}, \epsilon_{2i}, \epsilon_{3i})' \sim N_3(0, \sum)$ such that

$$\Sigma = \begin{pmatrix} 1 & \theta_{12} & \theta_{12} \\ \theta_{21} & 1 & \theta_{23} \\ \theta_{31} & \theta_{32} & 1 \end{pmatrix}.$$

It has been noted that constructing trivariate copula can be challenging, however, Song, Li and Yuan (2009) pointed out that Gaussian copula is an exception because of its flexible dependence structure. Copula dependence and marginal parameters can easily be estimated through Gaussian method. Each parameter depends on the joint model predictor including different type of covariates effects such as linear, non-linear, random and spatial effects (Marra & Radice, 2017a). A joint cumulative distribution function (CDF) of three discrete random variables Y_1, Y_2, Y_3 can be expressed as

$$F(y_1, y_2, y_3) = C(F_1(y_1|u_1, \sigma_1, v_1), F_2(y_2|u_2, \sigma_2, v_2), F_3(y_3|u_3, \sigma_3, v_3); \xi, \theta), \qquad (6)$$

where $(u_1, \sigma_1, v_1, u_2, \sigma_2, v_2, u_3, \sigma_3, v_3, \xi, \theta)^T, F_1(y_1|u_{1,}\sigma_1, v_1), F_2(y_2|u_2, \sigma_2, v_2)$ and $F_3(y_3|u_3, \sigma_3, v_3)$ are marginal CDFs of Y_1, Y_2, Y_3 and u_m, σ_m and v_m for $m = 1,2,3$ are marginal distribution parameters, *C* is a defined 3-place copula function with dependence coefficient θ and ξ is the number of degree of freedom (Marra & Radice, 2017a). In terms of joint modelling, a maximum likelihood approach is used as it can combine random effects.

DATA AND MODEL FITTING

Data

We apply GJRM to study the spatial distribution of WASH indicators in Namibia. The study used secondary data acquired through a cross-sectional design done on the 2013 NDHS data by MoHSS. A stratified sampling method was employed at a two-stage cluster sampling. The first stage selected the Primary Sampling Units (PSUs) and the second stage, a fixed number of households were systematically selected, covering all the 13 administrative regions in Namibia (MoHSS& Macro International Inc, 2014). A total of 9 849 households were successfully interviewed, giving a response rate of 92%. Full details of the survey can be found in the report by MoHSS (MoHSS & Macro International Inc, 2014).

The three binary response variables of interest were: household has improved drinking water source, improved sanitation, and improved hygiene. According to UNICEF (2016), an improved drinking water source refers to those that adequately protects the source from outside contamination, and include water piped into a dwelling, piped to yard or plot, public tap/standpipe, tube well or borehole, protected well, protected spring, rainwater and bottle water. A non-improved water source is not protected from outside contamination, such as an unprotected well, an unprotected spring and tanker or cart with a drum.

Table 9.2. Description of key variables

Variable	Description
Response variables	
Water	Source of drinking water: 1=improved drinking water source; 0=non-improved drinking water source
Sanitation	Sanitation facilities: 1= improved shared and not shared toilet facility; 0= non-improved toilet facility
Hygiene	Hand washing: 1= Place where household members wash their hands and presence of water and soap; 0= No place for washing hands, no water or no soap
Covariates	
Sex of household head	1 = Male; 2= Female
Household size	Number of people living in the house, fitted as continuous
Age of household head	The age of the head of the household, fitted as continuous
Place of residence	1= Urban, 2 = Rural
Household Wealth index	1= Poor, 2= Middle, 3= Rich
Women literacy	0= Cannot read at all 1= Able to read only part of sentence 2= Able to read whole language 3= No card with required language 4= Blind/visually impaired
Frequency of reading newspaper or magazine	0= Not at all; 1= Less than once a week; 2-At least once a week
Frequency of listening to radio	0= Not at all; 1= Less than once a week 2-At least once a week
Frequency of watching television	0= Not at all; 1= Less than once a week 2-At least once a week
Highest educational level attained by household head	0= No education; 1= Primary 2= Secondary 3= Higher
Current Marital Status of household head	0= Never married; 1= Married; 2= Living together; 3= Widowed; 4= Divorced
Respondent Occupation	0= Not working/unemployed/student/retired; 1= Unskilled; 2= Skilled 3= Low Skilled; 4= High Skilled
Respondent working status	0= No 1= Yes
Spatial Effect	
Region	Administrative boundaries. Data were collected from 13 regions during the 2013 NDHS.

Improved sanitation mean facilities that hygienically separate human excretion from human contact (UNICEF, 2016). These facilities are flush/pour flush to piped sewer system, flush/pour flush to a septic tank, flush/pour flush to pit latrine, ventilated improved pit (VIP) latrine, pit latrine with slab, and composting toilet. The non-improved sanitation are the flush/pour flush not to sewer/septic tank latrine, pit latrine without slab/open pit, bucket, hanging toilet/hanging latrine and no facility/bush/field. Meanwhile, improved hygiene is the availability of water and soap for handwashing.

Table 9.2 shows the description of the variables that were used in the study. The choice of the covariates was based on studies by Tiwari and Nayak (2013), Yang et al.(2013), Koskei, Koskei, Koske, and Koech (2013) and Adam, Boateng, and Amoyaw (2015) as they were found to predict access to improved drinking water, sanitation and hygiene.

Model Fitting

This study applied a copula regression to joint analysis of the three WASH indicators, quantify the determinants of improved water, sanitation and hygiene, and capture spatial dependency. A Gaussian copula was assumed. The MRF smoother was applied to model the spatial dependence based on the geographic location of survey respondents, as introduced in the mgcv package by Wood (2017). The variable region was fitted as an unstructured random effect. This method is common when the geographic areas or countries of interest are divided into discrete connecting geographic units or regions (Marra, Radice, Bärnighausen, Wood, & McGovern, 2017).

Based on the 2013 NDHS data, there were 13 geographic regions and this allowed us to use the information contained in the neighbouring observation that is located in the same country. Spatial pattern suggests that areal observations close to each other have more similar values than those that are far from each other. Thus,

$$N_{ij} = \begin{cases} 1 & if\ i\ and\ j\ are\ neighbours \\ 0 & otherwise \end{cases}$$

For continuous variables such as age and size of household, the smooth function was applied using the regression spline approach by Eilers and Marx (1996). These splines are used to model non-linear effects for variables such as age and household size.

Model fitting used a SemiParTRIV () function in the GJRM package by Marra and Radice (2017b) in R to fit the model. A SemiParTRIV fits a flexible trivariate binary model with several types of covariate effects. One joint model was fitted for WASH indicators. The model was fitted with spatial effects to capture regional spatial variation using a penalized likelihood estimation. The non-linear effects were included in the model to capture non-linearity for discrete variables such as household size and age. All the trivariate response used were discrete. The WASH indicators in the trivariate were assumed to be correlated as an unrelated regression model.

Considering a trivariate model and let y be the binary indicator at location i. Thus the joint (trivariate) model is given by

$$y_i = V'\alpha + s(nonlinear) + s(spatial)$$

where $V'\alpha$ are the fixed effects fitted and s is the smooth function applied.

In this study, the equation of GJRM fitted for Table 9.2 was as follow:

$$\begin{aligned}
y_{1i} &= \text{logit}\, p(improved\ hygiene_i) \\
&= v_{1i}^T\alpha_{11} + s_{11}(Members) + s_{12}(Age) + s_{13}(Region) + \varepsilon_{1i} \\
y_{2i} &= \text{logit}\, p(improved\ water_i) \\
&= v_{2i}^T\alpha_{21} + s_{21}(Members) + s_{22}(Age) + s_{23}(Region) + \varepsilon_{2i} \\
y_{3i} &= \text{logit}\, p(improved\ sanitation_i) \\
&= v_{3i}^T\alpha_{31} + s_{31}(Members) + s_{32}(Age) + s_{33}(Region) + \varepsilon_{3i}
\end{aligned}$$

where $y_{1i,}$ y_{2i} and y_{3i} are binary and categorical variables, α_{11}, α_{21} and α_{31} represent the effects of the variables in y_i, while s_u for $u = 1, 2, 3$ are smooth functions of members, age and region and ε_i is the error terms. The variables included in $y_{1i,}$ y_{2i} and y_{3i} were: sex, marital status, wealth, literacy, newspaper, radio, TV, respondent occupation and working status (Table 9.2).

RESULTS

Descriptive Summaries

The distribution of percentages of households with improved drinking water and sanitation in Namibia at regional level is shown in Figure 9.1. Figure 9.1(a) shows that many regions have households with improved drinking water, except for Kavango, Kunene and Ohangwena regions. For improved sanitation, Figure 9.1(b) indicates that the households with a moderate percentages of improved sanitation (60-80%) were in the Khomas, !Karas, Hardap, Erongo and Otjozondjupa regions. Households with low percentages of improved sanitation were found in Kunene, Zambezi, Kavango, Ohangwena, Omusati and Oshana regions. The distribution of improved hygiene in Figure 9.1 (c) shows that households with low improved hygiene (0-20%) were found in !Karas, Hardap, Kavango, Zambezi, Ohangwena, Omusati, Oshikoto and Oshana regions. The point maps in Figure 9.1 present results based on regional distribution. However, it is helpful to provide mean percentage per region as it gives an overall indication of household having access to improved drinking water, sanitation and hygiene. Table 9.3 presents results of the mean percentage, minimum percentage and maximum percentage of improved drinking water, sanitation and hygiene per region.

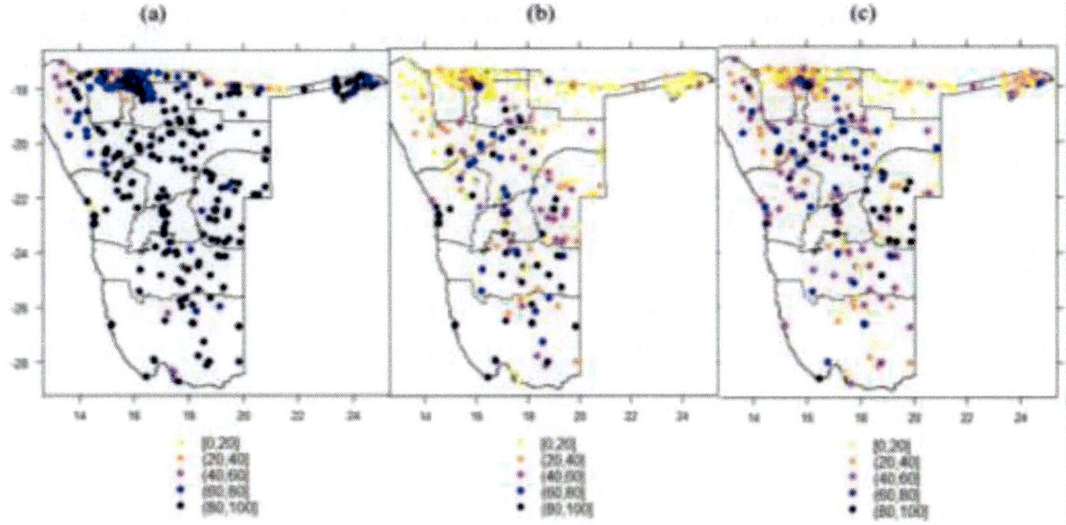

Figure 9.1. Point maps showing percentages of households with improved drinking water, sanitation and hygiene in Namibia.

Table 9.3. Summaries of percentage coverage of improved drinking water, sanitation and hygiene

Region	Improved drinking water					Improved sanitation					Improved hygiene				
	Mean	Median	SD	Min	Max	Mean	Median	SD	Min	Max	Mean	Median	SD	Min	Max
Zambezi	87.7	100	20.8	25.0	100	20.9	8.1	30.9	0	100	36.5	30	23.3	5.0	85.7
Erongo	97.9	100	11.9	10.0	100	85.3	100	29.2	0	100	60.6	55.3	20.9	11.1	100
Hardap	92.7	100	16.9	12.5	100	64.6	78.9	36.2	0	100	54.6	52.9	25.9	0	95.0
!Karas	92.8	100	13.1	45.0	100	66.5	71.4	32.5	0	100	49.2	55	27.0	0	100
Kavango	70.9	94.4	37.2	0	100	22.2	10	28.2	0	100	23.2	16.7	23.9	0	85.0
Khomas	99.1	100	2.8	86.7	100	80.1	94.7	29.6	0	100	66.3	72.2	26.3	6.7	100
Kunene	84.5	97.1	20.4	23.1	100	33.4	23.5	31.9	0	100	57.5	54.2	21.0	17.6	100
Ohangwena	65.4	68.3	25.6	5.3	100	15.2	10.3	20.9	0	100	20.4	16.7	18.7	0	82.4
Omaheke	96.5	100	6.5	68.7	100	39.1	36.8	30.6	0	100	54.8	60	35.4	0	100
Omusati	68.7	73.7	23.8	10.0	100	19.1	10	23.8	0	100	25.8	22.9	16.0	0	85.7
Oshana	88.2	94.9	16.9	27.8	100	49.3	50	29.0	0	100	56.8	59.4	17.0	23.5	84.2
Oshikoto	83.9	89.3	16.5	35.0	100	30.1	15.4	32.8	0	100	41.1	39.4	28.8	0	100
Otjozondjupa	98.1	100	3.7	83.3	100	60.9	62.8	33.4	0	100	70.3	72.6	22.9	12.5	100

Min = Minimum, Max = Maximum, sd= Standard deviation.

Table 9.3 shows that most households in the fourteen (14) regions have access to improved drinking water, with mean percentage value above 65%. Households with the lowest minimum percentage were found in Kavango region with 0%, while all regions were reported to have a maximum percentage of 100%. It is worth noting that the minimum percentage of households with improved drinking water was quite high in Khomas and Otjozondjupa regions with 86.67% and 83.33% respectively. With regard to improved sanitation, Ohangwena and Omusati regions had the lowest mean percentage of

15.15% and 19.14% respectively. In addition, the results show that the minimum percentage of improved sanitation in all region was 0%, while the maximum was 100%.

Furthermore, households with the lowest mean percentage of improved hygiene were found in Ohangwena region with 20.43%, while the region with the highest mean percentage of improved hygiene was Otjozondjupa with 70.33%. Hardap, !Karas, Kavango, Ohangwena, Omaheke, Omusati and Oshikoto regions had a minimum percentage of households with improved hygiene of 0%. The results showed that Erongo, !Karas, Khomas, Kunene, Omaheke, Oshikoto and Otjozondjupa regions had households with maximum percentage of improved hygiene of 100%.

Joint Model Results

Under this section, we present results of a joint spatial model of WASH indicators that was fitted to capture spatial variation and assess dependence. In the analysis, the continuous variables: age and household size were modeling as smooth functions using penalized splines, and region was the areal unit used to model spatial effects.

Correlates of Improved Hygiene

From Table 9.4, with regard to the covariates of improved hygiene, estimates indicate an association with residence, listening to the radio less than once a week and wealth index. However, there is no significance in covariates marital status, literacy, exposure to newspaper, exposure to TV, respondent occupation and working status. Figure 9.2 shows a high spatial variation of improved hygiene in Khomas, Omaheke, Erongo, Otjozondjupa and Kavango regions, while !Karas and Oshana regions have the lowest. Hardap, Kunene, Omusati and Oshikoto regions shows a moderate spatial variation.

Figure 9.3 reports the smooth function estimates for improved hygiene. The effects of household members and age show different degree of non-linearity. The intervals of the smooth for household members and age ranged from -1.0 to 0.2. This suggests that household size and age are good predictors of improved hygiene. There is a negative association with increase to household members as shown in Figure 9.3 (left panel).

Table 9.4. Results from the GJRM with spatial, fixed and non-linear effects of region to capture spatial variation and spline to model non-linear effect for variables age and household size, for improved drinking water, hygiene and sanitation

Variables	Improved Hygiene		Improved drinking water		Improved sanitation	
	Estimate	P-value	Estimate	P-value	Estimate	P-value
Residence (ref. Urban)						
Rural	-0.291	<0.001***	-0.530	<0.001***	-0.271	<0.001***
Sex (ref. Male)						
Female	-0.004	0.904	-0.092	0.039*	-0.043	0.303
Marital status (ref. Never married)						
Married	0.023	0.521	-0.024	0.633	-0.029	0.519
Widowed	-0.112	0.048*	-0.159	0.027*	-0.074	0.315
Divorced	-0.141	0.049*	-0.147	0.105	-0.044	0.648
Wealth index (ref. Poor)						
Middle	0.379	<0.001***	0.532	<0.001***	1.089	<0.001***
Rich	1.057	<0.001***	1.148	<0.001***	2.587	<0.001***
Literacy (ref. Cannot read at all)						
Able to read only part of sentence	0.043	0.484	0.011	0.895	0.039	0.619
Able to read whole language	0.084	0.067.	0.093	0.143	-0.060	0.307
No card with required language	-0.006	0.957	0.040	0.796	-0.097	0.434
Blind/visually impaired	0.676	0.136	-0.332	0.564	1.194	0.092 .
Frequency of reading Newspaper/Magazine (ref. Not at all)						
Less than once a week	0.004	0.896	-0.037	0.481	0.054	0.270
At least once a week	-0.074	0.085.	-0.034	0.573	-0.017	0.755
Frequency of listening to the Radio (ref. Not at all)						
Less than once a week	-0.087	0.041*	-0.022	0.714	-0.078	0.165
At least once a week	-0.042	0.251	-0.062	0.226	0.037	0.432
Frequency of watching Television (ref. Not at all)						
Less than once a week	-0.076	0.113	-0.019	0.773	-0.106	0.089.
At least once a week	0.017	0.649	-0.027	0.594	-0.017	0.727
Occupation (ref. Unskilled)						
Skilled	-0.126	0.189	0.009	0.945	-0.412	<0.001**
Low skilled	-0.138	0.108	-0.018	0.878	-0.411	<0.001***
High skilled	-0.081	0.421	-0.125	0.349	0.436	<0.001***
Working status (ref. No)						
Yes	0.084	0.318	-0.070	0.537	0.383	<0.001***

Significant. codes: 0 '***' 0.001 '**' 0.01 '*'0.05 .''0.1 '' 1.

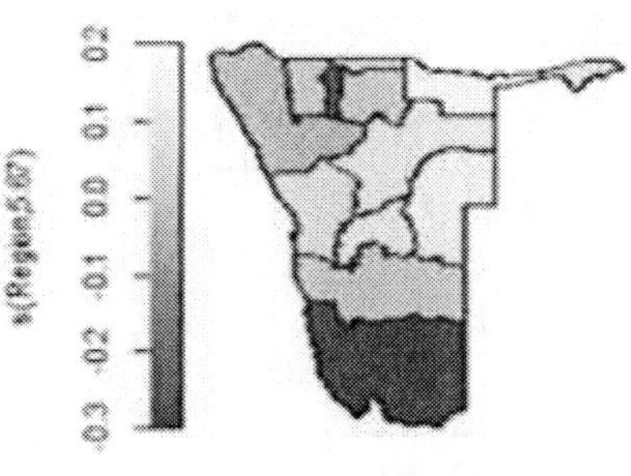

Figure 9.2.The region spatial variation of improved hygiene.

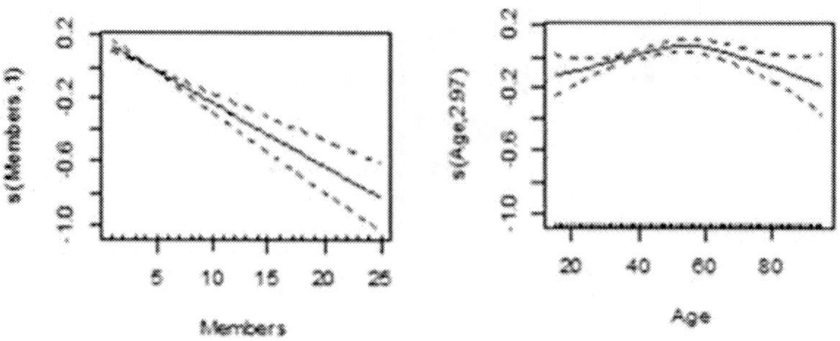

Figure 9.3. The smooth function of improved hygiene for: household size (left panel) and age (right panel), with corresponding confidence bands.

Correlates of Improved Drinking Water

Table 9.4 further shows that improved drinking water is associated with residence, sex and wealth. Figure 9.4 shows a high spatial variation of improved drinking water in Zambezi, Oshana, Ohangwena, Omaheke and Omusati regions, with low spatial variation in Erongo, Khomas, Hardap, Kunene, Otjozondjupa and !Karas regions. Moreover improved drinking water was negatively associated with household members as shown (Figure 9.5-left panel). The intervals of the smooth for household members and age ranged from -0.8 to 0.0. Similarly, Figure 9.5 (right panel), shows that smooth estimate of association between improved drinking water and age are significant at p<0.001.

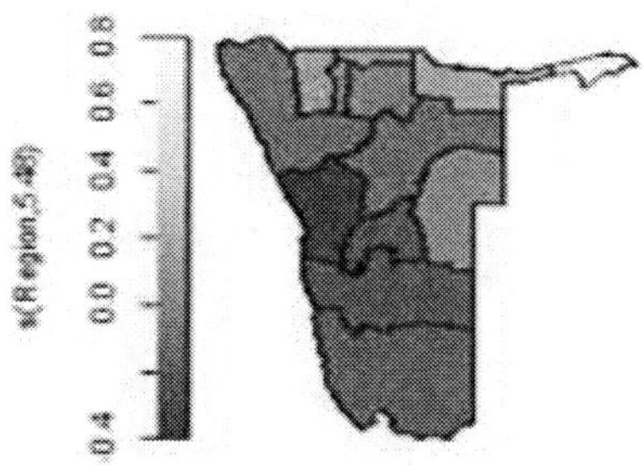

Figure 9.4. The region spatial variation of improved drinking water.

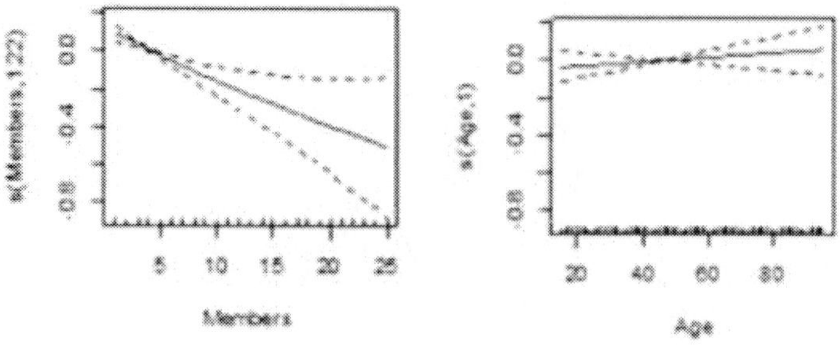

Figure 9.5. The smooth function on improved drinking water for: household size (left panel) and age (right panel), with corresponding confidence bands.

Correlates of Improved Sanitation

Furthermore, from Table 9.4, improved sanitation increased with wealth index and it was associated with occupation and working status. Marital status, literacy, sex and exposure to the media show no significant to improved sanitation. Poor sanitation was observed in rural areas. High region variation (Figure 9.6), in improved sanitation was captured in !Karas, Oshikoto and Kunene regions, while a low variation was captured in Omusati and Oshana regions. Khomas, Hardap, Erongo, Otjozondjupa, Ohangwena, Kavango regions had a moderate spatial variation.

The smooth function estimate of the the effects of household size and age on improved sanitation are shown in Figure 9.7. The intervals of the smooth for household members and age ranged from -2 to 3, while for region, the range was between -0.6 to 0.6. This suggests that household size and age are good predictors of improved sanitation. The p-value showed that region was an important determinant of improved sanitation. It has been observed that the probability of improved sanitation decreases with an increase in household members.

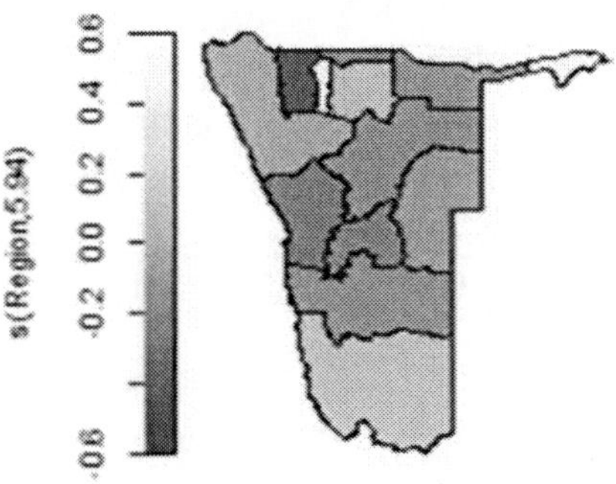

Figure 9.6. The region spatial variation of improved sanitation.

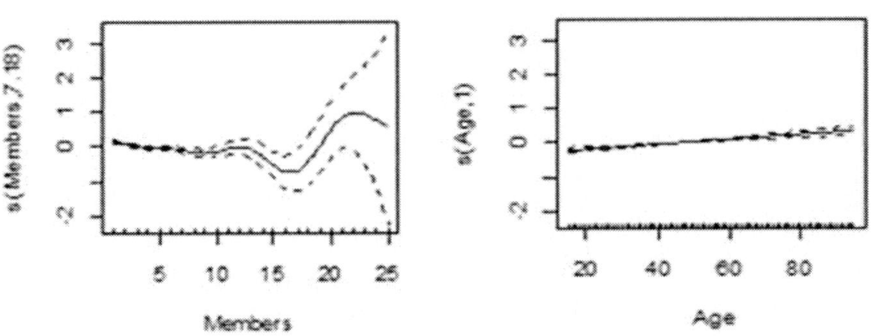

Figure 9.7. The smooth function estimates on improved sanitation: household size (left panel) and age (right panel), with corresponding confidence bands.

Table 9.5. Theta (θ) results from model fitted with spatial, fixed and non-linear effects

Theta (θ)	Estimated correlation, with CI
Improved hygiene and drinking water ($\widehat{\theta}_{12}$)	0.21 (0.163, 0.256)
Improved hygiene and sanitation ($\widehat{\theta}_{13}$)	0.217 (0.173, 0.251)
Improved drinking water and sanitation ($\widehat{\theta}_{23}$)	0.229 (0.173, 0.286)

The estimated correlations and confidence Intervals (CIs) for model fitted with spatial, fixed and non-linear effects using equation (5) are reported in Table 9.5. The results show that the indicators depend on each other because of the positive correlation between WASH indicators. It also indicates an increasing relationship between improved drinking water and sanitation.

Conclusion

Joint analysis of WASH indicators is of importance because these indicators depend on each other. Unsafe drinking water is additionally confounded by poor sanitation, which can make water become contaminated and incredibly increases the need for hygiene. In addition, improved drinking water has no meaning on health if improved sanitation is not concurrent. This study has shown that improved drinking water, sanitation and hygiene are of importance in improving the health and livelihood of the Namibian population.

References

Adams, E. A., Boateng, G. O., &Amoyaw, J. A. (2015). Socioeconomic and demographic predictors of potable water and sanitation access in Ghana. *Social Indicators Research*, 126:673-687. doi:10.1007/s11205-015-0912-y.

Baek, S. (2010). *A Copula-based method for analyzing bivariate binary longitudinal data. A dissertation in Epidemiology and Biostatistics*. University of Pennsylvania. Retrieved from http://repository.upenn.edu/edissertations/1564.

Bárdossy, A. (2006). Copula-based geostatistical models for groundwater quality parameters. *Water Resources Research*, 42(11),1-12, doi:10.1029/ 2005wr004754.

Besag, J., York, J., & Mollie, A. (1991). Bayesian image restoration, with two applications in spatial statistics. *Annals of the Institute of Statistical Mathematics*, 43(1), 1-20. doi:10.1007/BF00116466.

Bhat, C. R., &Sener, I. N. (2009). A Copula- based close form binary logit choice model for accommodating spatial correlation accross observational unit. *Journal of Geographical System*, 11(3), 243-272.

Blanchet, J., & Davison, A. C. (2011). Spatial modeling of extreme snow depth. The *Annals of Applied Statistics*, 5(3), 1699-1725.

Bohdalova, M., &Nanasiova, O. (2006). *Note to Copula Fuctions*. Retrieved from www.g-casa.co/download/Bohdalova_Copula_fuctions.pdf.

Costello, D. H. (2013). *An evaluation of a Water, Sanitation, and Hygiene (WASH) program for rural communities in Northern Afghanistan*. Retrieved from https://ir.library.oregonstate.edu/xmlui/bitstream/handle/.../CostelloDeniseH2013.pdf ?.Oregon State University.

Danaher, P. J., & Smith, M. S. (2011). Modelling multivariate distribution using Copulas. *An Application in Marketing Science*, 30(1), 4-21.

De Oliveira, V. (2012). Bayesian analysis of conditional autoregressive models. *Annals of the Institute of Statistical Mathematics*, 64(1), 107-133. doi: 10.1007/s10463-010-0298-1.

Eilers, P. H., & Marx, B. D. (1996). *Flexible smoothing with b-splines and penalties Statistical Science*, 11(2), 89-102.

Filippou, P., Marra, G., &Radice, R. (2017). Penalized likelihood estimation of a trivariate additive probit model. *Biostatistics,* 18(3), 569-585. doi:10.1093/ biostatistis/kxx008.

Garriga, R. G., &Foguet, A. P. (2013). Unravelling the linkages between water, sanitation,hygiene and rural poverty: The WASH poverty index. *Water Resources Management,* 27(5), 1501-1515. doi: 10.org/1007/s11269-012-0251-6.

Hoff, P. D. (2007). Extending the rank likelihood for semiparametric Copula estiamate. *The Annals for Applied Statistics*, 1, 265-283. doi: 10.1214/07-AOAS107.

Huard, D., Evin, G., & Favre, A. C. (2006). Bayesian Copula selection. *Computational Statistics and Data Analysis*, 51(2), 809- 822.doi: 10. 1016/j.cascla.2005.08.010.

Hughes, J., & Haran, M. (2013). Dimension reduction and alleviation of confounding for spatial generalized linear mixed models. *Journal of the Royal Statistical Society, Series B,* 75(1), 139-159.

Hutton, G., & Chase, C. (2016). The knowledge base for achieving the sustainable development goal targets on water supply, sanitation and hygiene. *International Journal of Environmental Research and Public* Health, 353-399. doi:10. 3390/ijerph13060530.

Irianti, S., Prasetyoputra, P., &Sasimartoyo, T. P. (2016). Determinants of household drinking-water source in Indonesia: An analysis of the 2007 Indonesian family life survey. *Cogent Medicine*, 3(1151143), 1-13. doi.org/10.1080/2331205X.2016. 1151143.

Kazianka, H., &Pilz, J. (2009). Spatial Interpolation using Copula based geostatistical models. In P. M. Atkinson, & C. Lloyd, *GeoENV VII Geostatistics for Environmental Application* (pp. 307-320). Berlin: Springer.

Koskei, E. C., Koskei, R. C., Koske, M. C., &Koech, H. K. (2013). Effect of socio-economic factors on access toiImproved water sources and basic sanitation in Bomet municipality, Kenya. *Research Journal of Environmental and Earth Sciences*, 5(12), 714-719.

Kuethe, T. H., Hubbs, T., &Waldof, B. (2009). *Copula models for spatial point patterns and processes*. Retrived from: http://www.ub.edu/sea2009.com/Paper/162.pdf.

Magalhaes, R., Barnett, A. G., & Clements, A. (2011). Geographical analysis of the role of water supply and sanitation in the risk of helminth infections of children in West Africa. *PNAS,* 108(50), 20084-20089.

Marra, G., &Radice, R. (2017). *Generalized Joint Regression Modelling* (GJRM). R package version 3.4.1. https://CRAN.R-project.org/package=GJRM.

Marra, G., Radice, R., Bärnighausen, T., Wood, S. N., & McGovern, M. E. (2017). A simultaneous equation approach to estimating hiv prevalence with non-ignorable missing responses. *Journal of the American Statistical Association*, 112(518), 484-496. doi: 10.1080/01621459.2016.1224713.

Masarotto, G., &Varin, C. (2012). Gaussian copula marginal regression. *ElectronicJournal of Statistics*, 6(2012), 1517-1549. doi: 10.1214/12-EJS721.

Masarotto, G., &Varin, C. (2017). Gaussian Copula Regression in R. *Journal of Statistical Software,* 8(77), 1-26. doi: 10.18637/jss.v077.i08.

Ministry of Health and Social Social Services (MoHSS) [Namibia] and Macro International Inc. (2014). *Namibia Demography and Health Survey 2013*. Windhoek, Namibia and Claverton, Maryland: MoHSS and Macro.

Mulenga, J. N., Bwalya, B. B., &Kaliba- Chishimna, J. (2017). Determinants and inequalities in access to improved water sources and sanitation among the Zambian households. *International Journal of Development and Sustainability*, 6(8), 746-762.

National Planning Commission (NPC). (2004). *Namibia Vision 2030*. Windhoek, Namibia: NPC.

Nelsen, R. B. (2006). *An Introduction to Copula.* New York: Springer.

Nketiah-Amponsah, E., Aidam, P. W., &Senadza, B. (2009). Socio-economic determinants of sources of drinking. Paper presented at the *Conference on International Research on Food Security, Natural Resource Management and Rural Development*, University of Hamburg, Germany. Retrieved from http://www. tropentag.de/2009/abstracts/full/185.pdf.

Ntozini, R., Marks, S. J., Mangwada, G., Mbuya, M. N., Gerema, G., &Mutasa, B. (2015). Using Geographic Information Systems and spatial analysis methods to assess household water access and sanitation coverage in the SHINE Trial. *ClinicalInfectious Diseases*, 61(7), 716-725. doi.org/10.1093/cid/civ847.

Paciorek, C. J. (2013). Spatial models for point and areal data using Markov random fields on a fine grid. *Electronic Journal of Statistics*, 7, 946-972. doi:10.1214/13-EJS791.

Perrone, E., & Müller, W. G. (2016). Optimal design for Copula models. *Statistics*, 50(4), 917-929.

Pitt, M., Chan, D., & Kohn, R. (2006). Efficient Bayesian inference for Gaussian Copula regression models. *Biometrika*, 93, 537-554. doi: 10.1093/biomet/93.3.537.

R Core Team (2016). *R: A language and environment for statistical computing. R Foundation for Statistical Computing*, Vienna, Austria. URL https://www.R-project.org/.

Rahut, D. B., Behera, B., & Ali, A. (2015). Household access to water and choice of treatment methods: Empirical evidence from Bhutan. *Water Resources and Rural Development*, 5, 1-16. doi:10.1016/j.wrr.2014.09.003.

Raso, G., Schur, N., Utzinger, J., Koudou, B. G., Tchicaya, E. S., &Rohner, F. (2012). Mapping malaria risk among children in Cote d'Ivoire using Bayesian geo-statistical models. *Malaria Journal*, 11(160), 1-11. doi.org/10.1186/1475-2875-11-160.

Sarmento, D. R. (2015). *An analysis of access to improved drinking water and sanitation and distance to the water source in a newly independent country*, Timor-leste: assessing geographical and socioeconomic disparities. Georgia State University. Retrieved from http://scholarworks.gsu.edu/iph_theses/391/.

Silva, R. S., & Lopes, H. F. (2008). Copula, marginal distribution and model selection. *Statistics and Computing*, 18, 313-320.

Sklar, A. (1959). *Distribution functions of n dimensions and margins. Publications of the Institute of Statistics*. University of Paris, Paris, pp. 229-231.

Smith, M. S. (2011). Bayesian approaches to Copula modelling. In P. Damien, P. Dellaportas, N. Polson, & D. Stephens, *Bayesian Theory and Applications*, (pp. 336–358).

Smith, M., Gan, Q., & Kohn, R. (2010). Modelling depedence using skew t Copulas: Bayesian inference and application. *Journal of Applied Econometrics (Forthcoming)*. doi: 10.1002/jae.1215.

Song, P. X.K., Li, M., & Yuan, Y. (2009). Joint regression analysis of correlated data using Gaussian Copula. *Biometrics*, 65, 60-69. doi: 10.1111/j.l541-0420.2008.01058.

Sudell, M., Kolamunnage-Dona, R., &Tudur-Smith, C. (2016). Joint models for longitudinal and time-to-event data: A review of reporting quality with a view to meta analysis. *BMC Medical Research Methodology*, 16(168), 1-11. doi: 10.1186/s12 874-016-0272-6.

Tiwari, R., &Nayak, S. (2013). Drinking water and sanitation in Uttar Pradesh: A regional analysis. *Journal of Rural Development*, 32, 61–74.

Trivedi, P. K., & Zimmer, D. M. (2005). Copula modeling: An introduction for practitioners. *Foundations and Trends in Econometrics*, 1(1), 1-111. doi: 10.1561/08 00000005.

UNICEF. (2010). *Retrieved from UNICEF Namibia*: https://www.unicef.org/namibia/hea lth_nutrition_13817.html.

UNICEF. (2016, April 10). Water, Sanitation and Hygiene. Retrieved from UN: https://www.unicef.org/wash/3942_3952.html.

Winkelmann, R. (2011). *Copula bivariate probit models: With an application to medical expenditures.* University of Zurich. Pp.29.

WHO & UNICEF. (2006). *Meeting the MDG Water and Sanitation Targets: The rural and urban challenges of the decad*e. WHO, Geneva and UNICEF, New York.

WHO & UNICEF. (2015). *Progress on sanitation and drinking water*. Geneva and New York.

Wood, S. N. (2017). *Mixed GAM computation vehicle with automatic smoothness estimation.* R package version 1.8-23. Retrieved from https://cran.r-project.org/web/ packages/mgcv/mgcv.pdf.

Yang, H., Bain, R., Bartram, J., Gundry, S., Pedley, S., & Wright, J. (2013). Water safety and inequality in access to drinking-water between rich and poor households. *Environmental Science and Technology*, 47(3), 1222–1230.

Yu, W., Bain, R., Mansour, S., & Wright, J. A. (2014). A cross-sectional ecological study of spatial scale and geographic inequality in access to drinking-water and sanitation. *International of Equity in Health*, 113-127.

Zimmer, D. M., & Trivedi, P. K. (2006). Using trivariate Copula to model sample seletion and treatment effects: Application to family health care demand. *Journal Business Economic Statistics*, 24, 63-76.

In: Statistical Modelling …
Editors: N-B. Kandala and L. Kazembe
ISBN: 978-1-53615-981-3

Chapter 10

BIVARIATE COPULA-BASED REGRESSION TO MODEL TIMING AND FREQUENCY OF ANTENATAL CARE UTILIZATION

Lawrence N. Kazembe*
Department of Statistics and Population Studies, University of Namibia,
Windhoek, Namibia

ABSTRACT

Access to reproductive health services is crucial towards promoting and improving maternal health and newborn outcomes. Particular emphasis is placed on adequate utilization of antenatal care (ANC) beginning with a first visit within the first trimester and attaining four focussed ANC visits during pregnancy. In this study, we proposed the use of a copula-based regression model for a joint analysis of the timing of first visit for antenatal care and total number of visits, formulated to account for negative dependence in the two outcomes. A bivariate mixture model of continuous-discrete outcomes using a Gamma and Poisson generalized linear models for the timing of first ANC visit (continuous) and number of ANC visits (discrete), respectively, was specified via Gaussian and Frank copula correlation. We then analyzed data from the 2006/07 Namibia demographic and health survey, specifically of pregnant women's utilization of ANC at their last birth. The effects of socio-economic and demographic variables are estimated and compared on both outcomes and copula models.

Keywords: bivariate mixture models, copula-based regression, antenatal care, Namibia DHS, Gaussian copula, Frank copula

---

* Corresponding Author's Email: lkazembe@yahoo.com.

INTRODUCTION

Many developing countries have a huge burden of maternal and newborn adverse outcomes (WHO, 2010). Globally, more than 500,000 women die of complications due to pregnancy and child birth (World Bank, 2010), and about 85% of these deaths occur in the sub-Saharan Africa and South eastern Asia. Equally, one in five children in every 1,000 live birth die in their first month or in infancy, and pregnancy or birth complications have been implicated as major contributing factor exacerbated by lack or poor utilization of antenatal care (Pell et al., 2013). This burden is highlighted in the Millennium Development Goals (MDG) targets 4 and 5, and subsequently in the Sustainable Development Goal number 3, which respectively stipulates urgent need for reduction in child mortality, improved maternal health and increased access to reproductive health services (World Bank, 2010).

In a bid to stem the burden, a number of interventions have been put in place including antenatal care (ANC), along with family planning, skilled delivery care and emergency obstetric care, aimed towards improving maternal and newborn health (WHO, 2003). The ANC was first introduced in the early 20th century in Europe and North America and is now universal in the developed world. The ANC strategy as mostly implemented in many developing countries, is based on the 2001 model which is called a focused ANC, which recommends four ANC visits to the health facility during pregnancy (Pell et al., 2013). The ANC is a package that incorporates a range of interventions, as such timely visits and adequate number of visits are required to detect pregnancy complications as well as ensure delivery at a health facility. A recent systematic review demonstrated that early timing of ANC interventions reduces the risk of having adverse effects in maternal and newborn outcomes (Hawkes et al., 2013). Mostly, at least a single visit is required in each trimester or within the first 10 weeks of pregnancy. However, the total number of visits largely depends on the time timing of the first visit.

A number of studies have identified lack of ANC or delayed ANC as a risk factor of maternal and child mortality (World Bank, 2010; WHO, 2003). Use of ANC depends on several or multiple interacting socio-economic and demographic factors including parity, maternal age, education, religion and ethnicity, place of residence and social position, among others (Magadi et al., 2000; Sagna and Sunil, 2012; Gage, 2007; Hawkes et al., 2013; Babalola and Fatusi, 2009; Magadi et al., 2007). In most studies, interest has been to determine factors associated with whether at least four visits were made during pregnancy (Gage, 2007; Babalola and Fatusi, 2009), or whether the visit was timely (Hawkes et al., 2013; Kisuule et al., 2013), although a few studies have considered both (Magadi et al., 2000; Sagna and Sunil, 2012).

Modelling of ANC has received much attention in health policy planning, health economics and public health in general, increasingly often under multilevel statistical modelling (Magadi et al., 2000, 2007; Sagna and Sunil, 2012; Babalola and Fatusi, 2009).

A very common approach is based on independent models, a logit model for early ANC and another logit for whether at least four ANC visits have been made. In some instances, a discrete outcome model, e.g., Poisson has been used to analyze number of ANC visits (Gayawan, 2013). The main assumption in such analysis is that the two outcomes are independent, which largely maybe misleading or untenable. As it is, statistically, this presents an interesting research question because these are mixed dependent outcomes which are better handled using joint analysis.

Joint analyses are found in many applications in public health. A common case is that of a bivariate set of outcomes, often characterized by either two continuous variables, or one of discrete-continuous outcome, or in some cases of mixed discrete outcomes only. Joint analysis has several advantages, including avoiding multiple tests, increased power and better control of Type I error rates, as well as efficiency mostly gained where missing data are being handled (de Leon and Chough, 2010; de Leon and Wu, 2011). However, a challenge in joint analysis is the specification of the joint distributions of mixed outcomes especially if such are non-standard random variables.

Therefore, current research interest has focussed on finding flexible model specification for any joint outcome at hand. The earliest strategy is through direct model specification by factorization into conditional and marginal distributions, given the joint outcomes, in what are called general location model. Example models in this class are the logistic-normal or probit-normal models (de Leon and Chough, 2010). Other strategies, as reviewed by de Leon and Chough (2010), including indirect approaches through random effects (e.g., correlated random effects and shared random effects models), and copula have been proposed to analyze multiple non-commensurate outcomes. Further reviews on copula approach are found in Kolev and Paiva (2009); Song et al. (2009).

This paper proposes a joint model for the correlated discrete and continuous outcomes constructed using copulas. Dependence modelling using copula has become popular in recent years. See a review by Kolev and Paiva (2009), a discussion by Song et al. (2009), or reference books by Joe (1997) and Nelsen (2006). The copula approach relates an arbitrary joint distribution to its corresponding univariate marginal distributions via copula. Our proposal is to combine the marginal distributions of timing (continuous variable) and number of visits (count variable) via bivariate copula. Modelling of bivariate mixed distributions have been used in a variety of applications (Kolev and Paiva, 2009; Song et al., 2009). The copula approach to discrete-continuous is particularly suited to estimate flexible dependency structures between continuous variable (timing) and discrete (number of visits) distributions, by allowing one to test several different copulas for the joint distribution of the bivariate outcomes (Nelsen, 2006). Further, we develop a flexible class to model error distributions of generalized linear models (GLM) that permits covariates under each marginal model in what is called Vector GLMs (Song et al., 2009).

In health economics literature, copula-based methods have been used to deal with sample selection, endogeneity in regressors and structural simultaneous equations of two-part and latent class. The development of copula-based regression for mixed outcomes data is recent in this area of health care, and has basically expanded following a systematic considerations of non-normal copulas which started with Smith (2003), in sample selection models with normal/normal and normal/Gamma marginals. Zimmer and Trivedi (2006) developed copulas for a joint model of negative binomial and normal marginals to analyze family health care demand, Winkelmann (2009) used a bivariate copula based probit models to estimate the effect of an endogeneous binary regressor on binary outcome variable, while Murteira and Lourenco (2011) applied a multinomial and Poisson/ Negative binomial marginals for analysis of health care utilisation and self-reported health. Additional applications can be found in Deb et al. (2013) who used dynamic copula-based hurdle model to study cost-offsets of prescription drug expenditure, and in Cameron et al. (2004) who used bivariate copula to model differences in self-reported and actual number of doctor visits. In this paper, the copula approach uses Gamma/Poisson marginals within the discrete-continuous mixed data. Particularly in our application, we construct a joint regression model by linking a marginal Gamma GLM for the average time (in months) and a marginal Poisson GLM for the number of visits using a mixed copula as developed by Czado et al. (2012), and first introduced by Song et al. (2009). Estimation of parameters is achieved using maximum likelihood (Kramer et al., 2012).

The remainder of this paper is organized as follows. Section 2 outlines the bivariate copula-based regression. Section 3 presents the data, analysis and results of the application. This is followed by the discussion in Section 4 and we conclude.

BIVARIATE COPULA-BASED REGRESSION

In a bivariate case, an arbitrary joint distribution F_{Y_1,Y_2} is related to its corresponding univariate marginal distributions F_{Y_1} and F_{Y_2} via a copula C as

$$F_{Y_1,Y_2}(y_1, y_2) = C(u_1, u_2|\rho),$$

where u_1 and u_2 are respective realizations of probability integral transforms $U_1 = F_{Y1}(y_1) \sim Unif\ [0,1]\ and\ U_2 = F_{Y2}(y_2) \sim Unif\ [0,1]$, and ρ is the dependence parameter measuring correlation between marginals F_{Y_1} and F_{Y_2}. This concept of defining a multivariate dependence structure from independent marginal distributions via copula is based on Sklar's theorem (Joe, 1997; Nelsen, 2006; Song et al., 2009). For continuous marginal distributions C is unique, however this is not the case when one or more of the margins is discrete. Further details of properties of copula are presented in Joe (1997); Nelsen (2006).

Marginal Regression Models

Building a regression model with several covariates when constructing copula presents a useful extension, and on-going research. In this case study, we follow Czado et al. (2012); Kramer et al. (2012), to develop a bivariate model of mixed outcomes, with the marginals following a Gamma and Poisson GLMs. The Gamma GLM represents the average time it takes for a pregnant woman to visit ANC (Y_{i1}), while the a zero-adjusted Poisson GLM models the corresponding number of visits to ANC (Y_{i2}). First we specify the marginal GLMs.

Suppose Y_{i1} and Y_{i2} are independent random variables for woman $i = 1,\ldots,n$, then the marginal GLMs are given as

$$Y_{i1} \sim Gamma(\mu_{i1}, \nu^2) \text{ with } ln(\mu_{i1}) = X'\alpha \tag{1}$$

$$Y_{i2} \sim ZTPois(\mu_{i2}E_i) \text{ with } ln(\mu_{i2}) = ln(E_i) + Z'\beta \tag{2}$$

where X and Z are (distinct or similar) set of covariates, with α and β corresponding regression coefficients associated with the continuous Y_{i1} and discrete Y_{i2} outcomes respectively, modelled on the means $\mu_{i1} = E(Y_{i1})$ and $\mu_{i2} = E(Y_{i2})$. In the Zero-Truncated Poisson (ZTPois) model, we included the offset $ln(E_i)$ to account for the expected length of time a visit will occur (or exposure time) among a group of pregnant women.

The average length of time Y_{i1} has a density

$$f_{Y1}(y_{i1}|\mu_{i1}, \nu^2) = \frac{1}{\Gamma\left(\frac{1}{\nu^2}\right)}\left(\frac{1}{\mu_{i1}\nu^2}\right)^{\frac{1}{\nu^2}} . y_{i1}{}^{\frac{1}{\nu^2}-1} . exp\left\{-\frac{1}{\mu_{i1}\nu^2}y_{i1}\right\}$$

such that $Var[Y_{i1}] = \mu_{i1}\nu^2$ for some dispersion parameter ν which is estimated in the marginal Gamma GLM through $E[Y_{i1}]/\sqrt{Var[Y_{i1}]} = 1/\nu$.

Here it should be noted that a visit has to occur first, then cumulative visits follow. As such the total number of visits considered in this model are only positive counts, thus a zero-truncated Poisson (ZTPois). The zero-truncated Poisson has a density

$$f_{Y1}(y_{i2}|\mu_{i2}) = \frac{\mu_{i2}{}^{y_{i2}}}{y_{i2}!(1-e^{-\mu_{i2}})} . exp\{-\mu_{i2}\} \text{ for } y_{i2} = 1, 2, \ldots \infty$$

Joint Regression Model

Now consolidating the two marginal GLMs given in (1) and (2), we get a joint distribution defined as

$$F(y_{i1}, y_{i2} | \mu_{i1}, \nu, \mu_{i2}, \rho) = C(u_1, u_2 | \rho). \quad (3)$$

Several choices for copula C are possible. See Nelsen (2006) and Joe (2014) for examples.The choice of copula, however, will depend on the stochastic processes assumed on the outcome variables. For instance, the direction of dependence might be of importance to capture. In the current case-study, two outcomes: The timing of first visit (y_1) and number of visits (y_2) may move in opposite directions, thereby creating a negative dependence in the data. We therefore search for copula that will permit both positive and negative dependence. The most common choice is the Gaussian copula given by

$$C(u_1, u_2 | \rho) = \Phi_2(\Phi^{-1}(u_1), \Phi^{-1}(u_2))$$

The dependence here is captured using a bivariate Gaussian copula cumulative distribution function $C(u_1, u_2 | \rho)$, where Φ is the standard normal distribution, and Φ_2 is the standard bivariate normal distribution, with dependence parameter ρ, often interpreted as the Pearson correlation coefficient which is restricted with $|\rho| < 1$. The Gaussian copula is an important copula family with a variety of applications (Song et al., 2009; Czado et al., 2012; Kramer et al., 2012), flexible and analytically tractable (Song et al., 2009; Nelsen, 2006), as well as convenient because of its nice marginal and conditional properties (de Leon and Chough, 2010; de Leon and Wu, 2011).

Another most frequently encountered copula is the Frank copula. This is given by

$$C(u_1, u_2 | \rho) = \begin{cases} -\frac{1}{\rho} log\left(1 + \frac{(e^{-\rho u_1} - 1)(e^{-\rho u_2} - 1)}{(e^{-\rho} - 1)}\right) & \rho \neq 0 \\ u_1 u_2 & \rho = 0 \end{cases}$$

Negative and positive dependence occurs when $\rho < 0$ and $\rho > 0$ respectively, with independence copula defined at $\rho = 0$. The copula correlation parameter attains the Frechet-Hoeffding upper and lower bounds under $\rho \rightarrow +\infty$ and $\rho \rightarrow -\infty$ respectively.

Accordingly, once the copula function has been identified, the joint density function of equation (3) of the bivariate mixture outcome is given by (Song et al., 2009),

$$f(y_{i1}, y_{i2} | \mu_{i1}, \nu, \mu_{i2}, \rho) =$$
$$f_1(y_{i1} | \mu_{i1}, \nu, \rho)[D_1(u_1, u_2 | \rho) - D1(u_1, u_2 - 1 | \rho)], \quad (4)$$

where $D_1(u_1, u_2|\rho) = \frac{\partial}{\partial u_1} C(u_1, u_2)$. Or in terms of the conditional density of Y_{i2} given Y_{i1}: $f_{Yi2\,|Yi1}\,(y_{i2}|y_{i1}, \mu_{i1}, \nu, \mu_{i2}, \rho)$, equation (4) becomes

$$f(y_{i1}, y_{i2}|\mu_{i1}, \nu, \mu_{i2}, \rho) = \quad f_1(y_{i1}|\mu_{i1}, \nu, \rho).\, f_{Yi2\,|Yi1}\,(y_{i2}|y_{i1}, \mu_{i1}, \nu, \mu_{i2}, \rho) \tag{5}$$

Estimation

The model depends on four parameters: μ_1, ν of the Gamma random variable, μ_2 for the Zero-Truncated Poisson, and the copula correlation ρ, dropping i for brevity. However, the model parameters in turn depend on covariates. We therefore need to estimate the unknown parameter vector

$$W = (\alpha, \beta, \rho, \nu) \in R^{p+q+2} \tag{6}$$

based on n observation pairs (y_1, y_2), where p and q are the number of parameters estimated in the two marginal models. Again following Czado et al. (2012), we use maximum likelihood estimation techniques. By definition, the log-likelihood of the model parameters in Eqn. (6) is

$$\ell(W\,|y1, y2) = \sum_{i=1}^{n} ln(f_{y_1,y_2}\,(y_1, y_2)|W\,). \tag{7}$$

The maximum likelihood estimates are given by

$$\widehat{W} = arg \max_{W} \ell(W|\, y_1, y_2)\,.$$

There is no closed form solution, and the log-likelihood is maximized by applying Broyden-Fletcher-Goldfarb-Shanno (BFGS) optimization algorithm. For the copula parameter ρ, optimization is through a Fisher transform function $g(\rho) = \frac{1}{2} ln \left(\frac{1+\rho}{1-\rho}\right)$. Thus we optimize $W' = (\alpha, \beta, g(\rho), \nu)$.

Alternatively, one can use an approach called maximization by parts (MBP), which is presented in detail in de Leon and Wu (2011), and is applied in Czado et al. (2012). Briefly, at first step, one can estimate respective model parameters from the Gamma and Zero-Truncated Poisson GLMs, and then these estimates are used to transform the observations (y_1, y_2) to obtain F_{y_1}, F_{y_2}. At the second, one optimizes the copula parameter ρ by maximizing the likelihood, similar to that in Eqn. (4), till convergence. In this paper, however, we use direct maximum-likelihood as carried out in Kramer et al. (2012).

Model Selection

Where two competing copula families have been used, comparison of the models can make use of the Voung test (Voung, 1989), which is a variant of the likelihood-ratio test (LRT). As defined in Kramer et al. (2012), let us denote by $\ell^{(1)}, \ell^{(2)}$ the vectors of pointwise loglikelihoods for a model with copula family 1 and 2 respectively. Here, we assume that both models have the same degrees of freedom, i.e., the same number of parameters. We now compute the differences of the pointwise log-likelihood as

$$m_i = \ell_i^{(1)} - \ell_i^{(2)}, i = 1, \dots, n.$$

Suppose the mean of the differences is denoted by $\bar{m} = \frac{1}{n}\sum_{i=1}^{n} m_i$, then the test statistic is define by

$$T_V = \frac{\sqrt{n}.\bar{m}}{\sqrt{\sum_{i=1}^{n}(m_i - \bar{m})^2}}.$$

T_V is asymptotically normally distributed with zero mean and unit variance. Where $T_V > \Phi^{-1}\left(1 - \frac{\alpha}{2}\right)$, we prefer copula family 1 to copula family 2 at significant level of α based on the standard normal distribution function Φ. If $T_V < \Phi^{-1}\left(\frac{\alpha}{2}\right)$, we prefer copula family 2. Otherwise, no decision among the two copula families is possible.

Further, once a copula family has been decided, Akaike Information Criterion (AIC) can be used to select between an independent and joint model. The AIC is defined as

$$AIC = -2\ell(W \mid y_1, y_2) + 2df,$$

where the degrees of freedom (df) are the number of parameters estimated in the model. For the joint model $df = p + q + 2$, while for an independent model we have $df = p + q + 1$. The model with a smaller AIC is preferred. Note that AIC can be used, in lieu of Voung test, to choose between two copula families, with a copula family of smaller AIC preferred.

APPLICATION TO ANC UTILIZATION IN NAMIBIA

Recent reports in Namibia show an increase in maternal and under-five mortality rates (MoHSS and Macro, 2008). The delay and inadequate uptake of ANC have been blamed as some of the possible contributing factors for the increase (National Planning

Commission, 2008). Of interest in this article is to study these two outcomes. Specifically, we consider a regression problem of $n = 3416$ pregnant women in Namibia, with the goal of assessing the extent timing of ANC and number of ANC visits are related, then subsequently to estimate the effect of socio-economic and demographic covariates on the mixed outcome.

Data, Variables, Summaries

We use data from the 2006/2007 Namibia DHS (MoHSS and Macro, 2008). The DHS are periodic cross-sectional health surveys funded by the U.S. Agency for International Development (USAID) Bureau for Global Health. The DHS have been conducted in over 70 countries since 1980s, and the data are publicly available at MEASURE DHS website. The data are completely anonymized before they are released for general public use. The DHS follows a cross-sectional multi-stage stratified survey design. In the Namibian context, a two-stage stratified sampling design was implemented to collect the data and provide direct estimates of demographic and health indicators at national and regional level. Details on sampling are presented in the Namibian DHS report (MoHSS and Macro, 2008).

The questions concerned in this study were asked of each woman based on their most recent birth that has occurred within three years preceding the survey date. These were part of a suite of health related issues including immunization, health-seeking on most recent illnesses, source of care, and feeding practices and other nutritional behaviour.

The dependent variables in this study are time to first ANC visit, measured in months and the number of ANC visits made during last birth. Figure 10.1 shows the histograms of the observed (i) timing of first visit and (ii) number of visits. Table 10.1 presents summary statistics for the dependent variables. The mean waiting time was 4.36 (standard deviation 1.62), slightly skewed to the right. Once a first visit has been initiated, each individual pregnant woman consulted for ANC 5.84 times on average with standard deviation of 3.38.

Table 10.1. Summary statistics for dependent variables: timing of first antenatal care (ANC) visit and number of ANC visits

Variable	Minimum	1st Quartile	Median	Mean	3rd Quartile	Max
Timing of 1st ANC visit	1	3	4	4.36	6	9
Number of ANC visits	1	4	5	5.84	7	40

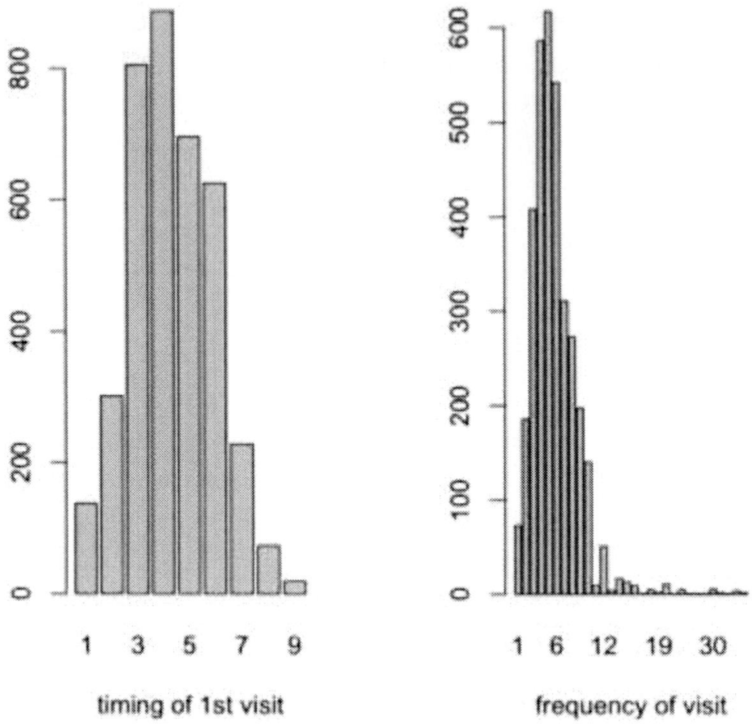

Figure 10.1. Histogram of antenatal care (ANC) utilization. Left panel: timing of care. Right panel: number of ANC visits.

The independent variables are given in Table 10.2, grouped into socio-economic and demographic variables. The selection of variables follow well-established theoretical and empirical frameworks developed by several authors (Magadi et al., 2000, 2007; Babalola and Fatusi, 2009; Sagna and Sunil, 2012; Pell et al., 2013; Hawkes et al., 2013). These are place of residence (rural or urban), region (13 categories), exposure to media (newspapers, radio or TV) measured in three categories, number of under-five children in the past three years preceding the survey (none, 1-2 children, or more), place where ANC was sought (public or private facility), sex of household head (female or otherwise), wealth index (poor, medium, rich), education level (none, primary, secondary or higher), whether the woman was working (yes or no), whether previous pregnancy was live birth (yes or no).

Basic summaries of the variables are also presented in Table 10.2. Individuals from female-headed household were 49%, those from urban were 41%. With regards education status, 9% had no formal education and 30% attained primary education. Working women were 45%, while 37% were from poor households, 26% were from medium wealth households. Most women had access to radio (61% daily, and another 26% occasionally listened to the radio). ANC services were mostly obtained from government facilities 270 (92% of sampled women). About 70% had previous birth experience, with under-five children between 1 and 2. Only 7% of the women had previous pregnancy terminated. Differences across regions were also evident.

Table 10.2. Explanatory variables: definitions and summaries*

Variable name	Variable definition	Mean	St.dev	Min.	Max.
fhead	=1 if female headed household	0.49	0.5	0	1
Urban	=1 if woman lives in urban area	0.41	0.49	0	1
Educnone	=1 if woman has no formal education	0.09	0.28	0	1
Educpri	=1 if woman attained up to primary education	0.30	0.46	0	1
Working	=1 if woman is working	0.45	0.50	0	1
Windpoor	=1 if woman is of low wealth ranking	0.37	0.48	0	1
Windmed	=1 if woman is of medium wealth ranking	0.26	0.44	0	1
Newsdaily	=1 if woman reads to newspapers/magazines daily	0.11	0.32	0	1
Newsoccas	=1 if woman reads to newspapers/magazines occasionally	0.47	0.5	0	1
Radiodaily	=1 if woman listens to the radio daily	0.61	0.49	0	1
Radiooccas	=1 if woman listens to radio occasionally	0.26	0.44	0	1
Tvdaily	=1 if woman watches TV daily	0.26	0.44	0	1
Tvoccas	=1 if woman watches TV occasionally	0.15	0.36	0	1
Govtfacility	=1 if ANC was done at public health facility	0.92	0.26	0	1
u5none	=1 if woman has no under-five year old children	0.13	0.33	0	1
u5b12	=1 if woman has between 1 and 2 under-five year old children	0.69	0.46	0	1
Pregterm	=1 if previous pregnancy gave live birth	0.07	0.26	0	1
Caprivi	=1 if woman is from Caprivi region	0.08	0.27	0	1
Erongo	=1 if woman is from Erongo region	0.06	0.24	0	1
Hardap	=1 if woman is from Hardap region	0.05	0.22	0	1
Karas	=1 if woman is from Karas region	0.04	0.29	0	1
Kavango	=1 if woman is from Kavango region	0.12	0.32	0	1
Khomas	khomas =1 if woman is from Khomas region	0.09	0.29	0	1
Kunene	=1 if woman is from Kunene region	0.05	0.22	0	1
Ohangwena	=1 if woman is from Ohangwena region	0.11	0.30	0	1
Omaheke	=1 if woman is from Omaheke region	0.06	0.23	0	1
Omusati	=1 if woman is from Omusati region	0.09	0.29	0	1
Oshana	=1 if woman is from Oshana region	0.08	0.26	0	1
Oshikoto	=1 if woman is from Oshikoto region	0.08	0.29	0	1

* Same variables were fitted to the joint regression model, i.e X = Z.

Analysis and Results

We consider the Gaussian and Frank copulas to combine the two marginals we considered in this analysis. The marginals are Gamma and Poisson corresponding to average time taken for first ANC visit, and average number of ANC visits during the period of pregnancy. The proposed model share similarities with that developed by Czado et al. (2012); Kramer et al. (2012) which they developed for the insurance data. In contrast, our model is adapted for health care analysis using national-wide cross-sectional survey data. All analysis were carried out in *CopulaRegression* package (Kramer et al., 2012) within R statistical software.

We first report on model selection criteria for the choice between copulas. Table 10.3 shows the copula dependence parameter ρ, Kendall's τ , the log-likelihood and AIC. Both the Gaussian and Frank copula shows a negative dependence between time of first visit and number of visits, $\rho_G = -0.54$ and $\rho_F = -4.54$ for the Gaussian and Frank copula parameter respectively. Additional insight into the nature of dependence implied by the copulas is captured by Kendall's τ, which is a transformation of the copula parameter ρ[1]. Kendall's τ measures the degree of concordance between the two outcomes, and varies between -1 and 1, such that a negative value shows negative dependence and vice-versa. Based on Kendall's τ, Frank copula ($\rho_F = -0.43$) seems to show higher dependency in the data as opposed to the Gaussian model ($\rho_G = -0.36$).

Table 10.3. Model comparison and related parameters under Gaussian and Frank copula

Component	Gaussian copula	Frank copula
Dependency parameter:		
ρ (joint)	-0.54	-4.55
τ (joint)	-0.36	-0.43
Model fit:		
Log-likelihood (independent)	14670.83	14670.83
Log-likelihood (joint)	14249.79	14124.88
Model selection:		
AIC (independence)	29,463.66	29,463.66
AIC (joint)	28,623.58	28,373.76
Δ*AIC* (joint vs independence)	840.08	1089.9
Δ*AIC* (Frank vs Gaussian)		249.82

The negative dependency estimated by both models imply that a delayed first visit leads to fewer subsequent visits for ANC services. Moreover, comparison of AIC shows the joint models should be preferred over those that assume independence ($\Delta AIC = 840.08$ in the Gaussian copula model and $\Delta AIC = 1089.90$ in Frank copula model). This shows that if the joint model is the true model, then we have to use the joint estimation, that is the dependence can not be ignored. Put differently, if dependency is not accounted for, there is likely to be bias in the model estimates. Furthermore, using Frank copula this dependence is well captured compared to the Gaussian copula ($\Delta AIC = 249.82$).

---

[1] Kendall's measure of dependence τ transforms ρ as follows: For the Frank copula, $\tau = 1 - \frac{4}{\rho}\left[1 - \frac{1}{\rho}\int_{t=0}^{\rho} \frac{t}{e^t - 1} dt\right]$, while for the Gaussian copula, $\tau = \frac{2}{\pi} sin^{-1}(\rho)$.

Table 10.4. Maximum likelihood estimates for timing of first visits model $[f\,(y_1|X)]$ under Gaussian and Frank copula-based joint model

Variable	Independent model		Gaussian model		Frank model	
	Coefficient	St. error	Coefficient	St. error	Coefficient	St. error
Fhead	-0.018	0.018	-0.013	0.017	-0.014	0.017
Urban	-0.021	0.024	-0.023	0.023	-0.02	0.022
Educnone	-0.047	0.019	-0.041	0.035	-0.015	0.034
Educpri	-0.059‡	0.023	-0.052†	0.022	-0.048†	0.021
Working	0.047‡	0.019	0.042†	0.018	0.044†	0.018
Windpoor	-0.079†	0.035	-0.078†	0.033	-0.065†	0.033
Windmedium	-0.068†	0.029	-0.071‡	0.028	-0.062†	0.027
Newsdaily	0.117‡	0.037	0.123‡	0.035	0.120‡	0.034
Newsoccas	-0.0013	0.023	0.00097	0.022	0.0059	0.021
Radiodaily	-0.0025	0.029	-0.0013	0.028	0.00073	0.027
radio occas	-0.018	0.032	-0.017	0.03	-0.015	0.03
Tvdaily	0.022	0.03	0.017	0.029	0.022	0.029
Tvoccas	0.025	0.029	0.029	0.027	0.013	0.027
Govtfacility	1.534‡	0.056	1.579‡	0.054	1.538‡	0.052
u5none	0.054	0.035	0.058	0.033	0.05	0.021
u5b12	0.061‡	0.024	0.056†	0.023	0.057†	0.023
Pregterminate	0.114‡	0.035	0.111‡	0.033	0.114‡	0.032
Caprivi	0.108†	0.047	0.104†	0.045	0.128‡	0.043
Erongo	0.268‡	0.048	0.263‡	0.046	0.272‡	0.044
Hardap	0.309‡	0.051	0.345‡	0.048	0.283‡	0.046
Karas	0.414‡	0.053	0.414‡	0.049	0.400‡	0.047
Kavango	0.189‡	0.043	0.189‡	0.042	0.178‡	0.041
Khomas	0.247‡	0.044	0.243‡	0.042	0.247‡	0.041
Kunene	0.155‡	0.051	0.177‡	0.048	0.184‡	0.047
ohangwena	0.225‡	0.046	0.128‡	0.044	0.227‡	0.042
Omaheke	0.137‡	0.049	0.143‡	0.046	0.162‡	0.045
Omusati	0.134‡	0.046	0.128‡	0.044	0.137‡	0.042
Oshana	0.200‡	0.047	0.202‡	0.044	0.217‡	0.042
Oshikoto	0.155‡	0.045	0.152‡	0.043	0.164‡	0.041

† Significant at $\alpha < 0.05$; ‡ Significant at $\alpha < 0.01$.

The full set of regression coefficients for the independence, Gaussian and Frank copula are displayed in Table 10.4 and 10.5. Independence and Gaussian model estimates are given for comparison. With regards the covariates on timing of first visit (Table 10.4), estimates indicated negative association with primary education, working status, and socio-economic status, while a positive association was observed with exposure to newspapers, use of government facility, having under-five chidren, and having previous pregnancy terminated. All regions coefficients were positively associated with timing of visit. This association was congruent across the independence, Gaussian and Frank copula, with reduced significance when dependency was considered.

Table 10.5. Maximum likelihood estimates for the number of visits model [$f\ (y_2|Z)$] under Gaussian and Frank copula-based joint model

Variable	Independent model		Gaussian model		Frank model	
	Coefficient	St. error	Coefficient	St. error	Coefficient	St. error
Fhead	0.025	0.017	0.027	0.017	0.022	0.016
Urban	-0.0029	0.022	0.0035	0.022	0.023	0.021
Educnone	-0.018	0.034	-0.029	0.034	-0.026	0.033
Educpri	0.041†	0.021	0.039	0.021	0.038	0.02
Working	-0.029	0.017	-0.027	0.018	-0.026	0.017
Wind poor	-0.013	0.033	-0.016	0.032	-0.0033	0.031
Wind medium	-0.0047	0.027	-0.0041	0.026	0.00039	0.026
Newsdaily	-0.132‡	0.036	-0.136‡	0.035	-0.135‡	0.035
Newsoccas	-0.012	0.021	-0.0064	0.021	-0.0035	0.021
Radiodaily	0.02	0.027	0.024	0.027	0.012	0.026
radio occas	0.012	0.029	0.01	0.026	-0.0037	0.028
Tvdaily	-0.032	0.028	-0.037	0.028	-0.033	0.028
Tvoccas	0.0035	0.026	-0.0041	0.025	0.0052	0.026
govt facility	1.563‡	0.051	1.553‡	0.051	1.571‡	0.049
u5none	-0.056	0.032	-0.065†	0.032	-0.068†	0.031
u5b12	-0.051†	0.025	-0.046†	0.022	-0.046†	0.021
pregterminate	-0.071†	0.034	-0.079†	0.033	-0.085‡	0.032
Caprivi	0.0046	0.043	0.03	0.042	0.024	0.041
Erongo	-0.120‡	0.046	-0.105†	0.046	-0.108†	0.045
Hardap	-0.014	0.046	-0.067	0.046	-0.092†	0.045
Karas	-0.056	0.049	-0.078	0.049	-0.084	0.047
Kavango	-0.0029	0.039	0.0079	0.039	-0.0073	0.037
Khomas	-0.01	0.041	-0.0086	0.04	-0.0098	0.038
Kunene	-0.0018	0.046	-0.051	0.046	-0.064	0.044
Ohangwena	-0.124‡	0.043	-1.001†	0.042	-0.116‡	0.041
Omaheke	-0.075	0.045	-0.075	0.044	-0.086†	0.043
Omusati	-0.071	0.042	-0.039	0.041	-0.038	0.041
Oshana	-0.148‡	0.049	-0.139‡	0.044	-0.149‡	0.042
Oshikoto	-0.083†	0.041	-0.068	0.041	-0.069	0.039

† Significant at $\alpha < 0.05$; ‡ Significant at $\alpha < 0.01$.

In Table 10.5, average number of visits were negatively related to daily access to a newspapers, having under-five children, and pregnancy termination. Positive association was observed with use of government facility, and marginally if a woman attained primary education. In five of the thirteen regions (i.e., Erongo, Hardap, Ohangwena, Omaheke and Oshana), we observed a negative association with average number of visits. Comparing estimates in the joint model to the independence model, we observed relatively more covariates significant in the independent model than in the Frank copula model. For those variables significant in both the independent and Frank copula model, we observed that significance was reduced under the joint model.

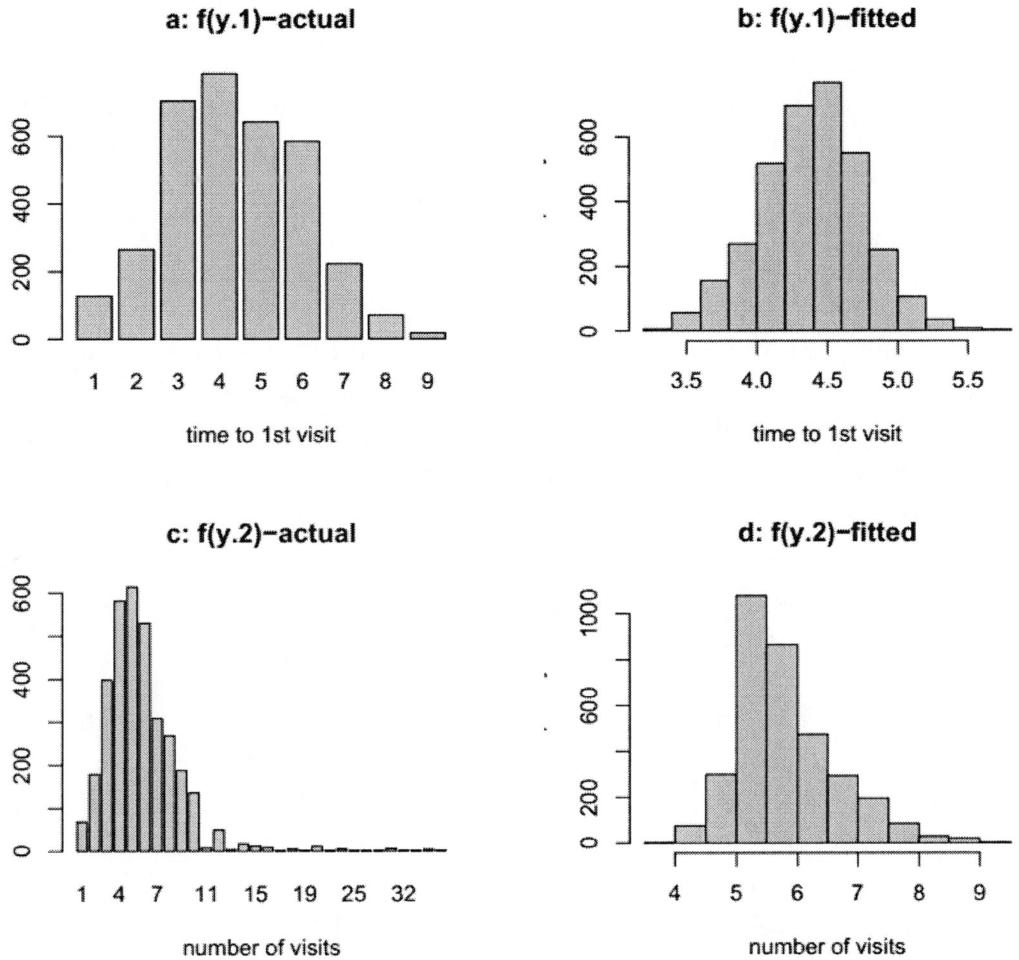

Figure 10.2. Histograms of empirical and fitted values of antenatal care (ANC) utilization. Top panel: timing of first ANC visit. Bottom panel: number of ANC visits.

We also compared the fitted versus empirical frequencies of the probability distribution/mass functions $f\ (y_1)$ and $f\ (y_2)$ based on the Frank copula-based model. Top panel of Figure 10.2 shows the pdf of $f\ (y_1)$, while the bottom panel shows the pmf of $f\ (y_2)$. Both model estimates seem to match well with actual distributions, however, the fit of the fitted count data is not as close to the actual distribution more especially at the upper tail.

DISCUSSION

Previous research on antenatal care has proposed independence of timing of first visit and number of visits. This has several limitations. First it ignores the bivariate dependence between the outcomes. Second, ignoring such dependence will lead to erroneous inferential conclusions. Third, the relationship between the outcomes and explanatory variables might relate differently once dependency is considered. This paper proposed a bivariate copula-based regression to account for dependency between the two outcomes. More generally, the paper adapted a Kramer et al. (2012) approach and fitted a Gamma and Poisson GLM, which used full information as opposed to an independency model (Table 10.3).

Results suggests that average number of visits in Namibia as of 2006/7 were well above the "focussed" four visits, however, this depended on when the first visit was initiated as evidenced in the copula dependency (Table 10.3). The corresponding model estimation results showed importance of demographic and socio-economic factors on ANC utilization. The results from the Frank copula-based model are more conservative indicating that assuming independence between the outcomes could be misleading, with significant covariates being mispecified.

The findings hold important implications for health policy planning. First, the extent of ANC utilization can only be enhanced since the average time to first visits is 4.36 ±1.62 months which fails short of the recommendation that the first ANC visit should be within first three months. Second, use of government facilities are important towards ANC uptake, therefore programmes aimed at promoting access to reproductive health should make public facilities within reach of communities in need. Third, while there is an increased use of ANC in all regions, the total number of visits is generally sub-optimal across regions. Therefore there is need to promote continuous and optimal use of ANC at all stages of pregnancy.

Our model might be limited in that other possible models for the two outcomes were not considered and compared. For example, for the waiting time for the first visit, log-normal or discrete-time duration models are possible candidates. For the count data, one can consider a negative binomial as an alternative to the Poisson model. However, our simulation of the observed data shows that Gamma and Poisson suffice for the data at hand. Another possible limitation is the choice of the copula model, in which only Frank and Gaussian copulas were considered, although other family of copulas (e.g., Clayton and Gumbel) are possible with *CopulaRegression* package. However, only Gaussian and Frank copula handle both negative and positive dependency. Comparing Gaussian and Frank copula, evidently Frank copula provides a better fit (Table 10.3), because Frank copula allow modeling dependence in arbitrarily high dimensions, governed by the strength of dependence (Joe, 1997).

Furthermore, we only included selected covariates in the models, other possible explanatory variables are possible as suggested in numerous literature. In additional, data is based on recall of recent birth (within three years preceding the survey date), and possibilities of recall-bias are pervasive. However, studies done on same data from 19 African countries show that women across all social strata (education, rural/urban, wealth ranking) were able to make correct judgement of events especially those pertaining to birth (Boerma et al., 2003). Be as it may, possibilities of heaping or rounding numbers is possible. DHS has resolved this by using health cards where these are available.

This paper makes an evaluative contribution towards formulating and estimating a mixed outcome model based on the copula-based regression. This, to our knowledge, is the first instance in the health services research to apply a joint Gamma and Poisson GLM as opposed to fitting separate logit model for "the time to first ANC visit" and another binary logit model for "whether four ANC visits have been reached". Examples of separate models are considered elsewhere (Magadi et al., 2000; Sagna and Sunil, 2012; Gayawan, 2013). As argued before, such separate models ignore the correlation that exist between outcomes, and may lead to erroneous small standard errors, thus resulting in committing Type I error. There are important further research questions from this study. First, there is need to explore other possible duration and count models particulary those that capture tail dependence, of which Vine copula are promising. Second, significant regional effects suggests substantial variation and heterogeneity still remain which can better be captured using random effects, therefore we hope an extension for correlated random effects would be worthwhile to explore. Third, we propose to explore the impact of ANC (whether early initiation of ANC or fulfilling focussed ANC) on the use of perinatal care or use of skilled delivery. This can further be modelled using multivariate sample selection models via copula methodology.

ACKNOWLEDGMENTS

We acknowledge permission granted by Macro International to use the 2006/2007 Namibia DHS data.This work was supported through the DELTAS Africa Initiative SSACAB [grant #DEL-15-005]. The DELTAS Africa Initiative is an independent funding scheme of the African Academy of Sciences (AAS)'s Alliance for Accelerating Excellence in Science in Africa (AESA) and supported by the New Partnership for Africas Development Planning and Coordinating Agency (NEPAD Agency) with funding from the Wellcome Trust [grant #107754/Z/15/Z] and the UK government. The views expressed in this publication are those of the author(s) and not necessarily those of AAS, NEPAD Agency, Wellcome Trust or the UK government.

REFERENCES

Babalola, S., Fatusi, A. (2009). Determinants of use of maternal health services in Nigeria- looking beyond individual and household factors. *BMC Pregnancy Childbirth* 9:43.

Boerma, J. T., Black, R. E., Sommerfelt, A. E., Rustein, S. O., Bicego, G. T. Accuracy and completeness of mother's recall of diarrhoea occurence in pre-school children in demographic and health surveys. *Int J Epidemiol* 1991;20:1073-1080.

Czado, C., Kastenmeier, R., Brechmann, E. C., Min, A. (2012). A mixed copula model for insurance claims and claim sizes. *Scandinavian Acturial Journal* 2012:278-305.

Cameron, A. C., Li, T., Trivedi, P. K., Zimmer, D. M. (2004). Modelling the differences in counted outcomes using bivariate copula models with application to mismeasured 411 counts. *Econometric Journal* 7:566-584.

Deb, P., Trivedi, P. K., Zimmer, D. M. (2013). Cost-offsets of presription drug expenditure: data analysis via copula-based dynamic hurdle models. *Health Economics. To appear.* Available as NBER Working Paper No. 15192.

Gage, A. (2007). Barriers to the utilization of maternal health care in rural Mali. *Social Science and Medicine,* 65:1666-1682.

Gayawan, E. (2013). A Poisson regression model to examine spatial patterns in antenatal care utilization in Nigeria. Population, Space and Place. *Early view.* DOI:10.1002/ psp.1775.

Hawkes, S. J., Gomez, G. B., Broutet, N. (2013). Early antenatal care: Does it make a difference to outcomes of pregnancy assocaited with syphilis? A systematic review and meta-analysis. *PLoS ONE*, 8(2):e56713.

Joe, H. (2014). *Dependence modelling with copulas.* Chapman and Hall, London.

Kisuule, I., Kaye, D. K., Najjuka, F., Ssematimba, S. K., Arinda, A., Nakitende, G., Otim, L. (2013). Timing and reasons for coming late for the first antenatal care vist by pregnant women at Mulago hospital, Kampala, Uganda. *BMC Pregnancy Childbirth* 13:121.

Kolev, N., Paiva, D. (2009). Copula-based regression models: A survey. *Journal of Statistical Planning and Inference* 139:3847-3856.

Kramer, N., Brechmann, E. C., Silvestrini, D., Czado, C. (2012). *Total losss estimation using copula-based regression models.* Preprint, Technische Universitat Munchen.

Leon, A. R. M. L., Chough, K. C. (2010). Mixed-outcome data. Encyclopedia of Biopharma-ceutical Statistics. Third Edition, vol 2. Chapter 132. In S.W Chow (ed). *Informa Healthcare, p*p: 817-822.

Leon, A. R. M. L., Wu, B. (2011). Copula-based regression models for a bivariate mixed discrete and continuous outcome. *Statistics in Medicine,* 30:175-185.

Magadi, M. A., Agwanda, A. O., Obare, F. O. (2007). A comparative analysis of the use of maternal health services between teenagers and older mothers in sub-Saharan Africa: evidence from demographic and health surveys (DHS). *Social Science and Medicine* 64:1311-1325.

Magadi, M. A., Madise, N. J., Rodrigues, R. N. (2000). Frequency and timing of antenatal care in Kenya: explaining the variations between women of different communities. *Social Science and Medicine,* 51:551-561.

Ministry of Health and Social Social Services (MoHSS) [Namibia] and Macro International Inc. (2008) *Namibia Demography and Health Survey 2006-2007.* Windhoek, Namibia and Claverton, Maryland: MoHSS and Macro.

Murteira, J. M. R., Lourenco, O. D. (2011). Health care utilization and self-assessed health: specification of bivariate models using copulas. *Empirical Economics* 41:447-472.

National Planning Commission (2008). *Second Millennium Development Goals Mid-Term Progress Report.* Ministry of Finance and Planning, National Planning Commission. Windhoek, Namibia: NPC.

Nelsen, R. B. (2006). *An introduction to Copulas.* Springer-Verlag, Berlin. 2nd Edition.

Pell, C., Menaca, A., Were, F., Afrah, N. A., Chatio, S., et al. (2013). Factors affecting antenatal care attendance: Results from qualitative studies in Ghana, Kenya and Malawi. *PloS ONE,* 8(2):e53747.

Sagna, M. L., Sunil, T. S. (2012). Effects of individual and neighbourhood factors on maternal care in Cambodia. *Health and Place* 18:415-423.

Smith, M. D. (2003). Modelling sample selection using Archimedian copulas. *Econometrics Journal* 6: 99-123.

Song, P. X. K., Li, M., Yuan, Y. (2009). Joint regression analysis of correlated data using Gaussian copula. *Biometrics* 65:60-68.

Vuong, Q. H. (1989). Likelihood ratio tests for model selection and non-nested hypotheses. *Econometrica.* 57:307-333.

Winkelmann, R. (2009). *Copula-based bivariate binary response models.* Working Paper No. 0913 Socioeconomic Institute, University of Zurich.

WHO, 2003. *Antenatal Care in Developing Countries: Promises, Achievements and Missed Opportunities: An Analysis of Trends, Levels and Differentials,* 1990-2001. WHO, 2010. World Health Statistics 2010. Geneva:WHO.

World Bank, 2010. *Safe Motherhood and Maternal Health.* Washington DC: World Bank.

Zimmer, D. M., Trivedi, P. K. (2006). Using trivariate copulas to model sample selection and treatment effects: Application to family health care demand. *Journal Business Economic Statistics* 24: 63-76.

In: Statistical Modelling …
Editors: N-B. Kandala and L. Kazembe
ISBN: 978-1-53615-981-3

Chapter 11

MULTISCALE SPATIAL MODELLING OF DIABETES AND HYPERTENSION IN NAMIBIA

Tommy R. Harris [*] ***and Lawrence N. Kazembe***
Division of Population and Housing Census and Demographic Surveys,
Namibia Statistics Agency, Windhoek, Namibia
Department of Statistics and Population Studies, University of Namibia,
Windhoek, Namibia

ABSTRACT

In Namibia, non-communicable diseases (NCDs) are alarmingly becoming an important cause of morbidity and mortality, and an emerging public health concern, although population based estimates in the area are lacking. To curb the rising trend of the burden of non-communicable diseases in Namibia, the Ministry of Health and Social Services embarked on several preventive initiatives such as raising awareness through preventive programmes, developing national promotion policies as well as health promoting school initiatives. The programmes, however, are implemented at national level but efficient targeting of such programmes requires the identification of high risk areas where the NCDs are most prevalent. This study demonstrates the use of disease maps to highlight areas of elevated risk for diabetes and hypertension, to inform better planning, monitoring and evaluation and assisting in targeting of resources. We used 15,462 cases of diabetes and 30,620 cases of hypertension, reported from 444 health facilities over a period of seven years (2008-2014), to fit a multiscale model at various geographic levels (region, health districts, and constituencies) in Namibia. A fully Bayesian inference based on Markov Chain Monte Carlo simulation techniques was used to derive results. Results show elevated risk for diabetes in Kavango, Caprivi, Khomas and Otjozondjupa regions, while risk for hypertension was hightest in Khomas region. At health district and constituency level, elevated risk for diabetes were found in Rundu

[*] Corresponding Author's Email: tharris@nsa.org.na.

health district and Rundu Urban constituency respectively. With respect to hypertension, highest risk were observed in the health districts of Andara, Rundu, Swakopmund and Windhoek, whereas at constituency level, Rundu Urban constituency exhibited the highest risk of hypertension. It is hoped that the study will assist policy makers especially those involved in health planning to develop targeted interventions in areas that were found to have elevated risk of diabetes and hypertension. Moreover, further studies should consider spatial factors related to diabetes and hypertension that could explain variability in risk.

Keywords: multiscale spatial modelling, Bayesian inference, diabetes, hypertension, Namibia

INTRODUCTION

Non-communicable diseases (NCDs) are increasingly becoming a challenge towards achieving global progress in improving population health (Ministry of Health and Social Services [MoHSS]) & ICF International, 2014). The World Health Organization (WHO, 2014) states that NCDs are becoming prevalent in low and middle income countries, where nearly three quarters of NCD deaths (28 million deaths) occur. Globally, they make the largest contribution to mortality and account for 60 percent (35 million) of global deaths (MoHSS & ICF International, 2014). It is projected that the NCD burden will increase globally by 17 percent in the next ten years, and in the African region by 27 percent (Islam et al., 2014).

In sub-Sahara Africa (SSA), NCDs are already a significant health problem, already the leading cause of death, and causing almost a third of deaths in the region (Marquez &Farrington, 2013). Moreover, the largest relative increase in NCD deaths globally in the next decade is expected to occur in Africa (Marquez & Farrington, 2013).

In Namibia, NCDs are equally on the increase. Statistics on NCDs (cancer, diabetes, cardiovascular diseases, hypertension etc.) as a cause of morbidity and mortality show that it is a growing public health concern even though population based data estimates in the area are rudimentary (MoHSS & ICF International, 2014). The WHO (2010) stated that between 2004 and 2008, in Namibia, institutional mortality due to cancer (all types) rose from 3.2% to 54.7%, cardiovascular diseases (all types) rose from 5.3% to 21.2%, and diabetes mellitus rose from 1.0% to 14.6%. The most prevalent NCDs in Namibia are cardiovascular diseases which accounted for 38 percent of total deaths across all age groups in 2008 while cancers non-communicable variants of respiratory diseases, and diabetes contributed 3% to total mortality (MoHSS & Macro International, 2008).

To curb the rising trend of burden of NCDs in Namibia, the MoHSS has embarked on a number of preventive initiatives including raising awareness through preventive programmes, passing laws related to the use of tobacco products as well as national promotion policy and health promoting school initiatives. These programmes are implemented at national level but an effective fight against NCDs will require the identification of areas with elevated disease risks. The use of disease maps to identify areas with elevated risk of NCDs burden therefore becomes important with available limited resources in Namibia.

There is, however, scant literature available on mapping of NCDs in Namibia which could assist to identify, at regional and local level, where the disease risk is raised for evidence-based deployment of limited resources, and targeting areas of high risk. With the rising trend in the burden of NCDs in Namibia, more efforts are needed to mitigate the burden of NCDs. Recent developments of epidemiological methods to study diseases risk remain of vital importance. For instance, knowledge of disease burden in populations, using spatial modeling, is essential for health authorities with limited resources as priority health programmes for prevention and care may effectively be identified. Disease mapping is one such approach that can serve as a basic tool in planning to optimize the reduction of NCDs (Lawson, 2008). The study is therefore significant as it will assist the MoHSS to derive health metrics, guide intervention strategies and advance epidemiological understanding using spatial modelling and mapping of diseases approaches.

The study of spatial variations in disease rates (disease mapping) is a classic epidemiological technique, where location is used as a surrogate for the mix of lifestyle, environmental and possibly genetic factors that may underline geographical differences in risk (Held & Best, 2001). According to Held and Best (2001) and Held, Natário, Fenton, Rue and Becker (2005), the purpose of spatial modeling is to describe such variations so that areas of elevated risk can be identified. Bayesian hierarchical models are typically used in this context (Lawson, 2018), which represent the risk surface using a combination of available covariate data and a set of spatial random effects. The random effects are typically modelled by a conditional autoregressive (CAR) prior distribution.

In modelling diabetes and hypertension in Namibia, the study applied a multiscale model to identify spatial variations of diabetes and hypertension at various geographic levels in Namibia to guide better planning, monitoring and evaluation and assist in targeted interventions at a finer scale and local level. The use of multiscale disease mapping has increased recently, building on the earlier work by Kolaczyk and Haung (2001), and expanded in Louie and Kolaczyk (2006). Aregay and others in particular have extended multiscale models using various approaches. For example, by introducing a shared random effect that the finer level inherits from the coarser level, or by assuming two independent convolution models at the finer and coarser levels, or by applying a convolution model at the finer level, and allow the relative risk at the coarser level to be obtained by aggregating the estimates at the finer level (Aregay et al., 2017a). Further extensions applied measurement error models in the framework of multiscale approach (Aregay et al., 2016), or compared multilevel and multiscale convolution models for small area aggregated health data (Aregay et al., 2017b). In another study, Aregay et al. (2016) investigated different spatial mixture multiscale models by mixing spatially varying probability weights of correlated (CH) and uncorrelated heterogeneities (UH), with the goal of capturing the predominance of the correlated or uncorrelated component for multiple scales. In this study, we adopted the approach that assumed two independent convolution models at the finer and coarser levels.

The organization of the rest of this chapter is as follows. Section 2 provides a review of the different methods used in disease mapping. Section 3 presents data collection methods, descriptive statistics, and the multiscale models fitted for diabetes and hypertension. The results are reported in section 4 and finally discussion of the results and drawing of conclusions are presented in section 5.

METHODS

Study Area and Context

This study covered the entire country of Namibia. As of 2013, the country was divided into 13 administrative regions, which were further sub-divided into 107 constituencies. Within the region, there were a total of 34 health districts offering health care services and each health district has several health facilities. In some instances, a region is equal to a health district, e.g., Khomas region is the same as Windhoek health district and Zambezi region is the same as Katima Mulilo health district. Each constituency has at least one health facility.

Morbidity Data

Diabetes and hypertension data were obtained from an administrative register called the Health Information System (HIS), maintained by MoHSS. We used data collected between January 2008 and December 2014. The data included both inpatient and outpatient observed cases of diabetes and hypertension from 444 health facilities across Namibia. Cases were based on raw data from individual patients encountered each day at the facility level using paper based registers and tally sheets. Reporting health facilities include all those managed by public, private and mission or not-for-profit non-governmental organisations (NGOs). At the end of each month, the data are transferred onto monthly summary report forms, which are submitted to the district level, and entered into HIS. Taken together, the indicators in HIS provide a comprehensive summary of the current status of the health of the nation, which allow to produce health service coverage, generate cause-specific mortality and morbidity on selected diseases or summaries of high burden diseases, and their corresponding risk factors.

Cases were disaggregated by health district, region and constituency level to perform spatial analysis. Estimates of population at risk at region, heath district and constituency level were linearly interpolated from the 2011 population data available at the Namibia Statistics Agency (NSA). The expected cases, for both diabetes and hypertension, using the population at risk were then calculated. Our analysis was restricted to those aged 15-64 years.

OVERVIEW OF DISEASE MAPPING

The most basic method for disease mapping is the standardized morbidity ratio (SMR), defined as $\frac{y_i}{e_i}$, which is an improvement of the crude ratio, $\frac{y_i}{n_i}$, where y_i and e_i denote the observed and expected number of cases respectively, with the population at risk, n_i, in a given area ($i = 1,2, \dots m$). The expected number of cases is defined as $e_i = n_i \left(\frac{\sum_i y_i}{\sum_i n_i}\right)$.

A common assumption with SMR is that the observed counts are assumed to be independent and drawn from a Poisson distribution, i.e., $y_i|\theta_i \sim_{iid} Poisson(e_i\theta_i)$ with mean $e_i\theta_i$, where θ_i is the true unknown relative risk (RR). Under this assumption, the maximum likelihood (ML) estimator of θ_i is the SMR, $\widehat{\theta_i} = \frac{y_i}{e_i}$. The SMRs may provide unstable estimates when the data is sparse and does not accommodate for the correlation between neighboring areas (Aregay et al., 2015a). To overcome the limitations, Besag et al., (1991) proposed a spatial convolution model (BYM) that estimates RR as a function of spatially structured and unstructured random effects. Tu and Greenwood (2012) state that BYM approach improves the precision and stability of risk estimates. Furthermore, BYM method provides a framework to simultaneously model the spatial and non-spatial (or heterogeneity) effects on disease risk.

Accordingly, using the BYM model, the risk in area i is estimated as:

$$\log(\theta_i) = \alpha_o + \beta^T x_i + u_i + v_i \quad (1)$$

where α_o represents the intercept of the log e_i; x_i is a vector of covariates associated with parameter β in ith area; u_i represents the ith area-wide heterogeneity, which is synonymous with global extra-Poisson variation; v_i denote the spatially dependent random effects which are defined by a range of different structures describing adjacency or closeness in space (Clayton & Kaldor, 1987). Generally, one may define a normal prior distribution for the non-spatial heterogeneity effects, such that $u_i \sim Normal\ (0, \sigma_u^2)$ (Besag et al., 1991). Further, we may assume that spatially correlated random effects v_i arise through a combination of independent random effects errors r_i that are normally distributed i.e., $r_i \sim Normal\,(0, \sigma_v^2)$ (Tu & Greenwood, 2012). We assume that the components v_i may be written as

$$v_i = \frac{\sum_{j \in \Theta i} r_j}{n_i} \tag{2}$$

giving rise to intrinsic conditional autoregressive (ICAR). Here Θ_i represents a set of areas sharing a common boundary with area i, with n_i denoting the number of neighbours for area i. Therefore, through the averaging of the independent random effects, r_j defines the effect of area j on the disease risk in area i. The Besag model can be extended to include area-level covariates such as socio-economic status, whilst accounting for both the spatially structured and unstructured (heterogeneity) random effects for the RRs across areas.

Alternatively, correlation models are available. For instance, a proper CAR model has been proposed for the correlated heterogeneities (CH) component, leading to a proper multivariate Gaussian distribution with a full rank covariance matrix (Cressie and Chan, 1989; Stern and Cressie, 1999). Moraga and Lawson (2012) considered Gaussian component mixture models (GCM) and compared it to the intrinsic conditional autoregressive (ICAR) assumption in a simulation study.

Multiscale Spatial Modelling

Although the convolution model has been widely used in spatial epidemiology, it does not accommodate a spatial scaling effect associated with aggregation in the data space. For example, data that is often observed in the form of aggregation at different multiple scales (i.e., census block, block group, census tract, and state levels etc.) will cause a scale effect to occur due to the aggregation of data from a finer level to a coarser level (Cressie, 1996; Banerjee, Carlin & Gelfand, 2004; Wong, 2009, Ch. 7; Lawson, 2013, Ch. 9; Aregay et al., 2016a). A spatial scaling effect is induced from a lower geographical level (finer) to a higher geographical level (coarser) by smoothing out the variation (i.e., averaging) (Wong, 2009; Aregay et al., 2017a). Hence, in disease mapping, the scaling effect is of special interest when dealing with multiple/hierarchical or multiscale diseases reporting levels (Aregay et al., 2016a, 2017a).

According to Aregay et al. (2017a), scaling reduces the variability of the data and makes them more alike, and hence the results from the finer data may not be consistent with the results from the coarser data. For example, we may find a relationship between an outcome and a covariate at a higher level, which is not retained at a lower level. Furthermore, an atomistic fallacy will occur if we try to generalize directly from the lower level (e.g., census tract) to the higher aggregation level (e.g., county) without adjusting for the variability of individuals' response to the diseases (Aregay et al., 2017a).

To encompass scaling effects, Kolaczyk and Haung (2001) proposed a multiscale modelling approach by factorizing the likelihood into the individual components of local information. The model assumes that the hierarchical partitions correspond to the successive aggregation of an initial data space (Aregay et al., 2017a). The model is briefly described as follows.

Assume that D is an arbitrary spatial area and $B_{i,k}$ are subareas in region D at the kth spatial scale and relative positions i within a scale, such that $\cup_{i=1}^{N_k} B_{i,k} = D$ and $\cup_{i' \in ch(i,k)} B_{i',k+1} = B_{i,k}$, where for a given choice of $k = 0,1,\dots,K$, and $i = 1,\dots N_k$, $ch(i,k)$ represents the collection of spatial indices i' at scale level k for which $B_{i',k+1} \subseteq B_{i,k}$. Note that $k = 0$ denotes the coarsest level with $N_0 = 1$ while $k = K$ represents the finest level. Further, let $Y_k = (Y_{1,k}, \dots, Y_{N_k,k})^T$ be the vector of aggregated counts at the kth scale and assume $Y_{ch(i,k)k+1}$ is the vector of aggregated counts $Y_{i',k+1}$ for whom $B_{i',k+1} \subseteq B_{i,k}$. Similar to wavelet decomposition, Louie and Kolaczyk (2006) assume that the information in the original measurement at the finest scale K, i.e., Y_K decomposes into conditionally independent components as those measurements are aggregated across each successively coarser level.

Mathematically, it can be expressed as

$$\Pr(Y_K|u_K) = \Pr\left(Y_{1,0}|u_{1,0}\right) \prod_{k=0}^{K-1} \prod_{i=1}^{N_k} \Pr(Y_{ch(i,k),k+1}|Y_{i,k}, \omega_{i,k}) \tag{3}$$

where the count at the coarsest level follows a Poisson distribution $Y_{1,0} \sim Poisson\left(u_{1,0}\right)$ and conditioning on the count at the coarser scale, the aggregated counts at the finer level follow a multinomial distribution, $Y_{ch(i,k),k+1}|Y_{i,k}, \omega_{i,k} \sim Multinomial\left(Y_{i,k}; \omega_{i,k}\right)$, with $\omega_{i,k} = u_{i,k}^{-1} \times \mu_{ch(i,k),k+1}$.

Nevertheless, their approach is limited by the conditional assumption, denoted $Y_{ch(i,k),k+1}|Y_{i,k}$. This means that conditioning on the counts at the kth coarser level $(Y_{i,k})$ the counts at $k+1$ finer scale level $Y_{ch(i,k+,k+1)}$ follow a multinomial distribution. This is not true in reality because health data are collected at the finer level and not at the coarser level (Aregay et al., 2017a). Moreover, the data at the coarser level are an aggregation of the data at the finer level. In addition, their method does not help to assess whether a general clustering effect exists in the overall region because it does not contain the spatial component which captures the extra-Poisson variation.

To address this issue, Louie and Kolaczyk (2006) considered a Poisson log-normal model at each individual and independent scale k. However, this method does not accommodate for the individual information at the finest level and it is not flexible enough to estimate RR at the lowest and highest levels at the same time. Aregay et al.

(2017a) developed a multiscale log-normal modeling framework extension to the model proposed by Louie and Kolaczyk (2006). The framework can be used to make inference both at the higher (areas) and lower levels (subareas) simultaneously using Bayesian models. Aregay et al. (2017a) incorporated the scale effect using a convolution BYM multiscale modelling approach, whereby they proposed three different models to account for the scaling effect. For their study, the modelling was done at both the finer and coarser levels simultaneously by including a shared random effect to handle the scaling effect.

The models proposed by Aregay et al., (2017a) are briefly described as follows for data that is aggregated at the county and public health (PH) district levels. Then $y_i^c, i = 1, \ldots, N$, is the county level count of disease and $y_j^{ph} = \sum_{i \epsilon j} y_i^c, j = 1, \ldots, n,$ *is* the jth public health district level count of disease aggregated to the county level. In model (4), they considered a joint convolution model at the county and PH district levels. The linkage between the two levels was incorporated in the model by including a shared spatial structured random effect, u_j^{ph}. The model is then given by:

$$\begin{aligned} &y_i^c \sim Poisson(e_i^c \theta_i^c), \\ &log(\theta_i^c) = \alpha_0^c + v_i^c + u_j^{ph}, \\ &y_j^{ph} \sim Poisson\left(e_j^{ph} \theta_j^{ph}\right), \\ &log\left(\theta_j^{ph}\right) = \alpha_0^{ph} + v_j^{ph} + u_j^{ph} \end{aligned} \tag{4}$$

Here, e_i^c and e_j^{ph} are the expected rate at the county and PH district level respectively. For model (4), Aregay et al., (2017a) assumed the following prior distributions. The intercept parameters, α_0^c *and* α_0^{ph}, and the uncorrelated heterogeneity (UH) random effects v_j^{ph} and v_i^c were assumed to be normally distributed, i.e., $\alpha_0^c \sim N(0, sd_{\alpha c}^2)$ and $\alpha_0^{ph} \sim N\left(0, sd_{\alpha ph}^2\right)$ $v_j^{ph} \sim N\left(0, sd_{vph}^2\right)$ and $v_i^c \sim N(0, sd_{vc}^2)$. On the other hand CH random effect, u_j^{ph}, was assumed to have a CAR distribution.

Model (5), is similar to Model (4), except that a spatial structured random effect at the county level, u_i^c, is added to the model. The model can be written as:

$$\begin{aligned} &y_i^c \sim Poisson(e_i^c \theta_i^c), \\ &log(\theta_i^c) = \alpha_0^c + v_i^c + u_i^c + u_j^{ph}, \\ &y_j^{ph} \sim Poisson\left(e_j^{ph} \theta_j^{ph}\right), \\ &log\left(\theta_j^{ph}\right) = \alpha_0^{ph} + v_j^{ph} + u_j^{ph}. \end{aligned} \tag{5}$$

With respect to model (6), the model assumes two separate convolution models at both the county and PH levels. The model is of the form:

$$
\begin{aligned}
&y_i^c \sim Poisson(e_i^c \theta_i^c),\\
&log(\theta_i^c) = \alpha_0^c + v_i^c + u_i^c,\\
&y_j^{ph} \sim Poisson\left(e_j^{ph} \theta_j^{ph}\right),\\
&log\left(\theta_j^{ph}\right) = \alpha_0^{ph} + v_j^{ph} + u_j^{ph}.
\end{aligned} \tag{6}
$$

Multiscale modelling proposed by Aregay et al. (2017a) has many advantages: (i) it incorporates the scaling effect during modelling and hence helps to recover the lost information during aggregation, (ii) it provides diseases risk estimates at different geographical levels simultaneously. Instead of modelling the disease risks at each scale level separately, jointly modelling the disease risks at different geographical levels will save time and energy, and (iii) it has been shown that multiscale model provides consistent covariate effect on outcome of interest at both finer and coarser levels compared to modelling separately at different scale levels (Aregay et al., 2015).

Binomial data have also been used in the context of multiscale modelling. For instance, Aregay et al. (2015) focused on modelling binomial data at different scale by applying two models. For both models, used a Bayesian approach, which combines the current and previous information. The models are given as:

$$y_{i_1} \sim \text{binomial}(n_{i_1}, p_{i_1}); logit(p_{i_1}) = \beta_{01} + \beta_{11} \times x_{i_1} + v_{i_1} + \varepsilon_{i_1}$$

$$y_{i_2} \sim \text{binomial}(n_{i_2}, p_{i_2}); logit(p_{i_2}) = \beta_{02} + \beta_{12} \times c_{i_2} + v_{i_2} + \varepsilon_{i_2} \tag{7}$$

where y_{i_1}, x_{i_1} are the response and covariates at county level and y_{i_2}, c_{i_2} are variables associated with the PH level, respectively. Further, v_{i_1} and v_{i_2} are the CH at the county and PH levels, whereas ε_{i_1} and ε_{i_2} are the UH at the county and PH levels. Aregay et al., (2015) assumed an ICAR structure for the correlated random effects for model (7). Aregay et al. (2015) applied the shared multiscale model proposed in Aregay et al., (2015), the multiscale convolution model can be given as:

$$y_{i_1} \sim \text{binomial}(n_{i_1}, p_{i_1}); logit(p_{i_1}) = \beta_{01} + \beta_{11} \times x_{i_1} + v_{i_1} + \varepsilon_{i_1} + v_{i_2}, \quad \text{for } i_1 \in S_{i_2}$$

$$y_{i_2} \sim \text{binomial}(n_{i_2}, p_{i_2}); logit(p_{i_2}) = \beta_{02} + \beta_{12} \times c_{i_2} + v_{i_2} + \varepsilon_{i_2} \tag{8}$$

where S_{i_2} is the set of subregions at the county level within the PH district. The shared random effect v_{i_2} is common for all the counties nested within the PH district or for data at different scales (Aregay et al., 2017a). In their work, Aregay et al. (2016a) considered four different scenarios for mixture multiscale convolution models: (i) no linkage between the scales, (ii) linkage using a shared uncorrelated random effect, (iii) a shared correlated component, and (iv) both shared correlated and uncorrelated random effects. In brief, their methods are summarized as below:

Let y_{i_k} be the outcome of interest in region $i = 1, 2, \ldots, N_k$ at scale $k = 1,2,3 \ldots j, i = 1, 2, \ldots, K,$ where N_k is the number of regions at spatial scale k. We assume that k is ordered from lowest (finest) to highest (coarsest) levels, that is $k = 1$ represents the lowest scale level, while $k = K$ denotes the highest level. Further, we assume that we have hierarchical data in the sense that the data at the lower level are nested within the data at the higher level. In practice, data are collected at the lowest level. If we have observed a response $y_{l_{k-1}}$ for scale $k = 2, \ldots, K,$ the outcome at the higher scale level k, namely y_{i_k}, is an aggregation of the outcomes observed within region i_k. Mathematically, this can be expressed as $y_{i_k} = \sum_{l \in S_{i_k}} y_{l_{k-1}}$, where S_{i_k} is the set of spatial subregions at scale level $k-1$ of region i_k. Conditioning on the relative risk θ_{i_k}, the responses y_{i_k} follow an independent Poisson distribution, that is, $y_{i_k} | \theta_{i_k} \sim Poisson(u_{i_k} = e_{i_k}\theta_{i_k})$.

The spatial mixture version of the multiscale model can be expressed as follows:

$$\begin{cases} log\theta_{i_1} = a_{0_1} + p_{i_1}v_{i_1} + (1-p_{i_1})u_{i_1} + (1-p_{i_2})u_{i_2} + \ldots + (1-p_{i_K})u_{i_K} \; for \; i_1 \in S_{i_2}, \\ log\theta_{i_2} = a_{0_2} + p_{i_2}v_{i_2} + (1-p_{i_2})u_{i_2} + \ldots + (1-p_{i_K})u_{i_K} \; for \; i_2 \in S_{i_3}, \\ \vdots \\ log\theta_{i_{K-1}} = a_{0_{K-1}} + p_{i_{K-1}}v_{i_{K-1}} + (1-p_{i_{K-1}})u_{i_{K-1}} + (1-p_{i_K})u_{i_K} \; for \; i_{K-1} \in S_{i_K}, \\ log\theta_{i_K} = a_{0_K} + p_{i_K}v_{i_K} + (1-p_{i_K})u_{i_K}, \end{cases} \tag{9}$$

where p_{i_k} is the mixing probability weight for region i at scale level k. Aregay et al. (2016a) considered four scenarios: a multiscale model without a shared component, with only a shared CH component, only a shared UH component, and both shared CH and UH components. They demonstrate the four scenarios for two-level $k = 1, 2$ aggregate disease incidence data.

Multiscale Spatial Models of Diabetes and Hypertension

For this study, multiscale models were applied to model spatially referenced outcomes by taking into account the spatial scaling effect, and assumed two independent convolution models at the finer and coarser levels. Health facility data collected at health facility level were aggregated to the health district, region and constituency level to carry out multiscale analysis. A number of candidate models were fitted with the aim of generating RR maps at region, health district and constituency level. The models were estimated as follows:

- Fit a multiscale random effects model for each disease by considering only spatially structured effects.
- Fit a multiscale random effects model that considered both spatially structured and spatially unstructured effects.
- Fit a multiscale random effects model for each disease by considering three different spatial scales (i.e., region, health district, and constituency) into the model. For each fitted model consider both spatially structured and unstructured effects.
- For each fitted model, observe the deviance, P_D, DIC and calculate the ΔDIC. The ΔDIC gives change in the deviance with respect to the deviance of the preceding model with -ΔDIC indicating an improved fit on the current model.
- Select any multiscale model using the ΔDIC rule of thumb that states that a ΔDIC less than 4 implies no significant difference between the models and therefore any model can be selected as a best fit model for the data.

For each outcome variable, four models were fitted and summarized below:

Diabetes

$M_1 = f_{spatial(hd)} + f_{spatial(const)}$

$M_2 = f_{spatial(hd)} + f_{spatial(reg)}$

$M_3 = f_{spatial(hd)} + f_{random(hd)} + f_{spatial(const)}$

$M_4 = f_{spatial(const)} + f_{random(const)} + f_{spatial(hd)} + f_{spatial(reg)}$

Hypertension

$M_5 = f_{spatial(hd)} + f_{spatial(const)}$

$M_6 = f_{spatial(hd)} + f_{spatial(reg)}$

$M_7 = f_{spatial(hd)} + f_{random(hd)} + f_{spatial(const)}$

$M_8 = f_{spatial(const)} + f_{random(const)} + f_{spatial(hd)} + f_{spatial(reg)}$

The study applied a fully Bayesian (FB) approach based on Markov random field priors and used the Markov chain Monte Carlo (MCMC) techniques for inference and model checking. MCMC techniques assist to overcome difficulties in calculation of the posterior distribution. The techniques allows for drawing random samples from the posterior, whose characteristics (such as the mean, standard deviation and quantiles) can be easily estimated by their empirical analogues. The analysis was carried out using version 2.1 of the BayesX (Belitz, Brezger, Kneib & Lang, 2009) statistical software which permits Bayesian inference based on MCMC simulation techniques. For all the models, 52000 iterations were run with a burn of 2000 for each model.

RESULTS

Observed and Standardized Morbidity Ratios

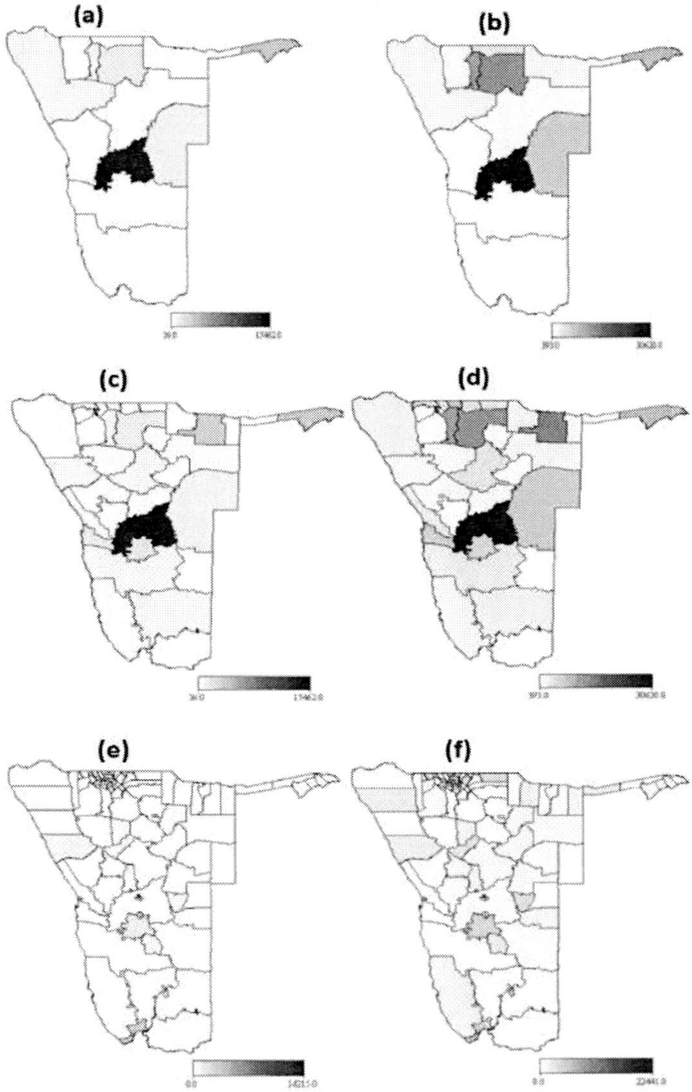

Figure 11.1. Total observed counts at regional level (top panel): (a) diabetes; (b) hypertension; at health district (middle panel): (c) diabetes; (d) hypertension; and at constituency level (bottom panel): (e) diabetes; (f) hypertension.

Figure 11.2. Standardized morbidity ratio (SMR) of diabetes (left panel) and hypertension (right panel) at region level (top), health district (middle panel) and constituency level (bottom panel).

Figure 11.1 and 11.2 present the observed and standardized morbidity ratios for diabetes and hypertension at regional, health district and constituency levels. Without smoothing the range of values is huge and outliers are evident in Figure 11.1. However, after standardizing the SMR values were reduced to reasonable range (Figure 11.2). The SMR associated with diabetes ranged between 0-3.1 at regional and health district level, and 0-15 at constituency level, while hypertension ranged between 0-5.1 at both regional and health district level, and 0-3 at constituency level.

Model Selection Criterion for Diabetes

Table 11.1 shows results for models fitted for the outcome variable diabetes. The change, ΔDIC, of all models was also calculated.

Table 11.1. Deviance, PD, DIC and ΔDIC for models of diabetes fitted in BayesX

	Model estimates for diabetes*	Deviance	P_D	DIC	ΔDIC
M_1:	diab = hd(spatial) + const(spatial)	73.20	5.17	78.91	
M_2:	diab = hd(spatial) + reg(spatial)	58.65	4.82	78.31	0.60
M_3:	diab = hd(spatial) + hd(random) + const(spatial)	57.32	6.39	70.11	8.80
M_4:	diab = const(spatial) + const(random) + hd(spatial) + reg(spatial)	55.01	11.36	67.75	11.16

*hd: health district; const: constituency; reg: region.

According to Table 11.1, M_1 was found to have a DIC equal to 78.91. On the other hand model M_2, which estimated spatial structured random effects at health district (hd) and regional (reg), showed a slight improvement with a difference in the DIC of 0.60 (DIC = 78.31). When unstructured random effects for HD were added in model M_3, the model improved much more (DIC = 70.11). Note that M_4 was obtained by adding both REG and HD structured spatial effects on model M_1, and it gave an overall better model, with DIC = 67.75 (Table 11.1). When random effects were added on model M_4, large DIC was obtained (results not shown), mostly driven by large deviance and effective number of parameters.

Model Selection Criterion for Hypertension

A similar approach used in fitting models for diabetes was used to fit and select the best model to examine the spatial variation of the outcome variable hypertension. Table 11.2 presents deviance, P_D, DIC and ΔDIC from the fitted models. All models were multiscale in nature. Model M_5 only considered spatially structured random effects only at CONST and HD, while model M_6 only considered spatially structured effects at HD and REG. Only model M_8 considered all three levels (HD, CONST, REG) and combined both spatially structured effects and spatially unstructured effects.

The model with the highest DIC was found to be M_5 (DIC = 56.5459), while model M_7 had the lowest DIC (= 54.2803). However, in our analysis we opted for model M_8 which had a better goodness-of-fit, as captured by the deviance of 44.27, and negligible ΔDIC compared to M_7. It also provides the estimates of all spatial components at all three levels: constituency, health district, and region. It should be noted that the differences in DIC between the various models fitted were relatively small.

Table 11.2. Deviance, PD, DIC and ΔDIC for models of hypertension fitted in BayesX

	Models for hypertension (ht)	Deviance	P_D	DIC	ΔDIC
M_5:	ht = hd(spatial) + const(spatial)	49.90	3.22	56.55	
M_6:	ht = hd(spatial) + reg(spatial)	49.95	3.20	56.75	-0.20
M_7:	ht = hd(spatial) + hd(random) + const(spatial)	44.87	4.71	54.28	2.27
M_8:	ht = const(spatial) + const(random) + hd(spatial) + reg(spatial)	44.27	5.65	54.57	1.98

*hd: health district; const: constituency; reg: region.

Model Estimates

Structured Spatial Variation of Diabetes and Hypertension at Regional Level

Based on models M_4 for diabetes and M_8 for hypertension, Figure 11.3 captures the spatial variation of diabetes a8n471d hypertension at regional level. For diabetes, the RR ranged from 0.969 to 1.027 while for hypertension the RR ranged from 0.976 to 1.019.

From the map in Figure 11.3(a), Kavango region had the highest RR for diabetes. Caprivi, Khomas and Otjozondjupa region also showed a relatively high RR of diabetes. The lowest RR of diabetes was found in Oshana region. Other regions with low RR were Omusati and Oshikoto with light grey colours, as shown in Figure 11.3(a).

The map of RR of hypertension at regional level is shown in Figure 11.3(b). Khomas region was found to have the highest RR of hypertension. Erongo, Hardap Otjozondupa, Kavango and Karas region also appeared to exhibit a high RR of hypertension. The regions with the lowest RR of hypertension was Oshikoto and Oshana region.

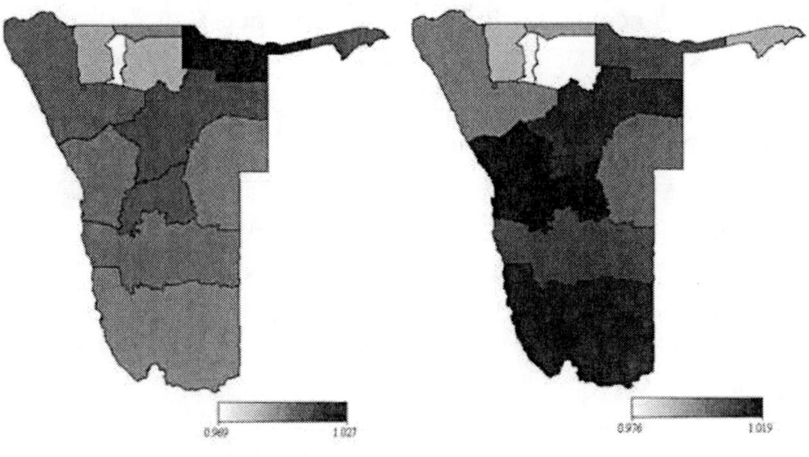

Figure 11.3. Spatial variation in risk (RR) at regional level for: (a) diabetes, and (b) hypertension.

Structured Spatial Variation of Diabetes and Hypertension at Health District

Figures 11.4(a) and 11.4(b) presents the structured spatial variation of diabetes and hypertension estimated from the multiscale models (M_4 and M_8) at health district level. Figure 11.4(a) shows that the RR for diabetes ranged from 0.967 to 1.037, while that of hypertension ranged from 0.975 to 1.012 as shown in Figure 11.4(b).

The highest RR of diabetes was found in Rundu health district as shown in Figure 11.4 (a). Andara, Nyangana, Nankudu, Grootfontein and Okakarara health districts also showed a high RR of diabetes. Aranos, Omaheke, Okahandja, Otjiwarongo, Omaruru, Khorixas, Kongo and Opuwo were also found to have a relatively high RR of diabetes.

On the other hand, Karasburg, Luderitz and Windhoek health districts had a reduced RR. There was a significant number of health districts with a low RR of diabetes. The lowest RR among health districts was found to be in Oshakati health district in Oshana region. Other health district with low RR included Eenhana, Engela and Oshikuku and among others.

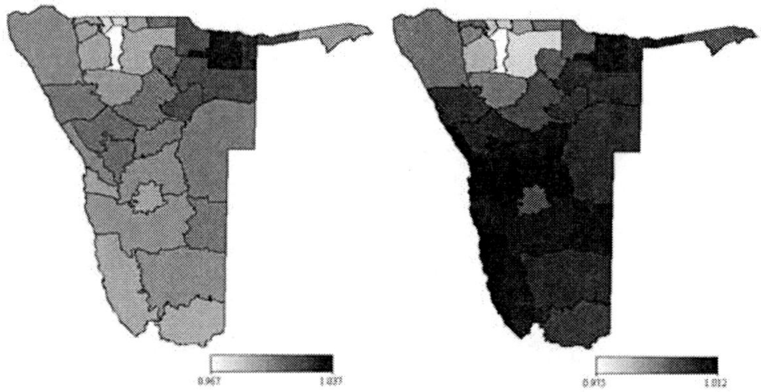

Figure 11.4. Spatial variation in risk (RR) at health district level for: (a) diabetes, and (b) hypertension.

Figure 11.4(b) also gives results of the RR of hypertension, the results showed that spatial variation exists amongst the different health districts in Namibia. The highest RR of hypertension was found in Andara, Rundu, Swakopmund and Windhoek health districts. There was also a relatively high RR of hypertension observed in some health districts. Health districts such as Aranos, Katima Mulilo, Luderitz, Okahandja and Usakos. Karasburg, Keetmashoop, Mariental, Gobabis, Omaruru, Khorixas, Okakarara, Grootfontein, Nyangana, Outapi and Nankudu also appeared to have a relatively high RR of hypertension.

The lowest RR of hypertension was evident in Oshakati health district. Other health districts that exhibited a relatively low RR of hypertension were Engela, Onandjokwe, Okahao and Oshikuku.

Structured Spatial Variation of Diabetes and Hypertension at Constituency Level

The RR of diabetes and hypertension at constituency level is given in Figures 11.5(a) and 11.5(b) respectively. The RR ranges of diabetes ranged from 0.972 to 1.064 while that of hypertension ranged from 0.967 to 1.03.

Rundu Urban was found to have the highest RR of diabetes at constituency level. Other areas that were found to have a high RR of diabetes at constituency level were Mashare, Kapako, Opuwo, Kahenge, Katima Mulilo Urban, Keetmanshoop Urban, Ndiyona, Khomasdal North, Mukwe, Otjimbinde, Okakarara, Khorixas, Tsumkwe, Tsandi, Otjiwarongo, Epupa, Aminuis, Okongo, Outjo, Katutura Central, Rehoboth East Urban, Walvis Bay Urban and Daures. Constituencies such as Oniipa, Okahandja, Omatako, Grootfontein, Ruacana, Okahao, Outapi, Engela, Sesfontein, Tobias Hainyeko, Berseba, Gibeon, Karibib and Tsumeb among others.

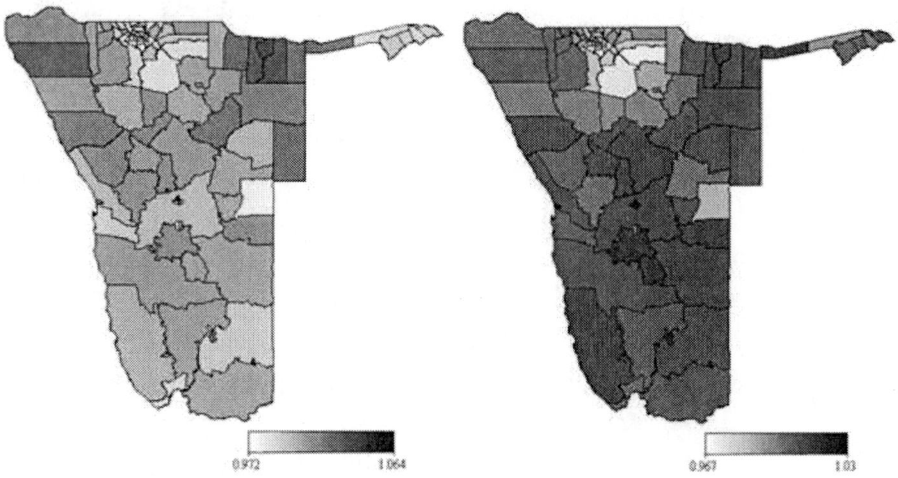

Figure 11.5. Spatial variation in risk (RR) at constituency level for: (a) diabetes, and (b) hypertension.

The study found the lowest RR of diabetes in Rehoboth West Urban constituency. Constituencies such as Oranjemund, Keetmanshoop Rural, Kalahari, Uuvudhiya, Omuthiyagwiipundi, Eengondi, Elim, Anamulenge, Ogongo, Okaku, Okatyali, Opundja, Uukwiyu, Onayena and Okankolo among others had a low RR of diabetes as shown in Figure 11.5(a).

With respect to hypertension, Rundu Urban was found to have the highest RR of hypertension at constituency level as shown in Figure 11.3(b). Other constituencies that appeared to have a high RR were Katima Mulilo Urban, Khomasdal North, Walvis Bay Urban, Swakopmund, Katutura Central, Rehoboth East Urban, Keetmanshoop Urban, Mukwe, Katutura East, Mariental Urban, Tobias Hainyeko, Windhoek West, Luderitz, Windhoek East, Aminuis, Samora Machel, Okahandja, Omaruru, Mariental Rural, Ndiyona, Rundu Rural West, Windhoek Rural, Khorixas, Omatako, Arandis, Kapako,

Gibeon, Keetmashoop Rural, Otjiwarongo, Tsumkwe, Kabe, Karasburg, Gobabis and Okakarara among others.

Constituencies such as Ruacana, Tsandi, Epupa, Karibib, Linyanti and Katima Mulilo Rural among others were found to have a reduced RR of hypertension as shown in Figure 11.5(b). The lowest RR of hypertension was however found in Ompundja constituency. Constituencies such as Okatyali, Eengondi, Omuthiyagwiipundi, Olukonda, Ohangwena, Okaku, Omuntele, Onyaanya, Onayena, Okankolo, Uuvudhiya, Uukwiyu, Etayi, Kalahari, Oshakati East, Okatana, Endola, Guinas and Ondobe also appeared to exhibit a low RR of hypertension.

Spatially Unstructured Random Effects for Diabetes and Hypertension at Constituency Level

The study presented the spatially unstructured random effects for diabetes and hypertension at constituency level, the results are shown in Figures 11.6(a) and 11.6(b) respectively. The RR for diabetes ranged from 0.98 to 1.017 while that of hypertension ranged from 0.985 to 1.011.

Opuwo constituency was found to have the highest RR of diabetes at constituency level. Other constituencies such as Khomasdal North, Keetmanshoop Urban, Katima Mulilo Urban, Oniipa, Oshakati East, Rundu Urban, Engela, Okakarara, Outjo, Otjombinde, Otjiwarongo, Rehoboth West Urban, Mukwe, Khorixas, Aminuis, Okahao, Tsandi, Daures, Mashare, Rundu Rural West, Okongo, Oshikuku, Outapi, Omuntele, Karibib, Walvis Bay Urban, Ndiyona, Rundu Rural West, Tobias Hainyeko and Eenhana were also found to have a high RR of diabetes as shown in Figure 11.6(a).

A reduced RR of diabetes at constituency level was also found in constituencies such as Karasburg, Katutura central, Epupa, Epembe, Guinas, Okahandja, Gibeon, Kapako, Oshikango, Steinhausen, Tsumeb, Ruacana, Gobabis and Berseba.

The RR of diabetes was found to be lowest in Kalahari constituency as shown in Figure 11.6(a). Other constituencies that showed a low RR of diabetes at constituency level include Rehoboth West Urban, Elim, Omuthiyagwiipundi, Eengondi, Ohangwena, Walvis Bay Rural, Katima Mulilo Rural, Oranjemund, Ongwediva, Okatyali, Anamulenge, Otjinene, Ondobe, Oshakati West, Keetmashoop Rural, Olukonda, Uuvudhiya, Ompundja, Samora Machel, Endola and Onayena among others.

With respect to hypertension, Figure 11.6 (b) shows that the RR was highest in Khomasdal North constituency. Other constituencies that appeared to have a high RR of hypertension were Katima Mulilo Urban, Engela, Oshakati East, Rehoboth East Urban, Mukwe, Eenhana, Oniipa, Rundu Urban, Swakopmund, Walvis Bay Urban, Khorixas, Aminuis, Outapi, Omatako, Otjiwarongo, Windhoek West, Oshikango, Okahao and Ondangwa.

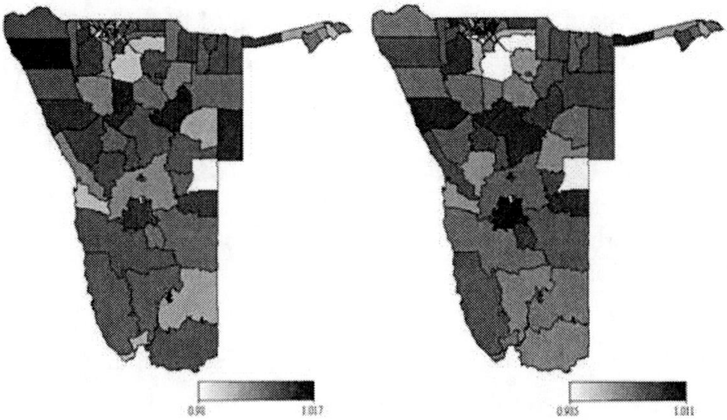

Figure 11.6. (a) Spatially unstructured effects at constituency level for: (a) diabetes and (b) hypertension.

A reduced RR of hypertension was also found in Ongwediva, Okakarara, Linyanti, Keetmashoop Urban, Katutura Central, Kahenge and Arandis among others. The lowest RR of hypertension was however found in Omuthiyagwiipundiconstituency. Constituencies such as Eengondi, Ompundja, Kalahari, Okatyali, Rehoboth West Urban, Ohangwena, Olukonda, Okankolo, Uuvudhiya, Kongola and Otavi Constituencies among others also showed a low RR of hypertension.

Variance Components of the Best Model for Diabetes and Hypertension

Table 11.3 presents the posterior means and 95% credible intervals of the variance components for the multiscale models M_4 and M_8. The variance component is important for the study as it provides information on how large the RR is in a particular area

For diabetes, the study found region has the variance component with a high posterior mean (1.0262) for model M_4 that considered health district, region and constituency as spatial effects.

With respect to health district and constituency, the study found a posterior mean of 1.0236 and 1.0155 respectively. The high posterior mean of region compared to the spatial effects health district and constituency suggests that the variability in RR for diabetes was large among the different regions and smaller in health districts and constituencies.

With respect to hypertension, the study also found region to have the highest variance component (posterior mean = 1.0168) as shown in Table 11.3. For health district and constituency, the posterior mean was 1.0125 and 1.0095 respectively. The high posterior mean for region compared to spatial effects health district and constituency considered in M_8 suggests that the variation in risk for hypertension was high in among regions compared to health districts and constituencies.

Table 11.3. Variance components of the best model of diabetes and hypertension. Given are the posterior means and 95% credible intervals in brackets

Disease	Variance component	Posterior mean (95% credible interval)
Diabetes(M_4)	Region	1.0262 (1.0005; 1.1705)
	Health district	1.0236 (1.0007; 1.1602)
	Constituency	1.0155 (1.0006; 1.0882)
Hypertension(M_8)	Region	1.0168 (1.0006; 1.1069)
	Health district	1.0125 (1.0005; 1.0702)
	Constituency	1.0095 (1.0006; 1.0491)

Discussion

The rising trend in NCDs calls for interventions that will assist in reducing the incidence and prevalence of diabetes and hypertension, yet many health funding agencies and policy makers are reluctant to divert scarce resources away from communicable diseases into other areas of disease burden such as NCDs (Kandala, Tigbe, Manda & Stranges, 2013). The failure to divert these scarce resources to NCDs to address the problems associated with NCDs may impose significant burden for health sectors and the economy of Sub Saharan African countries (including Namibia) because few people would seek treatment, which could lead to high morbidity and mortality rates from potentially preventable diseases (Belue et al., 2009).

Although, government and other relevant sectors such as civil society, faith based organizations, UN agencies, and individuals amongst others have made considerable efforts and spend resources on measures pertaining to the reduction of NCDs, in Namibia, less efforts on understanding the contributing factors to the increase in the RR of NCDs has been achieved. For example, of the studies found, the modelling of disease was primarily done at one level of analysis. However, there is limited literature on studies that have attempted to model diabetes and hypertension across different (multiscale) levels say at region, health district and constituency as demonstrated here.

The advantages of the multiscale spatial modelling approach are many. First, analysis at the regional, district and constituency suggest some underlying and unobserved factors that affect disease risk at those levels. Indeed, analysis of these data may allow to assess, compare and ultimately improve the care provided at all levels of health care. It may assist in monitoring and planning resource needs in a health system and designing appropriate interventions, tailored towards communities at high risk or lead to further investigations to identify important risk factors. The other major advantage of using a multiscale model concerns the scaling effect which is viewed as important in disease mapping. Aregay et al. (2015) argues that the scale effect reduces the variability of the data and makes them more alike, and hence the results from the finer data may not be consistent with the results from the coarser data.

Spatial Effects of Diabetes and Hypertension

Spatial effects for diabetes were presented as RR at different levels considered in the study. An RR greater than one implies that there is an increased risk for diabetes and hypertension in the area while an RR less than one implies that there is a decreased risk of diabetes and hypertension in an area. Regions such as Caprivi, Khomas and Otjozondjupa were also found to have an increased risk for diabetes. Although results also show that regions such as Omusati and Oshikoto appeared to have a low RR of diabetes, suggesting that there was a decreased risk for diabetes in those areas. The lowest RR was however found in Oshana region (RR = 0.969) suggesting that a reduced risk for diabetes was also evident in the region. The study found consistencies with what has been found by the MoHSS and ICF International (2014) where for instance, Khomas and Otjozondjupa region were among the regions with a high percentage of men and women informed by a health professional that they had diabetes while Oshana region was reported to be among the regions with a low percentage of men and women informed that they had diabetes which is consistent with the current study.

Results from the study also indicate that Erongo, Hardap, Otjozondjupa, Kavango and Karas region were among the regions with a high RR of hypertension, suggesting an increased risk of hypertension in those regions. The results for hypertension are not surprising since the MoHSS and ICF international (2014) also found Khomas, Erongo, Hardap and Kavango region among others to constitute regions where most men and women have been told by a health professional that they had hypertension which appears to be consistent with the current study. In terms of reduced risk for hypertension, there was some consistencies found with Oshana region that also appeared to have one of the lowest percentages of men and women being informed by a health professional that they had hypertension (MoHSS & ICF international, 2014).

At health district level, Rundu health district (RR = 1.037) was found to have the highest RR of diabetes suggesting that there was an increased risk of diabetes in the health district. Other health districts with an increased risk of diabetes were Andara, Grootfontein and Okakarara among others. Results further indicate that Oshakati health district (RR = 0.967) had the lowest RR of diabetes. Engela, Eenhana and Oshikuku were also found to have a low RR of diabetes suggesting that there is a reduced risk of diabetes in those health districts. For hypertension, Andara (RR = 1.012) had the highest RR. Other health districts with a high RR were Rundu, Swakopmund and Windhoek suggesting that an increased risk for hypertension were in these health districts. However,

Oshakati health district (RR = 0.975) was found to have the lowest RR indicating that a reduced risk for hypertension was in the health district.

Results further indicate that at constituency level, an increased risk for diabetes were mainly in Rundu Urban constituency (RR = 1.064) which was found to have the highest RR. Other constituencies that appeared to exhibit a high RR were Mashare, Kapako, Katima Mulilo Urban, Keetmanshoop Urban, Khomasdal North, Khorixas, Katutura Central and Okakarara among others. The results further indicate that Rehoboth West Urban was among the constituencies with the lowest RR of diabetes (RR = 0.972) indicating that a reduced risk for diabetes was in the constituency. Other constituencies such as Oranjemund, Keetmashoop Rural, Uuvudhiya, Eengondi and Onayena among others were also found to have a low RR of diabetes. Again, suggesting a reduced or decreased risk for diabetes.

With respect to hypertension at constituency level, an increased risk for hypertension were in constituencies such as Rundu Urban, Katima Mulilo Urban, Khomasdal North, Walvis Bay Urban, Swakopmund, Rehoboth East Urban, Katutura East, Keetmanshoop Urban, Mariental Urban and Tobias Hainyeko among others with Rundu Urban (RR = 1.03) found to have the highest RR of hypertension compared to the other constituencies mentioned. Results further indicate that there were also constituencies with low RR of hypertension, among others were Ompundja, Okatyali, Eengondi, Omuthiyagwiipundi, Olukonda, Ohangwena, Etayi, Oshakati East, Endola, Okaku and Kalahari with the lowest found in Ompundja (RR = 0.967). The low RR of hypertension in these constituencies suggests that there is a decreased risk for hypertension in the constituencies.

Limitations of the Study

The study had some limitations. Firstly, the data source only covered cases that were reported in the health facilities in Namibia implying that cases that were not reported at the health facilities were not part of the study. Therefore those who chose self-medication or used other means apart from public hospitals were not covered by the study. Hence, the study may not be representative to the situation on the ground. It would be of interest to match this study with further analysis using the 2013 Namibia Demographic Health Survey (NDHS) data, which collected self-reported data on prevalence of diabetes and hypertension. Examples, where DHS data has been used include Kandala et al. (2013). Secondly, the Namibia HIS did not collect information on behavior and habit factors such as smoking, alcohol consumption, health lifestyle including physical activity and eating habits which may be considered important in measuring the impacts of socio-economic factors on diabetes and hypertension. Lastly, there were no geo-coordinates available at health facility level, therefore the analysis could not be done at the level.

Conclusion and Recommendations

In terms of multiscale modelling of diseases risk in Namibia, the availability of limited literature for the study suggests that this is the first study that attempted to estimate the geographical variation of diabetes and hypertension at the region, health district and constituency level in Namibia.

As expected, the spatial analysis section shows that there is high risk across the different levels considered in the study. Furthermore, the multiscale model for diabetes and hypertension applied showed no clusters or spatial pattern. However, the areas found to have elevated or increased risk for diabetes and hypertension might need to be prioritized, without neglecting those areas found to have a reduced risk for diabetes and hypertension, probably through constant monitoring and evaluation.

It is therefore hoped that the study will assist policy makers particularly those involved in health planning to develop comprehensive programmes or targeted interventions particularly in areas that were found to have elevated risk of diabetes and hypertension. Moreover, the study hopes that considering spatial factors in planning of health programmes related to diabetes and hypertension could assist in the achievement of National development Plans (NDPs) goals such as those outlined in the NDP 5 in line with the progress towards the international goals such as UN Millennium Development Goals.

Because of the limitations of the data used, we recommended that a similar study should be conducted using population based data obtained from the recent 2013 NDHS survey undertaken in Namibia. Additionally measurement error models could also be considered to improve the current study. Moreover, the possibility of combining different survey datasets to generate better maps of RR could be explored. Furthermore, since recent studies have also been focusing on modelling disease outcomes jointly (Manda, 2011), the study recommends to consider jointly modelling diabetes and hypertension instead of modelling each disease separately.

References

Aregay, M., Lawson, A. B., Faes, C., Kirby, R., Carroll, R. and Watjou, K. (2015). Impact of income on small area low birth weight incidence using multiscale models. *AIMS Public Health* 2, 667–680.

Aregay, M., Lawson, A. B., Faes, C., Kirby, R. S., Carroll, R., & Watjou, K. (2016a). Spatial mixture multiscale modeling for aggregated health data. *Biometrical Journal,* 58(5), 1091-1112.

Aregay, M., Lawson, A. B., Faes, C., Kirby, R. S., Carroll, R., Watjou, K. (2016b). Multiscale measurement error models for aggregated small area health data. *Stat Methods Med Res.,* 25(4):1201-23. doi: 10.1177/0962280216661094.

Aregay, M., Lawson, A. B., Faes, C. and Kirby, R. (2017a). Bayesian multiscale modeling for aggregated disease mapping data. *Statistical Methods in Medical Research*, DOI: 10.1177/0962280215607546.

Aregay, M., Lawson, A. B., Faes, C., Kirby, R. S., Carroll, R., Watjou, K. (2017b). Comparing multilevel and multiscale convolution models for small area aggregated health data. *Spat Spatiotemporal Epidemiol.*, 22:39-49. doi: 10.1016/j.sste.2017.06.0 01.

Banerjee, S., Carlin, BP., Gelfand, AE. *Hierarchical modeling and analysis for spatial data*. London: Chapman and Hall/CRC Press; 2004.

Belitz, C., Brezger, A., Kneib, T., & Lang, S. (2009). *BayesX-software for Bayesian Inference in Structure Additive Regression Models,* version 2.01. Retrieved October 10, from http://www.dtst.uni-muechen.de/bayes2.01.

Belue, R., Okoror, T. A., Iwelunmor, J., Taylor, K. D., Degboe, A. N., Agyemang, C., & Ogedegbe, G. (2009). An overview of cardiovascular risk factor burden in Sub-Saharan African countries: a socio-cultural perspective. *Globalization and Health*, 5(1), 10.

Besag, J., York, J. C., and Mollié, A. (1991). Bayesian image restoration, with two applications in spatial statistics (with discussion). *Annals of the Institute of Statistical Mathematics* 43, 1-59.

Best, N., & Richardson, S. (2009). Introduction to Bayesian analysis using WinBUGS. Introduction to Bayesian Analysis and WinBUGS course, Imperial College.

Best, N., & Richardson, S. (2012). Introduction to Bayesian analysis using WinBUGS. In *Workshop-Hilton Glasgow Grosvenor* (Vol. 3, pp. 08-00).

Clayton, D. and Kaldor, J. (1987). Empirical Bayes estimates of age-standardised relative risks for use in disease mapping. *Biometics*, 43,671-681.

Cressie, N. A. C. (1996). Change of support and the modifiable areal unit problem. Journal *of Geographical Systems* 2, 83–101.

Cressie, N., & Chan, N. H. (1989). Spatial modeling of regional variables. *Journal of the American Statistical Association*, 84(406), 393-401.

Held, L., Natário, I., Fenton, S. E., Rue, H., & Becker, N. (2005). Towards joint disease mapping. *Statistical Methods in Medical Research*, 14(1), 61-82.

Hummer, R. A., & Lariscy, J. T. (2011). Educational attainment and adult mortality. In *International handbook of adult mortality*, 241-261, Springer Netherlands.

Islam, S. M. S., Purnat, T. D., Phuong, N. T. A., Mwingira, U., Schacht, K., & Fröschl, G. (2014). Non-Communicable Diseases (NCDs) in developing countries: A symposium report. *Globalization and Health*, 10(1), 1-8.

Kandala, N. B., Tigbe, W., Manda, S. O., & Stranges, S. (2013). Geographic variation of hypertension in Sub-Saharan Africa: A case study of South Africa. *American Journal of Hypertension*, 26(3), 382-391.

Kazembe, L. N., & Kandala, N. B. (2015). Estimating areas of common risk in low birth weight and infant mortality in Namibia: A joint spatial analysis at sub-regional level. *Spatial and Spatio-temporal epidemiology*, 12, 27-37.

Kazembe, L. N., & Mpeketula, P. M. (2009). Detecting geographical variability in risk of malaria-attributable morbidity using spatial models. *Analysis*, 4(5), 6.

Knorr-Held, L., & Best, N. G. (2001). A shared component model for detecting joint and selective clustering of two diseases. *Journal of the Royal Statistical Society: Series A (Statistics in Society),* 164(1), 73-85.

Kolaczyk, E. D., & Huang, H. (2001). Multiscale statistical models for hierarchical spatial aggregation. *Geographical Analysis*, 33(2), 95-118.

Lawson, A. (2008). *Bayesian Disease Mapping: Hierarchical modelling in spatial epidemiology*: CRC Press, Boca Raton, FL.

Lawson, AB. *Bayesian disease mapping: Hierarchical modeling in spatial epidemiology.* 2nd. New York: Chapman and Hall/CRC Press; 2013.

Louie MM, Kolaczyk ED. A multiscale method for disease mapping in spatial epidemiology. *Stat Med.* 2006; 25:1287–1306. [PubMed: 16217848].

Louie, M. M., & Kolaczyk, E. D. (2006). A multiscale method for disease mapping in spatial epidemiology. *Statistics in Medicine*, 25(8), 1287-1306.

Lunn, D., Jackson, C., Best, N., Thomas, A., & Spiegelhalter, D. (2012). *The BUGS book: A Practical Introduction to Bayesian Analysis.* CRC press. Boca Raton, FL, USA.

Manda, S. (2011). Joint Mapping Modelling for Multiple Health Problems in South Africa. Retrieved from http://sacemaquarterly.com/tag/joint-mapping-modelling.

Marquez, P. V., & Farrington, J. L. (2013). *The challenge of non-communicable diseases and road traffic injuries in sub-Saharan Africa: An overview*. Washington, DC: The World Bank.

Moraga P, Lawson AB. Gaussian component mixtures and CAR models in Bayesian disease mapping. *Comput Stat Data Anal*. 2012; 56:1417–1433.

Namibia Statistics Agency (2017). Namibia Inter-censal Demographic Survey 2016 report. *Namibia Statistics Agency*, Windhoek.

Neema, I., & Böhning, D. (2012). Monitoring murder crime in Namibia using Bayesian space-time models. *Journal of Probability and Statistics*, 2012.

R Core Team (2015). R: A language and environment for statistical computing. R Foundation for Statistical Computing, Vienna, Austria. *ISBN* 3-900051-07-0, URL http://www.R-project.org/.

Ross, C. E., & Wu, C. L. (1995). The links between education and health. *American sociological review*, 719-745.

Sacco, W. P., Bykowski, C. A., Mayhew, L. L., & White, K. E. (2012). Educational attainment moderates the effect of a brief diabetes self-care intervention. *Diabetes Research and Clinical Practice*, 95(1), 62-67.

Spiegelhalter, D. J., Best, N. G., Carlin, B. P., & Van Der Linde, A. (2002). Bayesian measures of model complexity and fit. *Journal of the Royal Statistical Society: Series B (Statistical Methodology),* 64(4), 583-639.

Stern, H. S., Cressie, N. A. Inference for extremes in disease mapping. In: Lawson, A. Biggeri, A. Bohning, D., et al., editors. *Disease mapping and risk assessment for public health.* Chichester: Wiley; 1999. p. 63-84.

The Ministry of Health and Social Services (MoHSS) & Macro International Inc (2008). Namibia Demographic and Health Survey 2006-07. Windhoek, Namibia and Calverton, Maryland, USA: MoHSS and Macro International Inc.

The Namibia Ministry of Health and Social Services (MoHSS) and ICF International. 2014. The Namibia Demographic and Health Survey 2013. Windhoek, Namibia, and Rockville, Maryland, USA: MoHSS and ICF International.

Tu, Y. K., & Greenwood, D. C. (Eds.). (2012). *Modern methods for epidemiology*. Springer Science & Business Media.

Whitaker, S. M., Bowie, J. V., McCleary, R., Gaskin, D. J., LaVeist, T. A., & Thorpe, R. J. (2014). The association between educational attainment and diabetes among men in the United States. *American journal of men's health*,8(4), 349-356.

Wong, D. (2009). The modifiable areal unit problem (MAUP). In: Fotheringham, S. and Rogerson, P. (Eds.), *The Sage Handbook of Spatial Analysis*. Sage, New York, pp. 108–123.

World Health Organization. (2010). *Levels & trends in child mortality*. Report 2010. Estimates developed by the UN Inter-agency Group for Child Mortality. United Nations Children's Fund.

World Health Organization-Namibia (2010). Non-Communicable Diseases (NCDs) Namibia's Silent Killers. Windhoek, *World Health Organization-Namibia.*

World Health Organization. (2014). Global Status Report on Noncommunicable Diseases 2013. Geneva: *World Health Organization.*

In: Statistical Modelling …
Editors: N-B. Kandala and L. Kazembe
ISBN: 978-1-53615-981-3

Chapter 12

MODELS FOR ANALYZING SPATIAL PATTERNS IN RISK OF URBAN MALARIA: A CASE STUDY OF BLANTYRE, MALAWI

Don P. Mathanga and Lawrence N. Kazembe*
Department of Community Health, College of Medicine, University of Malawi, Blantyre, Malawi
Department of Statistics and Population Studies, University of Namibia, Windhoek, Namibia

ABSTRACT

Malaria imposes a huge burden in most sub-Saharan countries, where the disease is endemic. Recently, there is increasing scientific interest towards eliminating malaria in areas of low transmission including highlands and urban areas. Motivated by two objectives: (i) the need to characterize the spatial distribution of malaria risk in Blantyre city, Malawi, and (ii) quantify the effect of travel to rural areas on malaria transmission, this chapter developed and applied two types of spatial models to analyze case-control data. The first model was a univariate Log-Gaussian Cox Process (LGCP) spatial model aimed to estimate the effect of travel and residual spatial structure, while the second model was a bivariate model aimed at examining the spatial segregation in risk between travelers and non-travellers. These models were estimated using the Bayesian inference. Through a hierarchy structure, the models incorporated correlated random effects estimated using bivariate conditional autoregressive model. Furthermore, the models were applied to a case-control data collected in 2007-2009, at Ndirande health centre, Blantyre, Malawi. Children who had travelled to rural areas were 7 times at higher risk than non-travellers. The spatial pattern associated with traveller and non-traveller showed

* Corresponding Author's Email: lkazembe@yahoo.com.

a strong dependence, suggesting a possible source-sink transmission dynamics, or common shared risk factors.

Keywords: Log-Gaussian Cox Process model, bivariate spatial analysis, case-control data, urban malaria, Malawi

INTRODUCTION

Malaria is a tropical disease caused by *Plasmodium* parasites, transmitted from human to human through an infected Anopheles mosquito during blood meals. Current estimates from the World Health Organization (WHO) in 2014 shows that the disease imposes a huge burden on the health system. In 2013, an estimated 198 million cases of malaria and 584 000 malaria-related deaths were reported, with the most affected being children under the age of five years (WHO, 2014). The disease is endemic in most Sub-Saharan African region, with isolated fringes where the disease transmission is of low intensity, for example in highlands, arid regions and urban areas.

The disease is preventable through vector-control using insecticide-treated nets and indoor spraying with residual spraying. Early diagnosis and treatment using anti-malarial medicines are used to cure the disease. Since the year 1998, there have been renewed and concerted efforts toward the reduction of malaria burden through the Roll-Back Malaria (RBM) initiatives (Nabarro and Tayler, 1998). Fast forward to 2010, the programme of malaria elimination in areas of low transmission settings was conceived, buoyed by the global gains made by the RBM (WHO, 2011). For this reason, this study was initiated in Blantyre city, Malawi, to investigate the feasibility of malaria elimination strategies.

In interventions such as malarial elimination, monitoring and evaluation became one of the central pieces of action. Spatial analysis was critical to monitoring and evaluating coverage of preventative measures, highlight clusters of disease and examine places of elevated risk. In this chapter, statistical models were developed to describe the spatial distribution of malaria risk in Blantyre city, thereby provide evidence to guide policy makers. The spatial structure inherent in the data made it suitable for the use of spatial statistics. To be precise, a model was developed from point processes framework, and a further set of multivariate spatial models were formulated to explore possible sources of infection in an area of relatively low transmission.

MOTIVATING EXAMPLE: BLANTYRE URBAN MALARIA PROJECT

One of the operational objectives of the Malaria Elimination Initiative (MEI) is Targeted Parasite Elimination (TPE) in areas of low transmission, e.g., in highlands and urban areas (WHO, 2017). Blantyre city, in Malawi, is both urban and highland with relatively low malaria risk. Since 2007, a study was established to investigate risk factors of urban malaria in Blantyre city to inform elimination efforts. The study was conducted by the Malaria Alert Centre, College of Medicine, University of Malawi. This study used data from 2007 to 2009 collected in Ndirande township, in Blantyre (Figure 12.1), through a nested case-cohort study design which was built on a cohort study on malaria drug efficacy.

Cases were defined as children aged below five years, who were definitively diagnosed of malaria at Ndirande health facility, in which controls were matched to cases for age (± 6 months), sex, and location. Controls consisted of two groups: clinic and community controls. For each case and control, geographical coordinates of homes from which the children came from were obtained (Figure 12.2). Also captured were the geo-coordinates of any potential putative malaria vector breeding sites including dam, river, standing/swamp water, or a garden nearby selected homes. In addition, using a structured questionnaire, information were collected on household assets, availability and use of insecticide treated nets (ITN), whether the child travelled to a rural area, and for how long was the visit, and whether a net was used when away. The motivation for this study was driven by the lack of local malaria transmitting vectors (Tambala et al., 1992), and it was then hypothesized that all malaria is imported through travel to rural areas.

The research questions were then structured as follows: How much of urban malaria is attributable to travel history? What are the spatial patterns of malaria in the study areas; and is there spatial segregation which may relate to the transmission mechanism (i.e., can the spatial patterns, thus observed, be conjectured to potential breeding sites in the area as well as travel history). Table 12.1 shows the summaries of variables used in the study, and from this table, it can be seen that travel and socio-economic status were the important risk factors of malaria based on chi-square test.

Fitting spatial models can be done using various approaches, including geostatistical, spatial point pattern and lattice models (Cressie 1993). However, the type of method to use is often driven by the research question. In this chapter, a unified approach by Diggle et al. (2013) was adapted, to examine the general pattern of malaria risk in Ndirande, and explore if there is any spatial segregation by fitting a bivariate model of two possible transmission processes (infections). Put differently, the first infection is due to travel to rural areas, which has been termed *the source* in this chapter, while the secondary infection in non-travellers, which is termed *the sink*, is attributed to those who travelled as a source of infection.

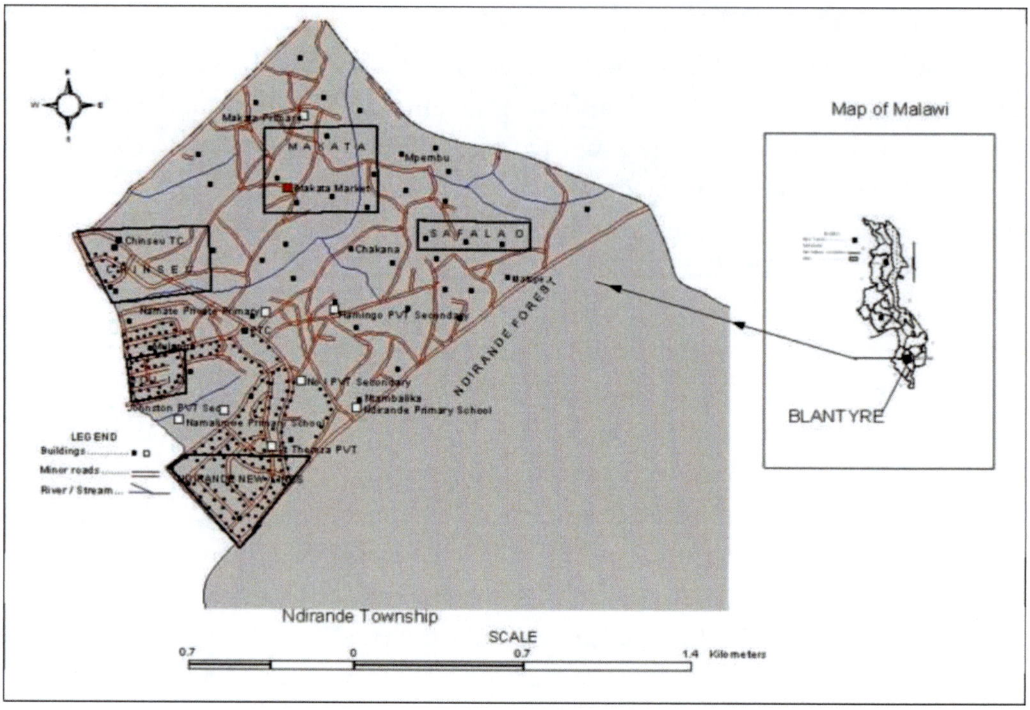

Figure 12.1. Map showing the study area- Ndirande Township in Blantyre city, Malawi.

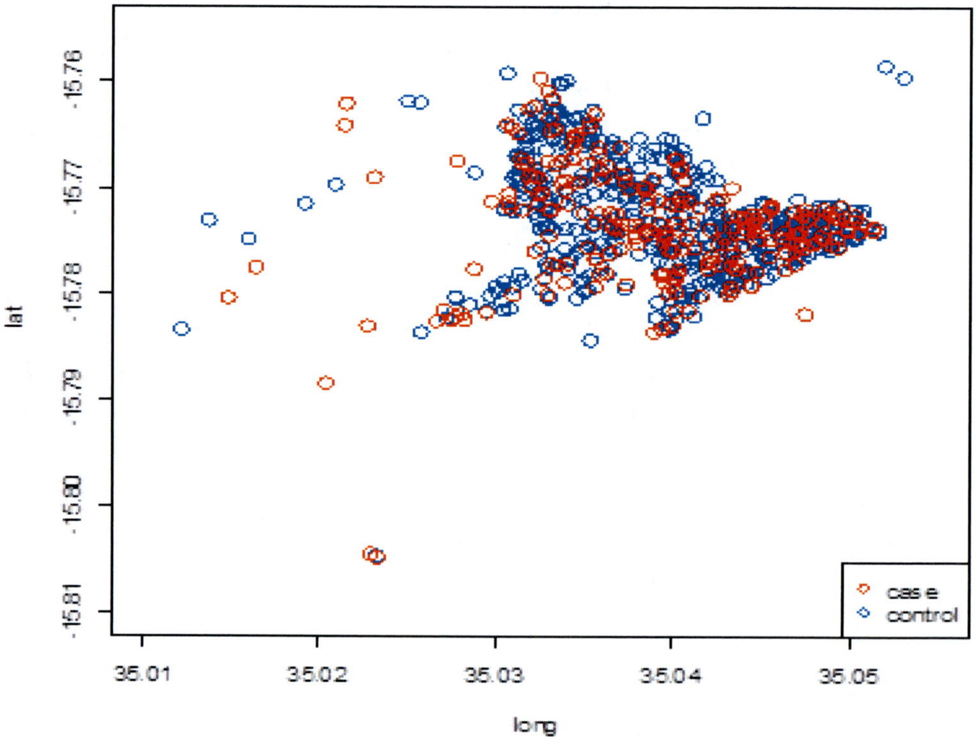

Figure 12.2. Locations of cases and controls in the study areas.

Table 12.1. Descriptive summaries of the variables used in the study

Variable	Category	n	Case (% with malaria)	p-value*
Travel	Yes	101	29.1	<0.001
	No	666	28.1	
Any net	Yes	492	34.2	0.26
	No	275	32	
Child used net	Yes	480	34.4	0.32
	No	287	32.4	
Garden (<500m)	No	430	32.3	0.18
	Yes	337	35.3	
Standing water (<500m)	No	344	34.9	0.14
	Yes	423	32.6	
River (<500m)	No	408	33.1	0.17
	Yes	359	34.1	
Swamp (<500m)	No	436	31.9	0.39
	Yes	331	36	
Socio-Economic status	Lowest	155	37.4	0.003
	Low	149	35.2	
	Medium	159	42.1	
	High	150	30	
	Highest	152	22.4	

*p-values based on chi-square test of association.

SPATIAL STATISTICAL MODELLING OF CASE-CONTROL DISEASE DATA

Statistical methodology for studying spatially referenced case-control data is now well established in Epidemiology (Chetwynd et al., 2001; Rogerson, 2006; Webster et al., 2006; Diggle 2013; Reeve et al., 2013). In this chapter, the analysis of such data using the Log-Gaussian Cox Processes (LGCP) was done. Spatial modelling using LGCP was introduced by Møller et al. (1998). The extension of the LGCP includes applications to analyze spatio-temporal processes (Brix and Diggle, 2001), as well as aggregated and multivariate data (Diggle et al, 2013; Waagepetersen et al., 2015). See also the third revision book by Diggle (2013), which provides further discussion and applications of the LGCP methodology.

Modelling Case-Control Data as Point Processes

The first approach of this paper used the LGCP, where the case-control data was considered as a point process. Diggle et al., (2013) defined a Gaussian process as a stochastic process $Y(s_i), s_i \in R^d$, where $Y(s_i) \sim MVN(\mu, \Sigma)$, s_i are the locations at $i = 1, \dots n$. A Cox process is an inhomogeneous Poisson process with stochastic intensity Λ. If Y is a Cox process, then conditional on Λ, the number of events in an observation window $W \subseteq S$ is defined by

$$number\ of\ Y\ in\ W \sim Poisson\left(\int_W \Lambda(s)\, ds\right) \tag{1}$$

As such Y is considered as continuous in space (Diggle et al., 2013), and for computational purposes, Y is piece-wise constant on finite grid. Therefore $Y(s)$ denotes the number of events in a grid cell containing s. That is,

$$\begin{aligned} &Y(s) \sim Poisson[\Lambda(s)] \\ &\Lambda(s) = C_A \lambda(s) \exp\{X(s)\beta + U(s)\} \end{aligned} \tag{2}$$

where C_A is the area of the grid cell containing s, often considered as a regular grid; $X(s)$ is a vector of covariates particular to the grid cell with s, β are the regression coefficients corresponding to $X(s)$, and $U(s)$ is a latent spatial process that follows a Gaussian process in the grid cell. The likelihood, conditional on the latent spatial process, is expressed as

$$L(\beta|U(s)) \propto exp\left(-\int_{C_A} \lambda(s)ds\right) \prod_{i=1}^{n} \lambda(U(s)). \tag{3}$$

Inference is based on Bayesian approach, with posterior density estimated through conditioning:

$$\pi(\beta, U, \omega|Y) \propto \pi(Y|\beta, U, \omega)\pi(\beta, U, \omega)\pi(\beta)\pi(U)\pi(\omega) \tag{4}$$

such that ω contains additional model parameters such as the variance σ and spatial dependence θ that control the latent spatial process. In a geostatistical context, these are called the sill and the range respectively. Estimation of the model parameters uses a Markov Chain Monte Carlo (MCMC) simulation technique. According to Diggle et al. (2013), the MCMC is evaluated using a mixture of Langevin kernels on the parameters U and β, and a random walk for ω.

Bivariate Modelling of Case-Control Data

The second approach of this chapter assumes that the transmission of malaria in Ndirande was driven by two correlated processes. Let y_{ij} be the observed event such that $y_{ij} \in \{0,1\}$; with 1 if a case and 0 if a control. The outcome can either be a first infection or secondary infection. The first infection occurred due to travel, and the secondary infection is a case within 500m from the traveller. Correlated outcomes were introduced to the predictor as follows:

$$I_{ij} = \begin{cases} 1 & if\ y_{ij} = 1 \\ 0 & if\ y_{ij} = 0 \end{cases}$$

and

$$y'_{ij} = \begin{cases} y_{ij} & if\ y_{ij} = 1 \\ otherwise & if\ y_{ij} = 0 \end{cases},$$

then

$$\pi_{ij} = \Pr(I_{ij} = 1 | x_{ij}, u_i, s_i) \tag{5}$$

where x_{ij} are the fixed effects, and u_i *and* s_i are the unstructured and structured spatial random effects in the grid respectively. The nested bivariate model for the two processes can then be defined as:

$$\begin{gathered} P(y_{ij} = 1 | Travel) \sim Bern(p_1) \\ P(y_{ij} = 1 | No\ Travel) \sim Bern(p_2) \\ p_i = P(y_i = 1 | \eta_i) = \frac{\exp(\eta_i)}{1+\exp(\eta_i)} \end{gathered} \tag{6}$$

where η_i is the predictor for travel related cases ($i = 1$) or secondary infection ($i = 2$). Thus, the two regression models are:

$$\begin{aligned} \eta_1 = \ & log\left(\frac{p_1}{1-p_1}\right) = \alpha_1 + \beta_{1j} x_{1j} + u_1 + s_1 \\ \eta_2 = \ & log\left(\frac{p_2}{1-p_2}\right) = \alpha_2 + \beta_{2j} x_{2j} + u_2 + s_2 \end{aligned} \tag{7}$$

with the random effects for the two model equations assumed to be correlated (Neelon et al., 2014).

The two parts (η_1 and η_2) in eqn. (7) are jointly normal through a bivariate Gaussian (biGauss) spatial process model. The predictor in eqn. (7) can be written as

$$\eta(\omega) = \mu(\omega) + \varphi(\omega)$$

where $\mu(\omega)$ provides the spatial mean surface, and $\varphi(\omega)$ is a zero-mean Gaussian process that capture unstructured random processes, e.g., error processes, while the term ω denotes the Eucledian distance between the two spatially dependent points. The error term is defined as $\varphi(\omega) \sim N(0, D(\varphi))$, such that $D(\varphi)_{jj} = \tau_j^2$. The spatial structured term $\mu(\omega)$ is measured through a cross-covariance matrix with respective to the correlation function

$$\rho_j(\omega) = \rho_j(\omega(i) - \omega(i')), j = 1, \dots p,$$

for any two given points of interest, i and i'. The cross-covariance matrix is given as

$$C(\omega) = \sum_{j=1}^{p} \rho_j(\omega(i) - \omega(i'))a_j\, a_j^T$$

with a_j being a non-singular matrix. This linear transformation induces dependence between the processes and also provides a distinct covariance function for each component (Banerjee et al., 2014; Gelfand and Banerjee, 2017). A common covariance function is an exponential function, which captures a decaying dependency with increasing distance,

$$C(\omega) = \sigma^2 \exp(-||\omega(i) - \omega(i')||; \theta).$$

where σ^2 is called the sill, and θ is the range in geostatistical terms.

The estimation of the model in eqn. (7) can be done using a Bayesian inference framework. In a Bayesian framework, all parameters are assumed to be random variables, and must be assigned prior distributions. This is easy to specify using a general hierarchical framework, which states three components: a data model, process model, and parameter model. A typical hierarchical spatial model is given by (Wikle, 2015):

$$\begin{aligned} &\text{DataModel: } Y |\, \beta, \eta, \sigma_\omega^2 \sim Gau(X\beta + H\eta, \sigma_\omega^2 I) \\ &\text{ProcessModel: } \eta |\, \theta \sim Gau(0, \Sigma(\theta)) \\ &\text{Parameter Model: } [\beta, \sigma_\omega^2, \theta] \end{aligned} \quad (8)$$

where X is an $m \times p$ matrix of covariates, often associated with the large-scale spatial mean (or "trend" as discussed above), β is the associated $p \times 1$ vector of parameters, H is an $m \times n$ matrix that associates the observations Y with the latent process η, σ_ω^2 corresponds to an independent, small-scale (nugget) spatial effect and/or measurement error process ω, and θ is a vector of parameters used for the spatial covariance function associated with the spatial latent process (e.g., a variance, spatial range parameter, smoothness parameter, etc.). The spatial covariance can be written as

$$\Sigma(\theta) = \sigma_\eta^2 R(\varphi, \kappa),$$

where $R(\varphi, \kappa)$ is the spatial correlation matrix, σ_η^2 is the variance, and φ and κ are spatial range and smoothness parameters, respectively.

Given suitable choices for the prior distributions of the parameters in eqn. (8), the posterior distribution, which is proportional to the product of the components in Eqn. (8) can be defined as:

$$[\eta, \beta, \theta, \sigma_\omega^2 | Y] \propto [Y | \eta, \beta, \sigma_\omega^2][\eta|\theta][\beta, \sigma_\omega^2, \theta].$$

IMPLEMENTATION

In this analysis, an exponential correlation function was assumed, with a spatial dependence parameter, φ (i.e., $r(d, \varphi) = exp(-d/\varphi)$, where d is the distance between two spatial locations). A relatively non-informative prior for β was chosen, i.e, $\beta \sim N(0, \Sigma_\beta)$, where Σ_β is a diagonal matrix, with each variance equal to 1000. A non-informative uniform prior for the β parameters was also considered, however, the posteriors were not sensitive to this choice. For σ_η^2 the non-informative prior $p(\sigma_\eta^2) \propto 1/\sigma_\eta^2$ was chosen, but the posteriors were not very sensitive to these choices. The prior for φ was assumed to be discrete uniform from 0.1 to 4.0 with increments of 0.1. This choice was based on the fact that the practical range (the smallest distance between points that gives negligible spatial correlation) for the exponential model was defined to be 3φ. Similarly, the prior for τ^2 was given by a discrete uniform distribution from 0.05 to 2.0 with increments of 0.05. All fitted models in this paper were analyzed in R software, using the INLA package (Rue et al., 2009).

ANALYSIS OF BLANTYRE MALARIA DATA

Four different models were explored for the analysis of the Blantyre malaria data, as shown in Table 12.2. For the univariate fittings, a model with only fixed effects was fitted (M0) first, then random effects were introduced to the model to capture the spatial structure (M1). Likewise, for the multivariate fittings, two types of models were considered. The first was the model with unstructured random effects only (M2), and the second was a model that included both the unstructured and structured random effects (M3). Table 12.2 also shows the Deviance Information Criteria (DIC) value per model. A smaller value of DIC signifies a preferred model. Using the DIC, for the univariate fittings, model M1, which combines both the fixed and spatial effects, is preferred over M0. Likewise, for the bivariate fittings, model M3 is preferred.

Table 12.2. Model selection summaries based on the deviance Information criteria (DIC)

Type of Model	$\bar{D}$	P_D	DIC
M0: Univariate (Fixed effects only)	1049	10	1085.64
M1: Univariate (Fixed + spatial effects)	1054	28	1025.44
M2: Bivariate model (unstructured effects only)	791	53	850.07
M3: Bivariate model (unstructured + structured effects)	747	110	852.85

Risk Factors of Urban Malaria

Table 12.3. Fixed effects of malaria given as OR with corresponding 95% credible intervals (CI)

Variable	Category	Odds ratio (OR)	95% CI
Travel	Yes	6.66	(4.79, 9.61)
	No	1.00	
Age of child	in months	1.01	(1.003,1.02)
Garden	(<500m)	1.14	(0.85, 2.50)
	Otherwise	1.00	
Standing water	(<500m)	0.90	(0.67, 1.22)
	Otherwise	1.00	
River	(<500m)	1.05	(0.78, 1.42)
	Otherwise	1.00	
Socio-Economic status	Lowest	1.00	
	Low	0.94	(0.67, 1.30)
	Medium	0.88	(0.64, 1.21)
	High	0.67	(0.47, 0.94)
	Highest	0.38	(0.25, 0.54)
Variance Components			
Sill (σ)		0.09	(0.02, 0.15)
Range (θ)		0.33	(0.09, 0.88)

As part of the objectives of this study, the effects of various factors of malaria were quantified. Table 12.3 presents a summary of Odds Ratios (OR) and their corresponding 95% credible intervals (CI) values obtained from fitting model M1.

From Table 12.3, travel to rural areas was an important risk factor, with OR= 6.66 (95% CI: 4.79, 9.61). Moreover, being within a distance from a garden or river was positively associated with malaria, although their effects were not significant based on CI. Futhermore, malaria infection was likely to be associated with low Socio-Economic Status (SES). Those having a highest SES had a lower risk of getting malaria infection, compared to the lowest SES (OR = 0.38, 95% CI: 0.25-0.54), and similar patterns were observed at other levels of SES (Table 12.3).

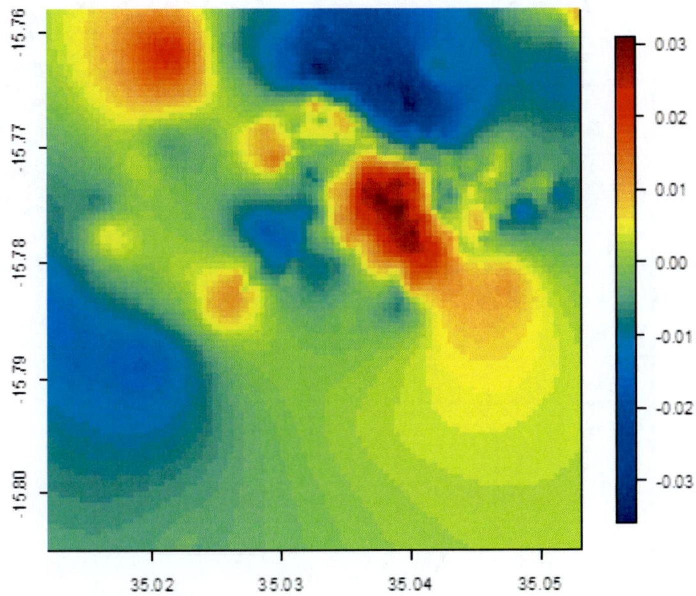

Figure 12.3. Residual random effects (on log odds scale) of malaria in Blantyre, Malawi.

Figure 12.3 presents the residual spatial effects after adjusting for fixed effects. The risk was relatively high in the centre of the area, and most of the high risk areas laid along the north-west and south-east transect. The low risk areas are depicted at northern corner and lower left corner. The sill and range were estimated as 0.09 and 0.33 respectively (Table 12.3).

Source and Sink Transmission Dynamics: Travel as Source of Primary and Secondary Infection

Next we intended to answer whether travelers, who were at about 7 times at high risk of malaria than non-travellers, were a source of infection? In other words, can it be argued that the non-travellers are *the sink*; and what is the spatial pattern associated with the source-sink transmission dynamics captured by the traveler- and non-travellers risk pattern? Figure 12.4 provides a segregation of cases according to travel and secondary cases (not travelled but in close proximity to travel cases).

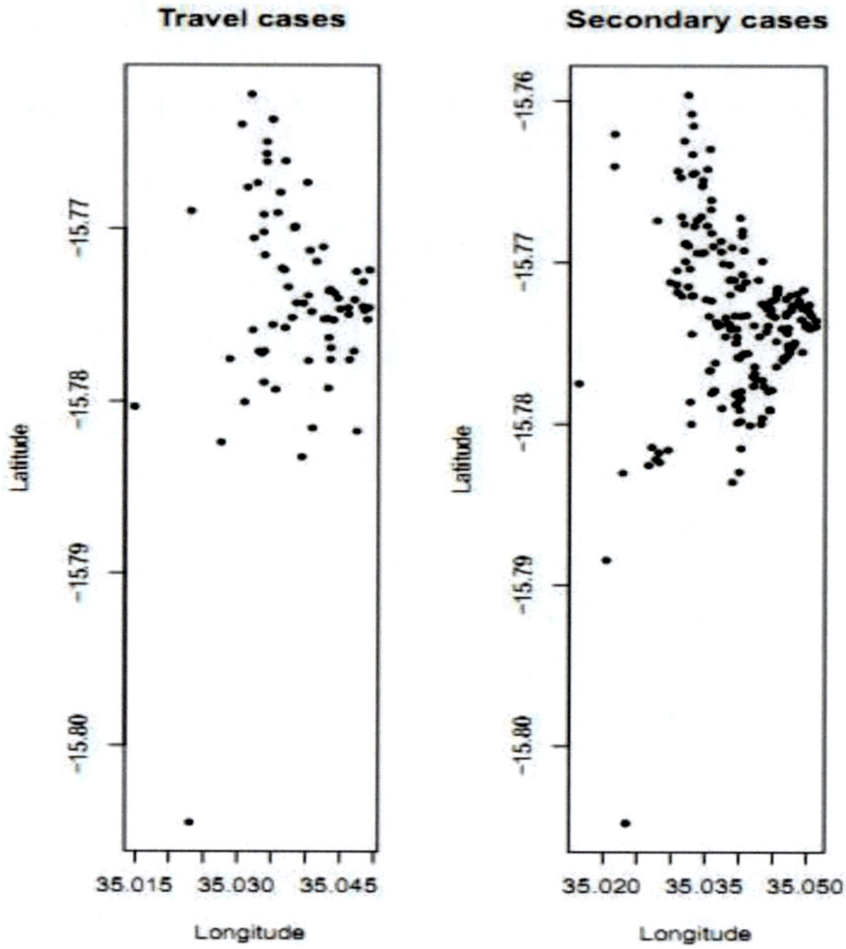

Figure 12.4. Segregating cases by travel, secondary cases (infected by nearness to travel cases), breeding site.

A bivariate model results, obtained from fitting model M3, is provided in Figures 12.5 and 12.6. The predicted risk of malaria prevalence is provided in Figure 12.5. The prevalence malaria among travellers show strong clustering of risk, in some cases rising to about 80%, and as low as 20% in others. For the non-travellers, prevalence ranged between 15-40%. There was evidence of correlation in risk between travellers and non-travellers, in that areas predicted with high risk for travellers also showed a higher risk among non-travellers (lower corner in both panels of Figure 12.5), and low prevalence among travellers correspond to lower risk for non-travellers (top corner in Figure 12.5).

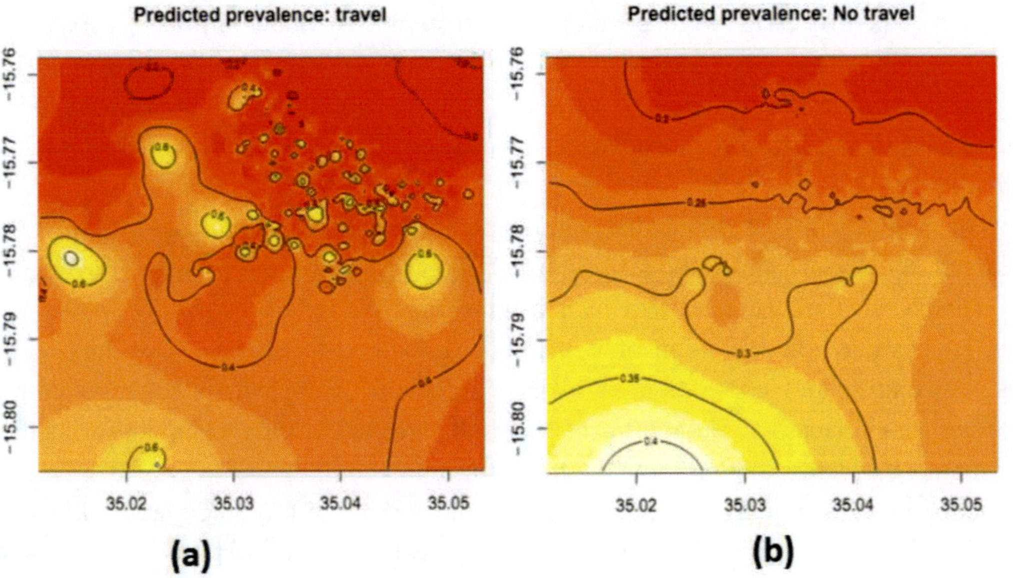

Figure 12.5. Predicted risk surface of malaria risk for travellers (panel a) and non-travellers (panel b).

Figure 6 displays the residual spatial effects. In both displays, the pattern shows a strong spatial dependence in the risk, which provides evidence that the underlying factors governing malaria transmission in both travelers and non-travelers is the same.

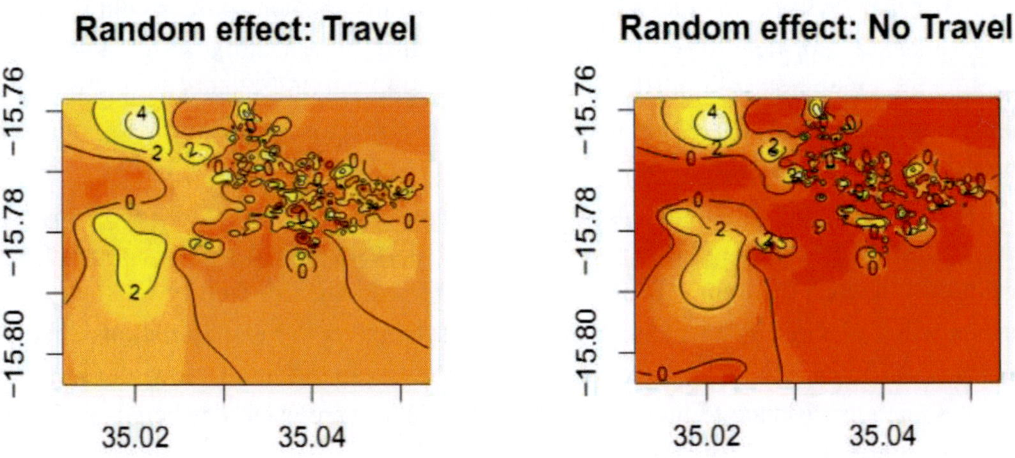

Figure 12.6. Total spatial random effects for travellers (left panel) and non-travellers (right panel).

Discussion and Conclusion

This study is among the first to look at the spatial structure related to travel history, and the effect travel imposes on further transmission of malaria in urban areas of Malawi, where the disease risk is low. For the analysis, flexible models within the class of Log-

Gaussian Cox Processes were developed than permitted under ordinary geostatistical models (Gelfand and Banerjee, 2017) to exploit both structured and unstructured latent (spatial) random effects inherent in the data. The methodology can be extended to study similar problems of in infection diseases dynamics that are at increased risk associated with travel.

The analysis revealed that the spatial pattern of risk associated with travel was clustered, mostly driven by the spatial distribution of cases (children) that travelled. On the other hand, the spatial pattern in non-travellers was relatively evenly spread in the area, suggesting little elevation of risk imposed for non-travellers. Nevertheless, there was a strong spatial correlation in malaria infection. This pattern suggests common unobserved explanatory variables, for example, health seeking behavior or common environmental factors that promote the transmission dynamics.

As part of the ongoing work of this study, the authors of this study are looking at the spatial pattern that emerge by considering the effect of being near a putative source of infection. In this initial analysis, a dichotomy approach was used (whether a household is within 500m of a potential breeding site or otherwise), which may create some arbitrariness in the associated effects. The exploration of using flexible modelling associated with actual distance using splines, as opposed to categorizing the metrical risk factors is currently being looked at. Furthermore, spatial segregation of risk by proximity to putative source of transmission will be investigated. However, there is a need to first consider if there will be any elevated risk with proximity to such sources, and a plausible cut-off distance will need to be determined. Currently, new techniques of insect vector sampling are being assessed in Blantyre (Harbison et al., 2006; Sikaala et al., 2013), with the hope to understand what type of Anopheles mosquitoes are transmitting malaria from travellers to non-travellers.

There were a few limitations to this study. Firstly, the cases were hospital-based, thus there could have been some other malaria-infected cases that were not reported to the hospital. Moreover, the data was only available from one health facility in Blantyre city. Further work planned is to expand the analysis to cover other populations that were reporting at other health facilities in Blantyre city. In addition, few environmental variables such as distance to breeding site, and altitude were used. Predicted surface may have been biased by the few choice of predictors.

Some concluding remarks can be reached from this study that may inform prevention strategies. For example, deliberate information and education campaigns to help those who travel to be (more) aware of malaria risk in highly endemic rural areas. Also, planning to treat all travel cases, as these are associated with secondary new cases among those who did not travel, would be an ideal targeted parasite elimination strategy for urban malaria. Furthermore, there is need for improved access and usage of preventive materials such as insecticide treated bednets and indoor residual spraying, especially, targeting the urban poor in areas of elevated risk.

REFERENCES

Banerjee, S., Carlin, B. P., Gelfand, A. E. (2014). *Hierarchical Modeling and Analysis for Spatial Da*ta. Boca Raton, FL: Chapman & Hall/CRC, second edition.

Brix, A., Diggle, P. J. (2001). Spatiotemporal prediction for log-Gaussian Cox processes. *J. R. Stat. Soc. Ser. B Stat. Methodol.* 63 823–841.

Chetwynd, A. G., Diggle, P. J., Marshall, A., Parslow, R. (2001). Investigation of spatial clustering from individually matched case-control studies. *Biostatistics* 2(3):277-93.

Cressie, N. A. C. (1993). Statistics for Spatial Data. New York: Wiley.

Diggle, P. J. (2013). Statistical Analysis of Spatial and Spatio-Temporal Point Patterns. Chapman and Hall. *Boca R*aton, Third Edition.

Diggle, P. J., Moraga, P., Rowlingson, B., Taylor, B. M. (2013). Spatial and Spatio-Temporal Log-Gaussian Cox Processes: Extending the Geostatistical Paradigm. *Statistical Science* 28(4):542–563. DOI: 10.1214/13-STS441

Gelfand, A. E., Banerjee, S. (2017). Bayesian Modeling and Analysis of Geostatistical Data. *Annual Review of Statistics and Its Applic*ation, 4(1):245-266.

Harbison, J. E., Mathenge, E. M., Misiani, G. O., Mukabana, W. R., Day, J. F. (2006). A simple method for sampling indoor-resting malaria mosquitoes *Anopheles gambiae* and *Anopheles funestus* (Diptera: Culicidae) in Africa. *J. Med. Entomol.*, 43: 473-479.

Møller J, Waagepetersen R, Syversveen AR. Log Gaussian Cox processes. *Scandinavian Journal of Statistics*, 25:451–482, 1998.

Nabarro DN, Tayler EM (1998). *The "Roll Back Malaria" Campaign. Science*, 280:2067-2068.

Neelon B, Anthopolos R, Miranda ML (2014). A spatial bivariate probit model for correlated binary data with application to adverse birth outcomes. *Statistical Methods in Medical Research*, 23(2), 119-133.

Reeve, N. F., Fanshawe, T. R., Keegan, T. J., Stewart, A. G., Diggle, P. J. (2013). Spatial analysis of health effects of large industrial incinerators in England, 1998-2008: a study using matched case-control areas. BMJ Open. 3(1). pii: e001847. doi: 10.1136/bmjopen-2012-001847.

Rogerson PA (2006). Statistical methods for the detection of spatial clustering in case-control data. *Stat. Med.* 25(5):811-23.

Rue, H., Martino, S., Chopin, N. (2009). Approximate Bayesian inference for latent Gaussian models by using integrated nested Laplace approximations. *J. R. Stat. Soc. Ser. B Stat. Methodol.* 71 319–392.

Sikaala, C. H., Killeen, G. F., Chanda, J., Chinula, D., Miller, J. M., Russell, T. L., Seyoum, A. (2013). Evaluation of alternative mosquito sampling methods for malaria vectors in lowland south-east Zambia. *Parasite and Vectors*, 6:91. Doi: 10.1186/1756-3305-6-91.

Tambala P, Macheso A, Ziba C (1992). *Malaria vector assessment.* Ministry of Health. Lilongwe Malawi.

Waagepetersen R, Guan Y, Jalilian A, Mateu J (2016). Analysis of multispecies point patterns by using multivariate log-Gaussian Cox processes. *Journal of the Royal Statistical Society: Series C (Applied Statistics)*, 65(1):77-96. DOI: 10.1111/rs sc.12108

Webster, T., Vieira, V., Weinberg, J., Aschengrau, A. (2006). Method for mapping population-based case-control studies: an application using generalized additive models. *International Journal of Health Geographics*, *5*, 26. http://doi.org/ 10.1186/ 1476-072X-5-26.

Wikle, C. K. (2015). Modern perspectives on statistics for spatio-temporal data. *Wiley Interdisciplinary Reviews: Computational Statistics* 7(1): 86–98, DOI:10.1002/ wics.1341.

World Health Organization (2011). A decade of partnership and results. Progress & Impact Series. *Roll Back Malaria Partnership*. Number 7 - September 2011.

World Health Organization (2014). *World Malaria Report 2014*. Geneva: WHO.

World Health Organization. *A framework for malaria elimination*. Geneva: World Health Organization; 2017. Licence: CC BY-NC-SA 3.0 IGO. (Accessed 24 July, 2017). Available at http://apps.who.int/iris/bitstream/10665/2547/61/ 1/9789241511988-eng.pdf.

In: Statistical Modelling …
Editors: N-B. Kandala and L. Kazembe

ISBN: 978-1-53615-981-3

Chapter 13

SPATIO-TEMPORAL MODELLING OF MALARIA RISK IN MALAWI: AN APPLICATION TO HEALTH MANAGEMENT INFORMATION SYSTEM DATA

Misheck Richard Luhanga** and Lawrence N. Kazembe
Chancellor College, University of Malawi, Zomba, Malawi
Department of Statistics and Population Studies,
University of Namibia, Windhoek, Namibia

ABSTRACT

Malaria is a major public health problem in Malawi with over 10 million new cases per year, particularly among young children. There has been a scale up of malaria control interventions since 2010, which necessitates the quantification of the spatial and temporal variation of malaria risk post scale up of interventions. Previous studies investigated spatial and not temporal variation in risk. The studies did not reveal how the risk has evolved following years of scale up of interventions, an important element for focusing interventions when resources are scarce. This study used malaria data from the Health Management Information System Data (HMIS), from 2010 to 2016, to explore the spatio-temporal variation of malaria risk using Bayesian spatio-temporal modelling approaches, implemented in R-INLA. Various models were fitted and the best model, based on the Deviance Information Criterion (DIC), was the spatially and temporally structured and unstructured random effects model. Results based on this model indicated that there was spatial variation in the risk of malaria across districts in the country, with the risk in Nkhotakota, Mwanza, Neno and Nkhata Bay being more than 60% higher than the national average. Overall results indicated that there has been a decreasing temporal trend in malaria cases from 2010 to 2016. Nevertheless, some districts such as Nkhotakota, Mwanza, Neno and Nkhata Bay showed an increasing trend of malaria risk. Given the

* Corresponding Author's Email:gomeluhanga@yahoo.co.uk.

fact that malaria control interventions are resource intensive, findings of this study should help the Ministry of Health (MoH) focus interventions in districts with higher than national average risk of malaria. Future studies should focus on understanding why Nkhotakota, Nkhata Bay and other districts still have a higher than average risk of malaria despite intensive and targeted malaria control interventions.

Keywords: spatial, temporal, spatio-temporal, Bayesian spatial generalized linear mixed models

INTRODUCTION

Malaria remains a major cause of morbidity and mortality in Malawi. With almost 3.5 million confirmed malaria cases and 3.8 thousand deaths in 2015, malaria accounts for 34% of all outpatient consultations and remains the leading cause of death among children under five (WHO, 2015). Prompt case management of malaria with an effective first-line drug is a cornerstone of malaria control policy in Malawi (WHO, 2015). Treatment of uncomplicated *Plasmodium falciparum* malaria with Artemisinin-based Combination Therapies (ACTs) reduces malaria-attributable mortality by 99% (WHO, 2015; Thwing, Eisele & Steketee, 2011).

The 2014 Malaria Indicator Survey results, in Malawi, showed that the prevalence of malaria in under-five children was at 33%. The prevalence rate was higher among children in poor households than those in the highest wealth quintile at 49 percent in the lowest quintile and 12 percent in the highest wealth quintile (Malawi National Malaria Control Programme (NMCP) and ICF International, 2014). The prevalence of malaria was also highest in rural areas at 37 percent compared to only 11 percent in urban areas. At 36 percent, the central region of Malawi had the highest prevalence rate followed by the southern region at 33 percent. The northern region had lower prevalence of malaria in under-five children at 29 percent (NMCP & ICF International, 2014). This suggests some disparities in the risk of malaria despite the huge burden at national level.

Over the past 12 years, the Ministry of Health, through the National Malaria Control Programme, has scaled up malaria prevention and control interventions. The current interventions include routine distribution of Insecticides Treated Nets (ITNs) to pregnant women and new born babies, universal coverage through mass distribution of ITNs where one net is given to every two members in a household, prompt access to effective antimalarial drugs, prevention of malaria in pregnancy and use of indoor residual spraying in selected districts.

Prior to 2012, ITNs were only provided through targeting the most vulnerable groups such as pregnant women and under-five children. In line with Rollback Malaria goal of universal access to malaria prevention and control interventions, the National Malaria Control Programme started mass distribution of ITNs to all households to ensure that every household member has access to an ITN. In Malawi, universal access to ITN is defined as one ITN for every two members of a household. According to the Malaria Strategic Plan 2011-2016, the mass of distribution of ITNs is expected to be done every 2 to 3 years. The first mass distribution of ITNs was in 2012 in all the districts in Malawi.

BACKGROUND INFORMATION

Currently, the MoH considers malaria hyper-endemic throughout Malawi, with districts along the lakeshore being considered to be of high malaria burden. Rolling out similar interventions across the country is very costly given the variation in risk among the districts and there are limited financial resources for malaria control interventions. Mapping the distribution of the disease risk may help to identify high risk districts (Bennett et al., 2013).

Disease mapping dates back to the 1800s in which maps were used to demonstrate the distribution of disease (Howe, 1989). The most popular example of disease mapping in epidemiology is the study by John Snow in 1854 where he used them to determine the spread and source of cholera. In his study, Snow was able to assess the spatial pattern of cholera cases in relation to potential risk factors which the location of water pumps (Snow, 1855).

The focus in disease mapping is the spatial or geographical distribution of the disease. The relative location of events is, therefore, important. The concepts of geographical information systems and spatial statistics are the key components of disease mapping. Given that there is geographic information, the fundamental question is how the incidence or prevalence of a disease can be analysed (Lawson, 2009).

Defining the Problem

There have been studies done in Malawi to quantify the risk and distribution of malaria. However, these studies either used data prior to 2011, before the intensive scale up of malaria control interventions, or only considered the spatial risk without looking at the distribution of risk in both space and time (Kazembe et al, 2006; Bennett et al., 2013). Therefore, there is still the need to understand the variation of risk in both space and time for policy makers and programme implementers to be able to address important epidemiologic questions on the stability of the estimated spatial patterns of disease, which could not be obvious if data is aggregated across time points or the analysis is done using data from a single time point.

Previous studies that have been done to quantify the risk and distribution of malaria in Malawi include one by Kazembe, Kleinschmidt, Holtz and Sharp (2006) which looked at mapping the risk of malaria using historical data. The study focused on children between the ages 1 and 10 years and used limited data points increasing possibilities of error and uncertainty. Another study (Chirombo, Lowe & Kazembe, 2014) estimated the risk and distribution of malaria using cross sectional data from the 2010 Malawi Malaria Indicator Survey. The study used structured additive regression models that allowed for spatial correlation to estimate the risk of malaria using point-referenced data from the 2010 Malaria Indicator Survey. This approach was appropriate for point-referenced data as opposed to area level data used in this study. Both Kazembe et al. (2006) and Chirombo et al. (2014) modelled the spatial distribution of malaria without considering the variation of risk in space and time. Understanding the risk of malaria, in both space and time, would enable planners to know areas that have an above average risk of malaria and say something about the temporal trend of risk (Blangiardo & Cameletti, 2015). As demonstrated by Abellan, Richardson and Best (2008), the disease risk may be constant across time and only vary spatially or it changes with time across all or some of the districts.

Understanding the full spatial and temporal profile of the risks of malaria would considerably strengthen the epidemiologic interpretation of overall patterns of risk, and particularly of high-risk areas (Abellan et al., 2008). Bennet et al. (2013) looked at the variation of malaria risk on space and time in which they mapped the transmission intensity of malaria using data from surveys spanning from 2000 - 2010 In their study they used space-time Bayesian geostatistical models for predicting *P. falciparum* distribution in Malawi from which the researchers estimated the population at risk of malaria.

As mentioned above, previous studies used approaches that have the limitation of not effectively accounting for space and time variation or used data prior to 2011 before the scale up of malaria control interventions to include universal access. Thus, this study used Bayesian approaches to model the risk of malaria both in space and time. More precisely, the main objective of this study was to use Bayesian approaches to model the Spatial and Spatio-temporal distribution of malaria risk in Malawi using passive routine health facility data from HMIS during the intervention scale up period from 2011 to 2016 in the 28 districts of the country. The goal was to provide the current spatial and temporal distribution of malaria incidence in Malawi and help estimate the impact of the current malaria interventions. Using Bayesian statistical modelling approaches, this study produced an updated risk map of the malaria incidence in Malawi, hence, making the significance of this study topical. The results of this study could assist the National Malaria Control Programme of the Ministry of Health, policy makers and stakeholders in targeting and selection of intervention districts based on the predicted risk.

METHODS

Data Source and Variable Description

In this study, data from the Malawi's Health Management Information System (HMIS) reported through a web-based system, DHIS 2, was used. To be precise, malaria data reported through the DHIS 2 system on a malaria specific reporting form was extracted. Reported malaria cases (presumed and confirmed) aggregated at the district level covering the period January 2010 to December 2016 was downloaded from the DHIS 2 system. A key indicator derived from reported malaria cases was the malaria incidence, which is the number of newly diagnosed malaria cases during a specified period per 1000 persons at risk of malaria. The catchment population was an estimate of the population at risk of malaria, an important variable in the estimation of the incidence of malaria in a defined catchment area. The estimated population at risk of malaria in each district for the period 2010 to 2016 was obtained from the National Statistical Office (National Statistical Office, n.d.). Moreover, the geographical coordinates for district boundaries were obtained from the Malawi Spatial Data Platform (http://www. masdap.mw). From the number of malaria cases and the estimated population at risk of malaria for each year, the malaria incidence (which is the number of new malaria cases per 1,000 persons per year) was calculated. This was then merged with the shape files for district boundaries. The merged data was transformed in an appropriate format and analysed in a Bayesian framework.

Using R software, data was organised in the appropriate space-time format. The data set had 28 areal units (districts) with observations taken at 7 times points (years). Four models viz: Besag-York-Mollie Model (BYM), linear time trend model and non-parametric trend model, were fitted using the R-INLA package of the R programming language. Each of the four models was implemented in R-INLA using the function, inla(). This function is similar to the lm() or glm() R functions.

Likelihoods for Spatial and Temporal Disease Modelling

The first step in defining a spatial model in a Bayesian framework is to identify the appropriate probability distribution function for data. In this study, the outcome of interest is the number of malaria cases in each district i and year j. Candidate distributions for this type of outcome data are Binomial, Poison or Negative Binomial distributions. As pointed out by Bolker (2009), the Binomial distribution is appropriate if there is an upper limit on the number of successes. The other conditions are that the probability of success is known and constant over all trials and that the number of trials is specified. Under the

Binomial likelihood, the formulation of the model is as follows. Let Y_{ij} indicate the observed number of malaria cases for district *i* and year *j*. Then Y_{ij} is assumed to follow a Binomial distribution with mean $\mu_{ij} = E_{ij}\,\lambda_{ij}$, that is

$$Y_{ij} \sim Bin\,(n_{ij},\, \mu_{ij}), \tag{1}$$

where E_{ij} is the mean expected number of malaria cases in district *i* and year *j* , λ_{ij} is the relative risk and n_{ij} is the number of individuals at risk of malaria in district *i* and year *j*. E_{ij} can be estimated as

$$E_{ij} = N_{ij}\left(\sum Y_{ij} / \sum N_{ij}\right)$$

where N_{ij} is the population at risk of malaria in each district and year. In Malawi, everyone is at risk of malaria. As such, N_{ij} , will be equal to the mid-year population for each district and year.

In Malawi, everyone is at risk of malaria and some people may have more than one episodes of the disease in a year, implying a possibility that there could be more malaria cases than the total population in a district. In addition, it is not possible to fix the probability of success. Thus, the Binomial distribution is not appropriate for modelling the number of malaria cases in a given year.

Based on the aforementioned, the next candidate distribution for this data would be the Poisson. Under the Poisson distribution, Y_{ij} represents the observed number of malaria cases for district *i* and year *j*. Then Y_{ij} is assumed to follow a Poisson distribution with mean $\mu_{ij} = E_{ij}\,\lambda_{ij}$, that is

$$Y_{ij} \sim Po\,(E_{ij}\,\lambda_{ij}), \tag{2}$$

where E_{ij} is the mean expected number of malaria cases in district *i*, is estimated as above and year j and λ_{ij} is the relative risk.

One of the most important assumptions of the Poisson distribution is that the mean and the variance are the same. Preliminary analysis of the data showed that the mean number of malaria cases per 1,000 persons over the five-year period was 413 and the variance was 31,935 indicating over dispersion of the data. This also showed that the Poisson distribution was not suitable for the given data.

This led to the choice of the Negative Binomial distribution as appropriate for the spatial temporal analysis of malaria risk in Malawi. Under this distribution, Y_{ij} is assumed to indicate the observed number of malaria cases for district i and year j. Then Y_{ij} is assumed to follow a Negative Binomial distribution with mean $\mu_{ij} = E_{ij}\,\lambda_{ij}$, that is

$$Y_{ij} \sim NB\,(\mu_{ij}, \gamma), \tag{3}$$

where E_{ij} is the mean expected number of malaria cases in district i and year j , λ_{ij} is the equivalent relative risk; γ the overdispersion parameter; and variance $V(Y_{ij}) = E_{ij}\,\lambda_{ij}(1 + \frac{E_{ij}\,\lambda_{ij}}{\gamma})$.

Modelling Bayesian Space Time Random Effects

There is a general agreement in literature that Bayesian approaches to spatio-temporal modelling of disease risk and mapping have advantages over other approaches such as maximum likelihood, moving window, and Kalman filter methods. This is so because Bayesian methods have the capacity to borrow strength over neighbouring regions and get more accurate region specific estimates.

The outcome variable in this study is the number of malaria cases in each district in Malawi and it is assumed that the country is divided into N non-overlapping districts. This study used a class of linear models called Bayesian Spatial Generalized Linear Mixed Models (GLMMs) proposed by Abellan, Richardson and Best (2008) and Besag, York and Mollié (1991). These approaches (GLMMs) were chosen because they were the most appropriate for modelling the spatial temporal variation of malaria risk. In these approaches, the fitted model estimates the common spatial and temporal trends based on the data and identifies districts whose residual temporal patterns show increased risks of malaria or are unstable.

Literature proposes various Bayesian models that can be appropriate for studying the spatiotemporal pattern of malaria data to account for spatial heterogeneity. These include the Besag-York-Mollie Model (BYM), linear time trend model (Bernardinelli et al., 1995) and non-parametric trend model (Knorr-Held, 2000). The models explained below were the only ones investigated because they were widely discussed in literature, provide less complexity than the other models (see Banerjee et al. 2004 for a comprehensive review) and were more relevant for the spatial and temporal modelling of malaria risk. This study explored four models in application to the incidence of malaria in Malawi for the period 2010-2016. These are discussed below.

The Besag-York-Mollie Model

The Besag-York-Mollie (BYM) developed by Besag et al. (1991) is formulated as follows (model 1):

$$\eta_{ij} = \beta_0 + \upsilon_i + \nu_i \,, \tag{4}$$

where η_{ij} is the log relative risk, β_0 is the overall level of disease risk, $\upsilon_i = f_1(i)$ and $\nu_i = f_2(i)$ are the two area specific effects, the spatially structured and unstructured residual effects respectively; $i = \{1, \ldots, n\}$ is the indicator for each spatial area.

A BYM specification (Besag et al., 1991) is assumed, so that υ_i, the spatially structured residual, is modelled using an intrinsic Conditional Autoregressive Structure (iCAR) specified as

$$\nu_i | \boldsymbol{\nu}_{j \neq i} \sim Normal(m_i, s_i^2);$$

with

$$s_i^2 = \frac{\delta_\nu^2}{\# N(i).} \; and \; m_i = \frac{\sum_{j \in N_i} \nu_j}{\# N(i).},$$

where ν_i is the mean for district i, $\# N(i)$ is the number of districts which share boundaries with the i-th district (i.e., its neighbours), s_i^2 is the variance for the same district which depends on the number of its neighbors ($N_i = \# N(i)$) and m_i is the mean of the areas bordering district i. The component υ_i represents the unstructured residual and was modelled using an exchangeable prior

$$\nu_i \sim Normal(0, \delta_\upsilon^2)$$

with $\boldsymbol{\delta}_\upsilon^2$ being the common variance (precision parameter)

Parametric Trend Model

The parametric trend model (model 2) as proposed by Bernardinelli *et al.* (1995) assumes that the linear predictor is written as:

$$\eta_{ij} = \beta_0 + \upsilon_i + \nu_i + (\beta + \delta_i) \times t \,, \tag{5}$$

where β_0 is the common intercept, υ_i and ν_i are the spatially uncorrelated and correlated random components over time respectively, β is the global time effect with a differential trend δ_t which identifies the interaction between year and district. This model is an extension of the BYM model presented by Besag et al. (1991).

Non Parametric Dynamic Trend Model

As proposed by Knorr-Held (2000), the linearity constraint imposed on the differential trend δ_t in (5) can be released to have a dynamic non parametric formulation (model 3:

$$\eta_{ij} = \beta_0 + \upsilon_i + \nu_i + \gamma_t + \phi_t , \tag{6}$$

where β_0 is the common intercept, υ_i and ν_i are as defined above, while γ_t and ϕ_t are the uncorrelated and correlated temporal components (model 3).

The spatial effects were modelled using intrinsic Conditional Autoregressive Structure (iCAR) priors (Knorr-Held, 2000) where ν_i , the spatially structured residual, was modelled using an iCAR structure where

$$\nu_i | \boldsymbol{\nu}_{j \neq i} \sim Normal(m_i, s_i^2$$
$$s_i^2 = \frac{\delta_\nu^2}{\# N(i).}\ and\ m_i = \frac{\sum_{j \epsilon N_i} \nu_j}{\# N(i).},$$

where ν_i is the mean for district *i*, $\# N(i)$) is the number of districts which share boundaries with the i-th district (i.e., its neighbours), s_i^2 is the variance for the same district which depends on the number of its neighbors ($N_i = \# N(i)$). The parameter υ_i represents the unstructured residual and was modelled using an exchangeable prior: $\upsilon_i \sim Normal(0, \delta_\upsilon^2)$; as presented in Banerjee (2004) and Blangiardo and Cameletti (2015).

The parameter ϕ_t in (6) represents the temporally structured effect, modelled dynamically using a random walk (RW) of order 1 or 2. The random walk of order 1 (RW1) is defined as:

$$\phi_t | \phi_{t-1} \sim Normal(\phi_{t-1}, \delta^2), \tag{7}$$

while the random walk of order 2 is defined as:

$$\phi_t | \phi_{t-1}, \phi_{t-1} \sim Normal(2\phi_{t-1} + \phi_{t-2}, \delta^2). \tag{8}$$

The parameter γ_t in (6) represents the temporally unstructured effect, modelled by using a Gaussian exchangeable prior: $\gamma_t \sim \text{Normal}(0, {}^1/_{\tau_\gamma})$ (Blangiardo & Cameletti, 2015).

Space-Time Interaction Model

The non-parametric dynamic trend model can easily be expanded to allow for interaction between district and year to explain the differences in the time trend in malaria cases for different districts. The model specification for the interaction (model 4) is as follows:

$$\eta_{it} = \beta_0 + \upsilon_i + v_i + \gamma_t + \phi_t + \delta_{it}. \quad (9)$$

In this model, the matrix of interaction terms $\{\delta_{it}\}$ defined by $\boldsymbol{\delta}$ is assumed to be multivariate normal. That is $\boldsymbol{\delta} \sim MVN(0, \delta_\phi{}^2 R_\phi)$ where $\delta_\phi{}^2$ is the variance parameter and R_ϕ is a structure matrix identifying the type of temporal and /or spatial dependence between the elements of $\boldsymbol{\delta}$. R_ϕ can be factorized as the Kronecker product of the structure matrix of the corresponding main effects that interact (Knorr-Held, 2000). Other studies have investigated four types of interactions in spatio temporal disease mapping which are type I, type II, type III and type IV (Knorr-Held, 2000; Lawson et al, 2016). In this study, type IV interaction which assumes that temporal trends differ between areas but are more likely to be similar for adjacent areas and are different for far apart districts was considered. This is considered as the most appropriate model for this data if interactions are present (Knorr-Held, 2000).

Model Selection and Validation

The statistics used in choosing the model that best fitted the data was the Deviance Information Criterion (DIC) and the effective number of parameters (p_D). The best fitting model is the model with a combination of lower DIC and small effective number of parameters to estimate (Spiegelhalter, Best, Carline and Van der Lindel. (2002)). The other statistic used in Bayesian modelling is the Watanabe Akaike Information Criterion (WAIC) which is another Bayesian version of the Akaike Information Criterion (AIC) and is asymptotically similar to the Bayesian cross-validation (Watanabe, 2010). In this criterion, the best fitting model is the model with a combination of lower DIC and small effective number of parameters (p_D) to estimate.

Sensitivity Analysis

Results from Bayesian spatio-temporal models need to be checked if they depend on the choice of priors. It is therefore recommended that sensitivity to prior specification should be examined (Lawson, 2013). This is an important step in Bayesian analysis to assess if the choices of the priors for the variance parameters result in different posterior distributions and changes in the estimates for relative risk. A sensitivity analysis was performed to assess if the posterior distribution depended on choice of the prior paratemeters.

Mapping of Malaria Risk

One of the objectives of this study was to identify districts that had a higher than national average risk of malaria. Thus, districts where malaria risk was clustered were identified as the districts that had a higher than national average risk of malaria. To identify districts with excess malaria risk, the posterior mean for the district specific relative risk of malaria, $\zeta_i = exp(\upsilon_i + v_i)$, compared to the whole of Malawi, was calculated and mapped. In addition, the uncertainties associated with posterior means, exceedance probabilities $Pr(\zeta_i > 1|data)$, were calculated and mapped (Lawson, 2013). An exceedance probability is defined as the probability that the risk θ exceeds some threshold c: $\Pr(\theta > c)$ which is often estimated from posterior sample values (Lawson, 2013).

RESULTS

We begin by exploring if the data displays the spatial and temporal variation by comparing the districts over the seven-year period (2010-2016). The first part of the analysis in this study was the descriptive statistics. The first plot was that of the incidence of malaria cases per 1,000 persons by year. This was plotted to demonstrate the trend of malaria cases over years. Another plot was a line plot demonstrating the trend of malaria cases by year in each district and finally a box plot of malaria cases by district.

Descriptive Results

Figures 13.1 to 13.3 present the crude incidence of malaria cases per 1000 persons at risk. As it can be noted, there was a general reduction in malaria cases per 1000 persons at risk from 2010 to 2016 across all the districts. Overall, at the national level (Figure 13.1), the incidence decreased from 541 malaria cases per 1,000 persons in 2010 to 389 cases per 1000 persons at risk in 2016. The results also show that the decrease was noticeable between 2010 and 2013. However, there was a slight increase in the incidence of malaria cases between 2013 and 2014.

The variation in the number of malaria cases can also be noted among the districts as shown in Figures 13.2 and 13.3. The crude results show that districts such as Neno, Mwanza, Nkhotakota and Nkhata Bay reported a higher incidence of malaria over the years compared to the other districts. Districts such as Chitipa, Thyolo, Blantyre and

Lilongwe reported a lower incidence of malaria over the past seven years compared to the rest of the districts. This analysis of the crude incidence indicates that there ~~are~~ were indeed variations in the risk of malaria in both space and time, thereby justifying the need for further analysis using methods that balance model complexity and robustness.

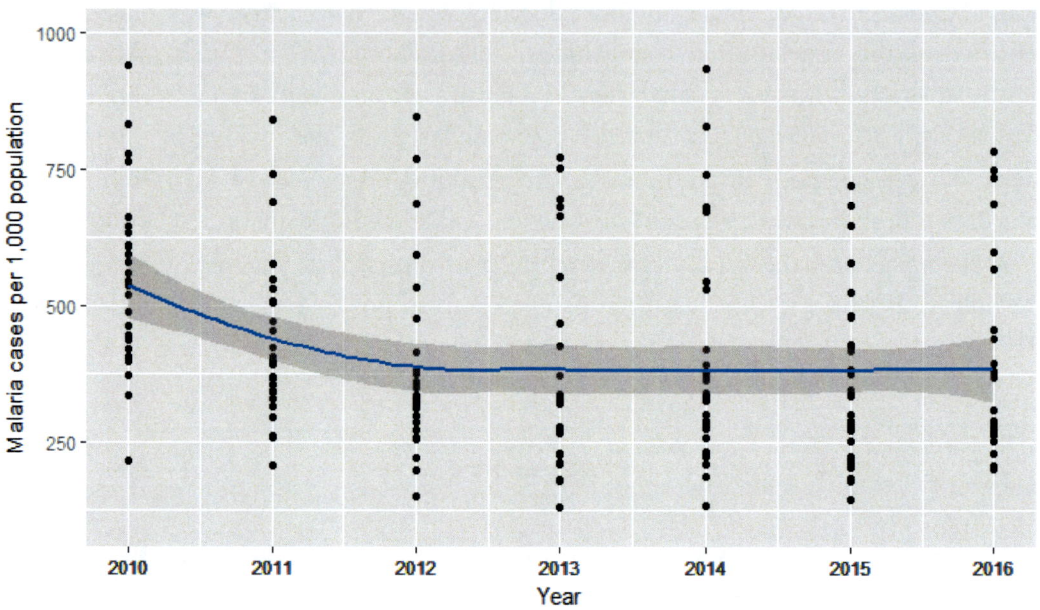

Figure 13.1. Number of malaria cases per 1000 persons per year, Malawi, 2010-2016.

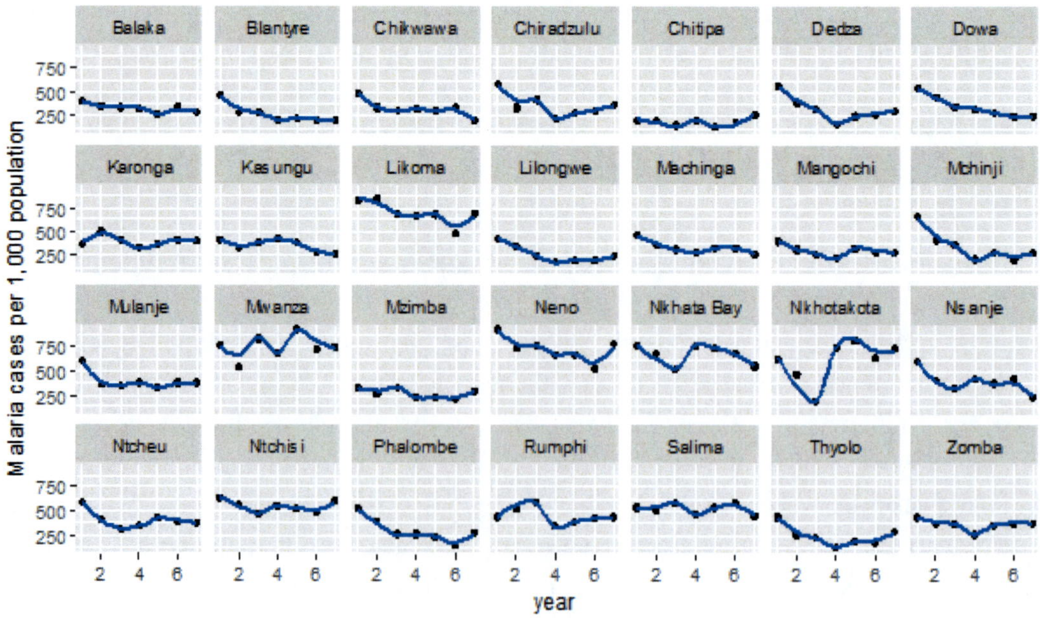

Figure 13.2. Number of malaria cases per 1000 persons per year by district, 2010-2016.

The box plot (Figure 13.3) also indicates that the distribution of malaria cases was highly skewed implying that there was great variation in the risk of malaria across the districts.

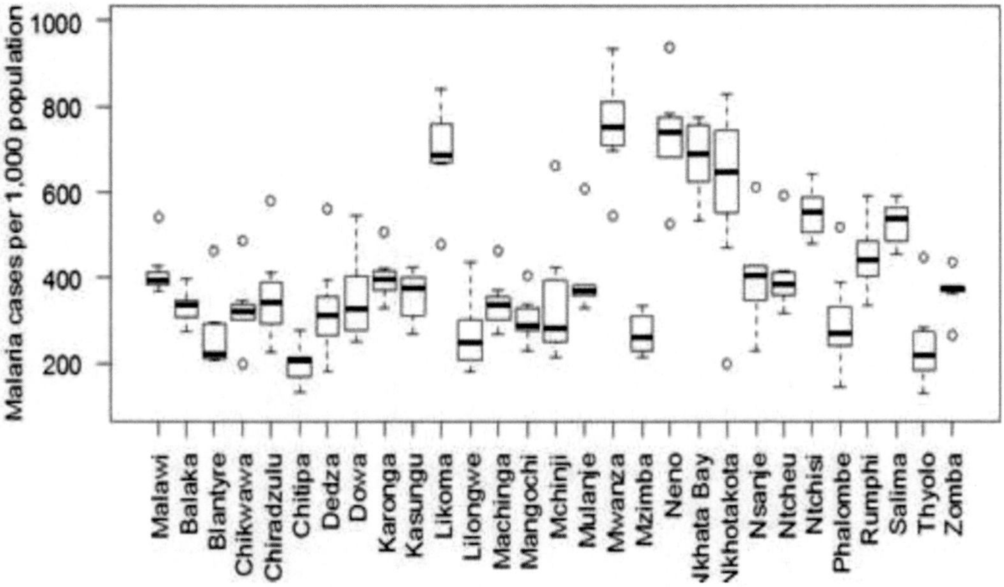

Figure 13.3. Box plot of malaria cases, 2010-2016.

Results of the Bayesian Models

Table 13.1 presents the DIC, P_D and WAIC values for the BYM, Parametric time trend, nonparametric time trend and space time interaction term model considered in this study. The best fitting model to use is the one with a combination lower DIC and small effective number of parameters (P_D) to estimate. As can be seen in Table 13.1, the BYM model is was far from the best model as it has the largest DIC value of 4723.75. Likewise the model proposed by Bernardinelli et al. (1995) and the type IV space-time interaction model proposed by Knorr-Held (2000) had higher DIC values than the non-parametric time trend model. Therefore, the Nonparametric time trend model, which is the model with uncorrelated and correlated spatial and temporal random effects had the smallest DIC value of 4,633.93 and small effective number of parameters. This model was therefore the best in terms of model fit and complexity. The posterior parameter estimates presented and discussed in the following sections were based on the best fitting model. Since the difference in DIC value between Nonparametric time trend and Model the Space time interarction model was less than 5 (Kass & Raftery, 1995), the posterior risk estimates for the spatial temporal model with interaction were also plotted to observe the spatial and temporal trend of malaria risk for each district over the years (as seen in Figures 13.7, 13.8 and 13.9).

Table 13.1. Deviance Information Criterion (DIC) and Watanabe Akaike Information Criteria (WAIC) values for the spatio-temporal models in this study

Model	Components	p_D	DIC	WAIC
BYM	$\beta_0 + \upsilon_i + \nu_i$	26.8	4723.75	4727.1
Parametric time trend	$\beta_0 + \upsilon_i + \nu_i + (\beta + \delta_t) \times t$	31.4	4668.09	4669.8
Nonparametric time trend	$\beta_0 + \upsilon_i + \nu_i + \gamma_t + \phi_t$	31.4	4633.93	4634.7
Space time interaction term	$\beta_0 + \upsilon_i + \nu_i + \gamma_t + \phi_t + \delta_{it}$	65.2	4639.26	4637.1

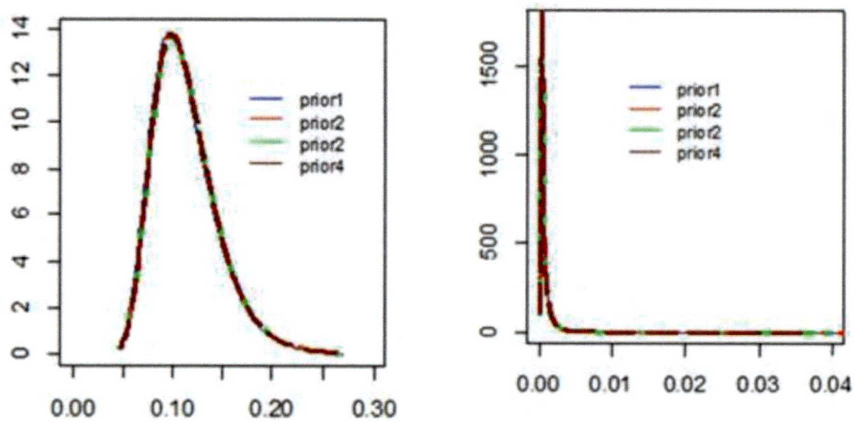

Figure 13.4. Marginal posterior densities of the variance of the spatially structured (left) and unstructured random effects (right).

Table 13.2. Posterior mean and 90% credible interval (CI) for the fixed effects and the variance parameters by prior

Parameter	Priors							
	Prior 1-INLA default		Prior 2 –logGamma (1,0.01)		Prior 3 logGamma (1,0.001)		Prior 4- logGamma (1,0.00005)	
	Mean	90% CI	Mean	90% CI	Mean	90% CI	Mean	90% CI
β_0	0.12	(-0.007, .249)	0.12	(-0.008, 0.247)	0.12	(-0.008, 0.248)	0.12	(-0.008, 0.2478)
δ_u	10	(5, 16)	10	(5, 16)	10	(5, 16)	10	(5, 16)
δ_v	2,887	(370, 8,956)	2,886	(370, 8,996)	2,940	(372, 9,037)	2,986	(376, 9,133)
δ_γ	470	(55, 1,781)	389	(51, 1,356)	386	(51, 1,360)	388	(52, 1,363)
δ_ϕ	17,945	(1,232, 5,015)	18,656	(1,200, 66,942)	18,679	(1,200, 66,987)	18,069	(1,193, 65,921)

The results of the sensitivity analysis to choice of hyper prior parameters are presented in Figure 13.4 and Table 13.2. The priors tested for the best fitting model ~~are~~ were the prior 2-log *Gamma* (1, 0.01), prior 3-log *Gamma* (1, 0.001) and prior 3-log *Gamma* (1,0.00005) which have been used previously in a spatial temporal Bayesian modelling (Ugarte, Goicoa, Ibanez, & Militino, 2009). These priors were compared against the INLA default priors, labelled as prior 1 in Figure 13.4. As noted in Figure 13.4 and Table 13.2, the results are quite stable to the choice of priors for the variance parameters. The posterior densities overlap for the different priors.

Spatial and Temporal Variation in Malaria

Spatial Variation in Malaria Risk

In line with the objectives of this study, the spatial risk $\zeta_i = \exp(\upsilon_i + \nu_i)$ and the spatial exceedance for a relative risk of 1 ($\Pr[\zeta_i > 1]$) was calculated and mapped for the best fitting model, the Space time interaction term. The spatial risk is interpreted as the residual relative risk of malaria for each district compared to the rest of Malawi. As such, exceedance probabilities for a district are the posterior probabilities whose spatial risk is greater than one. Exceedance probabilities have been proposed as a Bayesian approach to hotspot identification in both space and time by several authors (e.g., Lawson, 2013).

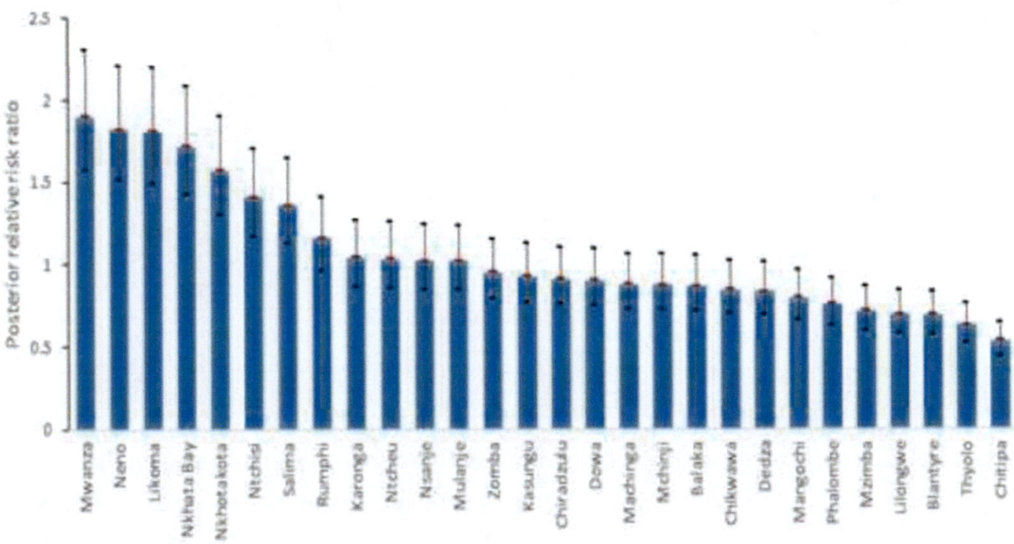

Figure 13.5. Posterior relative risk and corresponding 95% credible interval (CI) based on the best fitting model.

Figure 13.5 shows the posterior relative risks and their corresponding 95% credible intervals (CI) for each district in Malawi. The results indicate, in concordance with the descriptive results, that Mwanza and Neno had the highest Relative Risk (RR) of malaria compared to the rest of Malawi with an RR ratio of 1.9 (95% CI:1.58,2.30) and 1.83(95% CI:1.51,2.20), respectively. The results also showed that the risk of malaria ~~is~~ was lowest in Chitipa and Thyolo with an RR of 0.53 (95% CI: 0.44, 0.64) and 0.63 (95% CI: 0.52, 0.76), respectively, compared to the rest of Malawi.

Figure 13.6 shows district specific relative risk (RR) of malaria $\zeta_i = exp(\upsilon_i + v_i)$ compared to the whole of Malawi (left) and exceedance probability (RR > 1) (Right). The risk map divided the country into three regions of malaria risk, which were the low, medium and high risks. Districts with RR≤1 were classified as low risk, while districts with RR between 1 and 1.5 were classified as medium risk. On the other hand, ~~and~~ districts with RR≥1.5 were classified as high risk. Of the 28 districts in Malawi, districts with the high risk of malaria were Mwanza, Neno, Nkhotakota and Nkhata Bay.

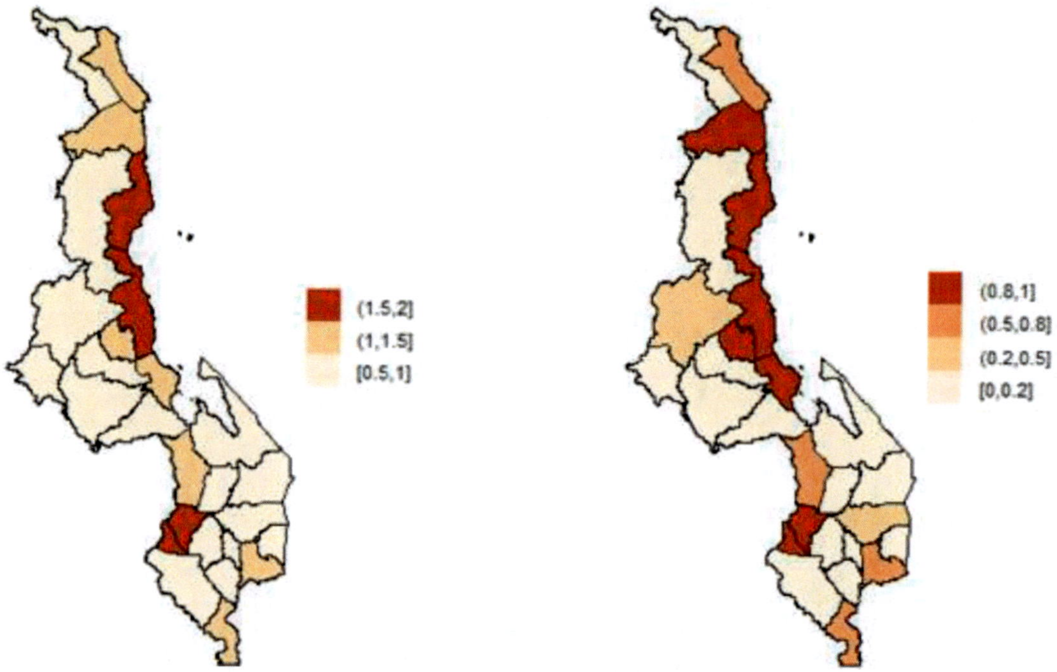

Figure 13.6. Posterior mean for the district specific Relative Risk (RR) of malaria and exceedance probability (RR >1).

The relative risk of malaria in these districts ~~is~~ was more than 60% higher than the average national risk (RR=1). Seven districts were in the medium risk zone of malaria, namely Ntchisi, Salima, Rumphi, Karonga, Ntcheu, Nsanje, and Mulanje in the order of decreasing risk. The remaining 16 districts were in the low risk zone with the lowest risk being in Blantyre, Lilongwe, Thyolo and Chitipa in order of decreasing risk. These results are consistent with the exceedance probability map that consistently indicated the same districts with excess risk of more than one given the data.

Temporal Variation in Malaria Risk

Figure 13.7 shows the posterior temporal trend for malaria cases in Malawi. The overall posterior risk is a combination of the structured temporal effect (an indication of the temporal correlated heterogeneity) and the unstructured temporal effect (an indication of temporal uncorrelated heterogeneity). As shown in Figure 13.7, there has been a decreasing structured temporal trend in the malaria cases over the years with a deep decrease from 2011 to 2013 and a slower decrease from 2013 to 2016.

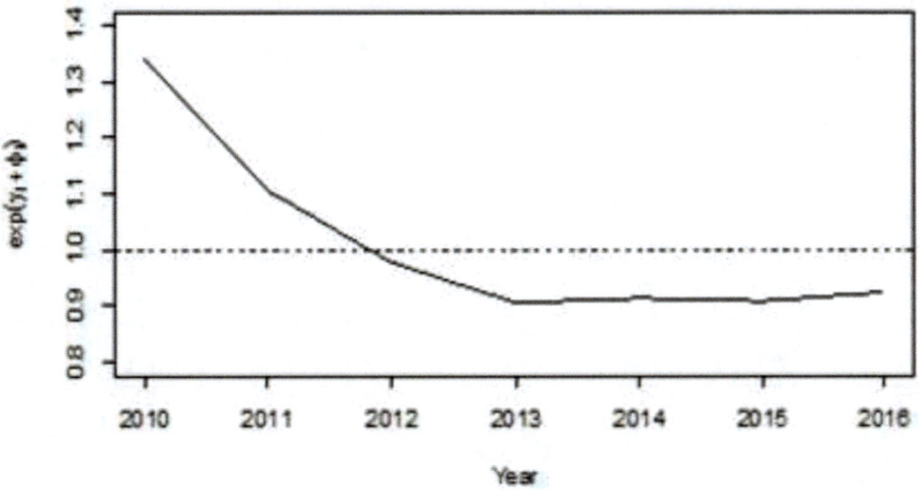

Figure 13.7. Average temporal posterior trend for malaria risk in Malawi, 2010-2016.

Spatial Temporal Variation in Malaria Risk

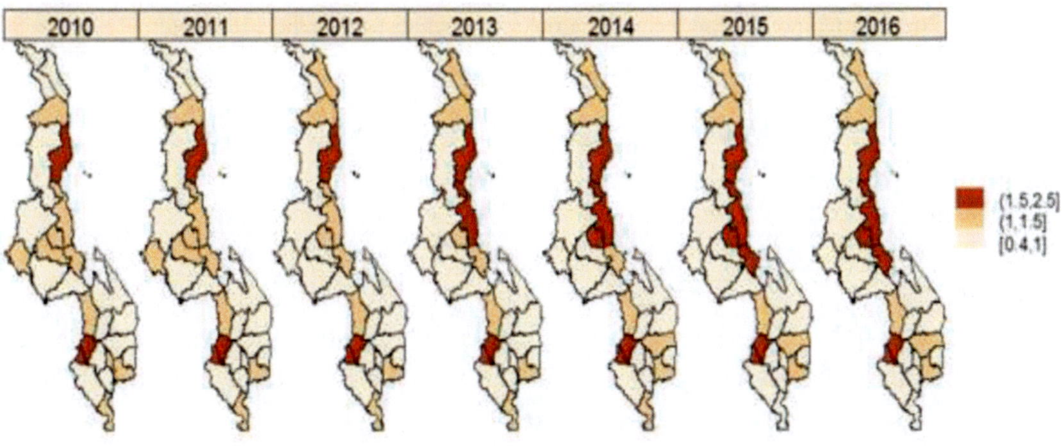

Figure 13.8. Spatial and spatio-temporal effects of malaria risk.

To demonstrate the spatial temporal variation of posterior risk of malaria, the trend in malaria risk by district over the years was also plotted (Figures 13.8 and 13.9). The results showed that while the overall trend of malaria cases indicateds a decline in malaria cases from 2010 to 2016, the district specific posterior trends showed that some districts have had an increase of malaria cases over the years. Districts such as Nkhotakota, Nkhata Bay, Mwanza, Neno and Salima had an increasing trend of malaria cases over the years. The results also revealed that some districts like Nkhata Bay and Neno maintained a higher risk than the national average risk (RR=1) of malaria over the years, while other districts moved from low to high risk. Nkhota kota moved from a medium to a high risk district over the years. These results are consistent with those of the descriptive analysis presented above. Again, some districts like Dowa, Blantyre, Lilongwe, Thyolo and Chitipa had a decreasing trend of malaria risk over the years.

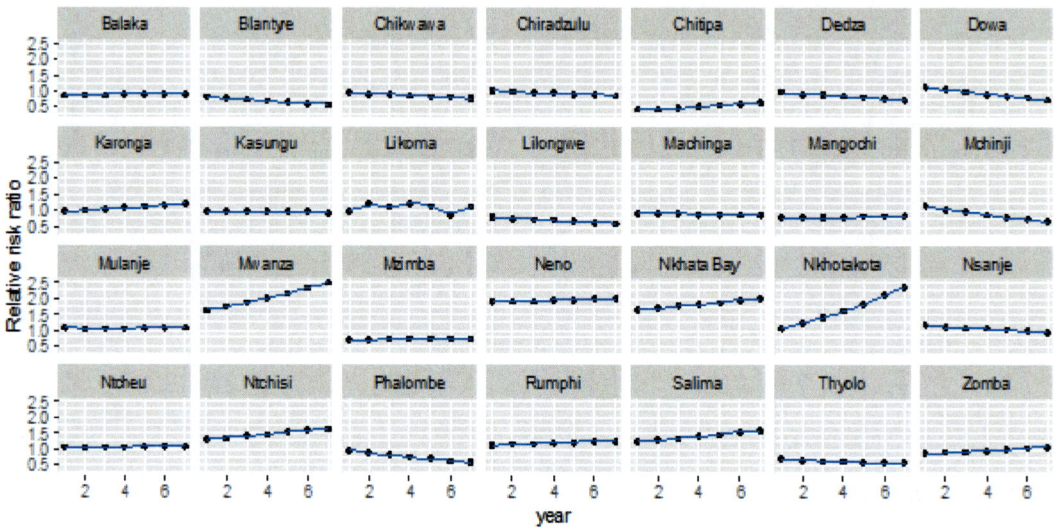

Figure 13.9. Posterior temporal trend for malaria risks for each district from 2010-2016.

Discussion and Conclusion

Discussion

In this study, a Bayesian spatiotemporal approach for modelling the risk of malaria in Malawi using R-INLA was presented. The results of this study help to understand the spatial, temporal and spatiotemporal risk of malaria and could help the Ministry of Health in planning malaria interventions. To achieve the study objectives, four models were fitted using the Bayesian Spatiotemporal approaches and the model with spatial and temporal correlated and uncorrelated random effects, nonparametric time trend, was the best model in terms of model fit and complexity.

The results as depicted on the spatial maps and plot of the posterior relative risk revealed that there ~~is~~ was indeed a spatial and temporal variation in the risk of malaria in Malawi. The results further demonstrated that the risk of malaria ~~is~~ was highest in the Mwanza, Neno, Nkhotakota and Nkhata Bay districts. Districts with medium risk of malaria were Salima, Rumphi, Karonga and Nsanje districts. The study also showed that the districts with the lowest risk of malaria were the Blantyre, Lilongwe, Thyolo and

Chitipa districts. Moreover, the results of the spatial variations of malaria risk were to a greater extent consistent with previous studies by Kazembe et al. (2006) and Bennett et al. (2013) in which they concluded that the highest risk of malaria ~~is~~ were in areas along the lakeshore, shire valley and central plain areas. These researchers also concluded that the lowest risk of malaria were in areas in the highlands of Rumphi, Mzimba and Chitipa (Kazembe et al., 2006; Bennett et al., 2013). Elevation has been considered as a very important factor in the spatial prediction of malaria risk (Kazembe et al., 2006) and as shown in the mapping of this study (Figure 13.6), all districts with the lowest risk of malaria were in the high elevation areas, while all districts with the highest risk of malaria were in the low altitude areas.

Additional insight revealed in this study was the temporal variation in malaria risk. The study by Kazembe et al. (2006). did not consider the temporal variations as they used a cross sectional point reference data. However, this study considered an additional variation to the spatial variation in the risk of malaria, which was a time dependent trend in the risk of malaria between the period 2010 and 2016. This trend was decreasing from 2010 to 2013. After 2013, the decrease was slower than the years before that. This decrease was expected as the Ministry of Health scaled~~s~~ up malaria interventions from 2010 to date. During the study period, the Ministry of Health had introduced Indoor Residual Spraying (IRS) in districts presumed to be of high malaria burden in Karonga, Nkhotakota, Salima, Chikwawa and Nsanje. In addition to the IRS, the Ministry also introduced mass distribution of insecticide treated nets to all households in the country. Over 6 million nets have since been distributed through this channel (NMCP and ICF International (2014). This scaled~~s~~ up ~~is~~ was noted in the results of the 2014 Malaria Indicator Survey where it was reported that 70% of the households owned an insecticide-treated mosquito net (ITN) and 73% of the of households had at least one ITN and/or IRS in the past 12 months. Ownership of ITN had increased from 58% in 2010 (NMCP and ICF International (2014). The increase in ownership of nets also impacted on the usage as the results of the 2014 Malaria Indicator Survey showed that usage of nets among under-five children increased from 55% in 2010 to 67% in 2014. These interventions were likely to be the contributing factors for the decrease in the incidence of malaria over the period under study.

The findings of this study also revealed that some districts like Nkhotakota, unexpectedly, did not register a decrease in the risk of malaria over the scaled up period. Results from the 2015-2016 Malawi Demographic and Health Survey showed that Nkhotakota district had a coverage higher than the national coverage of the proportion of households with at least one ITN and/or IRS in the past 12 months at 86% versus 59% nationally. In addition, Nkhotakota district had a higher than national proportion of household population using an ITN in a household with at least one ITN at 76% against 59% nationally (National Statistical Office/Malawi and ICF (2017).

This high coverage of prevention strategies was expected to result in a reduction in the risk of malaria than the national average. Moreover, Nkhotakota district had a combination of IRS and ITN and studies have shown that there is significant added protection from combining IRS and ITNs compared to ITNs alone (Kleinschmidt et al., 2009). Furthemore, Nkhotakota is the district where IRS has been implemented over a long period. Lack of expected decline in the risk of malaria in districts such as Nkhotakota, Nkhata Bay and Nsanje despite a combination of scaled up interventions can be attributed to the emergence of pyrethroid resistance that has been detected in *An. funestus* and *An. arabiensis* populations across Malawi and has increased over the years (Mzilahowa et al., 2016).

Pyrethroids are the insecticides that the World Health Organization recommends for use in ITNs. As such, the emergence resistance to pyrethroids in Anopheles mosquitoes, the main vector for malaria in Malawi and the region, threatens the efficacy of ITNs. The insecticides used for IRS in Malawi over the years are mainly the Alpha cypermethrin and Lambda cyhalothrin, which are pyrethroids (Mzilahowa et al., 2016), making it more plausible that the lack of decline of malaria risk in some districts despite the high coverage of malaria prevention interventions could be attributable to the presence of resistance to pyrethroids.

Conclusion

This study applied Bayesian spatial temporal approaches to model the variation in the risk of malaria. The best model for the analysis of malaria risk in both space and time for this data was the nonparametric time trend, with both spatial and temporal correlated and uncorrelated heterogeneity. The best model has shown that indeed there is variation in the risk of malaria across the districts in Malawi with some having a relative risk that was more than 50% higher than the national average relative risk. Districts with the highest risk were the Nkhotakota, Nkhata bay, Mwanza and Neno and those with the lowest risk of malaria were the Blantyre, Lilongwe, Thyolo and Chitipa.

Using the best model, we conclude that there has been a temporal variation of malaria risk, with a steep decrease noted between 2011 up to 2013 and a slower decrease from 2013 to 2016. In addition, while there was an overall decrease in the risk of malaria, this trend varied by district. Some districts did not register a decrease in the risk of malaria over the years. Furthermore, some districts that were expected to have a decrease in the risk of malaria over the period because of the package of interventions did not register any decrease, prompting the need for further investigation in addition to the fact that this could be due to the emergence of resistance to pyrethroids.

Implications and Recommendations

In line with the findings of the study, the Ministry of Health should focus interventions in districts with higher that national average risk of malaria instead of rolling out the same kind and level of interventions across the country. Nkhotakota, Nkhata Bay, Mwanza and Neno districts should be given the highest priorities in the wake of scarce resources. Resources permitting, Ntchisi, Karonga, Salima, Rumphi, Karonga and Nsanje should be the next set of priortized districts as they had medium risk of malaria compared to the national average. As Mzilahowa et al. (2016) recommend, and in line with recommendations in the Global Plan for Insecticide Resistance Management in malaria vectors (GPIRM) (WHO, 2012), new insecticides and new tools are necessary as part of the integrated vector management strategies. Changing interventions and insecticides would aid in understanding whether Nkhotakota, Salima and Nkhata Bay would still have a higher than average risk of malaria despite the malaria control interventions.

Limitations of the Study

R-INLA is very powerful as it results in great saving in time as a valid alternative to Markov chain Monte Carlo (MCMC) approaches to Bayesian Spatial temporal modeling of disease. The main limitation is that R-INLA only works for a limited type of models, the latent Gausian models and the type of likelihood. For example, R-INLA cannot be used for ordinal data and the multinomial distribution (Lawson, 2016). In addition, R-INLA cannot be efficient if the number of hyper-parameters is relatively small, less than 20. This study only used R-INLA without comparison with other alternative approaches such as the MCMC.

The findings obtained in this study were validated using different priors to assess the validity of the model in terms of complexity and robustness. Furthemore, only the main effects parameters were included in the model without individual and household level covariates. While they would have assisted in explaining the variation in risk, the covariates could not be included in this analysis because malaria cases are reported at an aggregate level and not at individual and household levels. Selection bias was also present given that people who get ill of malaria and seek medical attention may have characteristics that are different from those that do not seek medical attention and this pattern may vary by district.

REFERENCES

Abellan, J. J., Richardson, S., & Best, N. (2008). Use of Space–Time Models to Investigate the Stability of Patterns of Disease. *Environmental Health Perspectives,* 116(8), 1111–1119. http://doi.org/10.1289/ehp.10814.

Banerjee, B. Carlin, B. & Gelfand, A. (2004). Hierarchical Modeling and Analysis and Analysis of Spatial Data. New York: Chapman & Hill.

Bennett, A., Kazembe, L., Mathanga, D. P., Kinyoki, D., Ali, D., Snow, R. W., & Noor, A. M. (2013). Mapping Malaria Transmission Intensity in Malawi, 2000–2010. *The American Journal of Tropical Medicine and Hygiene,* 89(5), 840–849. http://doi.org/10.4269/ajtmh.13-0028.

Bernardinelli, L., Clayton, D.G., Pascutto, C., Montomoli, C., Ghislandi, M. & Songini, M. (1995). Bayesian Analysis of Space-Time Variation in Disease Risk. *Statistics in medicine,* 14, 2433-43. https://doi.org/10.1002/sim.4780142112.

Besag, J., York J.C., Mollié, A. (1991). Bayesian image restoration, with two applications in spatial statistics. Annals of the Institute of Statistical Mathematic, 43, 1–109. https://doi.org/10.1007/BF00116466.

Blangiardo, M. & Cameletti, M. (2015). Spatial and Spatio-temporal Bayesian Models with R – INLA.West Sussex: John Wiley and Son Ltd.

Bolker, B. (2009). Ecological Models and Data in R. New Jersey: Princeton University Press.

Chirombo, J., Lowe R. & Kazembe, L. (2014). Using Structured Additive Regression Models to Estimate Risk Factors of Malaria: Analysis of 2010 Malawi Malaria Indicator Survey Data. *PLoS ONE 9*(7): e101116. https://doi.org/10.1371/journal.pone.0101116.

Howe, G. M. (1989). Historical Evolution of Disease Mapping in General and Specifically of Cancer Mapping. In: P. Boyle., C. S. Muir, & E. Grundmann E. (Eds.), *Cancer Mapping. Recent Results in Cancer Research*. 114. Springer, Berlin, Heidelberg. https://doi.org/10.1007/978-3-642-83651-0_1.

Kass, R. E., & Raftery, A. E. (1995). Bayes factors. *Journal of the American Statistical Association*, 90, 773-795. https://doi.org/10.1080/01621459.1995. 10476572.

Kazembe, L. N., Kleinschmidt, I., Holtz, T. H., & Sharp, B. L. (2006). Spatial analysis and mapping of malaria risk in Malawi using point-referenced prevalence of infection data. *International Journal of Health Geographics*, 5, 41. http://doi.org/10.1186/1476-072X-5-41.

Kleinschmidt, I., Schwabe, C., Shiva, M., Segura, J. L., Sima, V., Mabunda, S. J. A., & Coleman, M. (2009). Combining Indoor Residual Spraying and Insecticide-Treated Net Interventions. *The American Journal of Tropical Medicine and Hygiene*, 81(3), 519–524.

Knorr-Held, L. (2000). Bayesian modelling of inseparable space-time variation in disease risk. Statistics in medicine. 19, 2555-2567. Lawson, A. (2009). *Bayesian Disease Mapping: Hierarchical Modeling in Spatial Epidemiology*. New York: Chapman and Hall.

Lawson, A. (2013). Bayesian Disease Mapping Hierarchical Modeling in Spatial Epidemiology (Second edition). New York: Chapman and Hall.

Lawson, A. B., & Choi, J. (2016). Spatiotemporal Disease Mapping. In A.B. Lawson, S. Banerjee, R.P. Haining & M.D. Ugarte (Eds.), *Handbook of Spatial Epidemiology* (pp. 345-348). CRS Press.

Mzilahowa, T. Chiumia, M. Mbewe, R.B., Uzalili, V.T., Luka-Banda, M. Kutengule, A., Gimnig.JE. (2016). Increasing insecticide resistance in Anopheles funestus and Anopheles arabiensis in Malawi, 2011–2015. Malar J., 15:563. https://doi.org/10.1186/s12936-016-1610-1.

National Malaria Control Programme (NMCP) [Malawi] and ICF International. (2014). Malawi Malaria Indicator Survey (MIS) 2014. Lilongwe, Malawi, and Rockville, Maryland, USA: NMCP and ICF International.

National Statistical Office/Malawi. ICF: Malawi demographic and health survey 2015–16. Zomba, Malawi: National Statistical Office and ICF; 2017. Snow, J. (1855). *On The Mode of Communication of Cholera*. London: Churchill. Available at https://collections.nlm.nih.gov/ext/cholera/PDF/0050707.pdf.

Spiegelhalter, D., Best, N., Carlin, B. & Van der Linde, A. (2002). Bayesian measures of model complexity and fit. *Journal of the Royal Statistical Society Series B 64*, 583–639. https://doi.org/10.1111/1467-9868.00353.

Thwing, J., Eisele, T. P., & Steketee, R. W. (2011). Protective efficacy of malaria case management for preventing malaria mortality in children: a systematic review for the Lives Saved Tool. *BMC Public Health,* 11(Suppl 3), S14. http://doi.org/10.1186/1471-2458-11-S3-S14.

Ugarte, M.D., Goicoa, T., Ibanez, B. & Militino, A.F. (2009). Evaluating the performance of spatio-temporal Bayesian models in disease mapping. *Environmetrics,* 20(6):647-665. https://doi.org/10.1002/env.969.

Watanabe, S. (2010). Asymptotic equivalence of Bayes cross validation and widely applicable information criterion in singular learning theory. *Journal of Machine Learning Research,* 11, 3571–3594.

World Health Organization. (2012). Global Plan for Insecticide Resistance Management in Malaria Vectors. Geneva, Switzerland: World Health Organization.

World Health Organization. (2015). Guidelines for the treatment of malaria (Third edition). Geneva, Switzerland: World Health Organization.

In: Statistical Modelling …
Editors: N-B. Kandala and L. Kazembe
ISBN: 978-1-53615-981-3

Chapter 14

MODELLING SPATIAL AND SPATIAL-TEMPORAL PATTERNS OF TB AND HIV MORTALITY IN NAMIBIA

Andreas I. Shipanga and Lawrence N. Kazembe*
Department of Statistics and Population Studies,
University of Namibia, Windhoek, Namibia

ABSTRACT

Namibia is among sub-Sahara African countries with high burden of TB and HIV co-infection in the world. The interaction between TB and HIV/AIDS makes the diagnosis and management of the co-infection difficult. The purpose of this study is to identify risk factors, assess spatial patterns and space-time clusters of mortality among TB and HIV in Erongo Region, in Namibia. A retrospective cohort study from 2003 to 2017 was carried out using data consisting of 3,145 subjects from all the 16 health facilities providing ART and TB management in the region. Global Moran's I and LISA statistics were used to help analyse the spatial distribution and clusters. Kaplan Meier and Cox proportional hazard model were used to model survival (time to death) and identify risk factors among TB and HIV co-infected patients, using a structured additive regression models via Bayesian inference approach. Of the 3145 patients, 424 (13.5%) defaulted treatment, 42(1.3%) stopped treatment, 542 (17.2%) were transferred out and 1673 (53.2%) were alive at the end of the study. About 464 (14.8%) patients experienced event of interest (death), during the period. Mortality was higher in Usakos district as compared to the other 3 health districts in the region. Global moran's I shows that the EAs tend cluster, 0.17 (p-value = 0.0001) and detected presence of hot spots in Erongo Region. Mortality worsened with increasing age and WHO clinical stage. The detection of space-time clustering was useful in identifying higher risk areas. From this study it is recommended that there is a need to have targeted intervention among these areas to ensure that Namibia strives to attain its NDP5, which is intended to reduce mortality of TB among HIV patients. Patients aged 50+ and those with bedridden functional status should be strictly followed to reduce mortality. This information can be used by the region for

* Corresponding Author's Email: lkazembe@yahoo.com.

planning purposes, allocation of resources and even dissemination of TB and HIV information.

Keywords: TB/HIV co-infection, spatial analysis, spatial-temporal modeling, Erongo-Namibia

INTRODUCTION

The Human Immunodeficiency Virus (HIV) pandemic presents a significant challenge to global tuberculosis (TB) control. The recent TB factsheet by the World Health Organization indicates that people who are HIV positive are 20 to 30 times more 27 likely to develop active TB (WHO, 2018). For HIV positive patients, being diagnosed with TB is a sign of progression to AIDS (CDC, 2017). TB is a leading preventable cause of death among people living with HIV, affecting more than one third of all AIDS-related deaths in 2015, and almost 60% of the estimated global HIV-related TB cases are not diagnosed and not treated (UNAIDS, 2017).

The World Health Organization (WHO) recommends 12 collaborative HIV/TB activities, including the "Three I's for HIV/TB" (isoniazid preventive treatment [IPT], intensified case finding, and infection control for TB), which should be seen as core prevention, care, and treatment services for HIV infection. Of the 12 activities, there has been progress in implementing testing for HIV infection, providing trimethoprim-sulfamethoxazole preventive therapy, and antiretroviral therapy (ART) (Getahun, Gunneberg, Granich, & Nunn, 2010). Globally, the WHO estimated that, in 2008, only 20% of people living with HIV knew their status (WHO, 2018). In 2017, 920 000 people living with HIV estimated to have fallen ill with TB; 22% of patients with TB were tested for HIV infection (WHO, 2018). Controlling and preventing HIV/AIDS, tuberculosis and other opportunistic infections are among the key global health priorities, particularly in developing countries such as Namibia.

Namibia is the fourth worst TB affected country in the world, with 9 882 patients diagnosed with the disease in 2014 (MoHSS, 2016a). It also remains one of the country with the highest adult HIV prevalence rate in the world, according to the Surveillance Report of the 2016 National HIV Sentinel Survey (MoHSS, 2016c) the prevalence rate was at 17.2%. The country has one of the highest HIV/TB co-infection rates in the world, meaning that on average, about 4 out of every 10 patients with tuberculosis are also HIV-positive (MoHSS, 2016a). Even though the government of Namibia have put up several strategies to reduce TB and HIV/AIDS incidence, the two diseases remain some of the highest causes of death in Namibia (WHO, 2018).

Despite progress made towards reducing the burden of tuberculosis (TB) in the country, the disease remains a major public health concern in the country. While there has been a consistent decrease in reported TB cases since 2004, it is noteworthy that the

2016 World Health Organisation (WHO) *Global Tuberculosis Report* included Namibia among the 30 countries with the highest TB burden globally, due to the country's high estimated per capita TB incidence. The country is also included among the countries with very high rates of TB/HIV coinfection, with an HIV prevalence of 38% among TB patients in 2016 (MoHSS, 2017b).

Coverage of HIV services for TB patients continues to improve, with 95% of TB patients having a known HIV status (MoHSS, 2016a). About 40% of the TB patients were co-infected with HIV, a marginal decrease from 44% in 2014 (MoHSS, 2016a). The majority (92%) of HIV infected TB patients were on Antiretroviral therapy (ART). The treatment success rate for TB/HIV patients was however relatively low (81% for the 2015 cohort), primarily due to a relatively high (11%) death rate (MoHSS, 2016a). Other contributing factors include poverty and associated risk factors (overcrowding, poor housing conditions, poor nutrition and delays in seeking health care), smoking, alcohol use and general community transmission due to the high prevalence of the two diseases in the country (MoHSS, 2017b).

In an effort to combat TB occurrence, several campaigns have been initiated all aiming at eliminating the TB disease. These campaigns include adoption of universal ART provision, WHO End TB strategy which targets 35% reduction in TB cases by 2020 and the UN zero campaigns targeting TB, AIDS, Malaria and poverty among others (WHO, 2016a). In Namibia the TB treatment success rate is estimated at 83% in 2015 which is lower than the WHO target of 90%, the main reason for this success rate is due to the reduction in funding the country has received from international donors such as the Global Fund and PEPFAR (Menges, 2018).

The main objective of the study is to identify risk factors, assess the spatial distribution, and examine the presence of the spatial-temporal clustering as well as analyse the survival of TB and HIV co-infections in Erongo Region, Namibia.

The use of survival models involving a random effect or "frailty" term is becoming more common (Banerjee, Carlin, & Gelfand, 2014). Usually the random effects are assumed to represent different clusters, and clusters are assumed to be independent. We consider random effects corresponding to clusters that are spatially arranged, such as clinical sites or geographical regions. That is, we might suspect that random effects corresponding to strata in closer proximity to each other might also be similar in magnitude. Very often, time-to-event data will be grouped into strata (or clusters), such as clinical sites, geographic regions, and so on. In this setting, a hierarchical modelling approach using stratum-specific parameters called frailties is often appropriate, this is a mixed model with random effects (the frailties) that correspond to a stratum's overall health status (Banerjee et al., 2014).

We review selected studies that have applied such methods in TB/HIV co-infection epidemiology. Uthman, Yahaya, Ashfaq, & Uthman (2009) showed that there was no decline in growth in the number of deaths due to TB among HIV positive in most Africa countries. The results of the study states that Eastern, Southern, Western, and Middle Africa experienced an upward trend in the number of reported TB-HIV deaths. They further showed that there was presence of 'hot-spots' and very large differences persist between sub-regions. The spatial distribution of TB cases was non-random and clustered, with a Moran's $I = 0.454$ ($p = .001$). Spatial clustering suggested that 13 countries were at increased risk of TB-HIV deaths, and six countries were grouped as "hot spots" (Uthman et al., 2009). They concluded that only by tackling TB and HIV together will progress be made in reversing the burden of both diseases. There is a great need for scale-up of preventive interventions such as the World Health Organization '3I's strategy' (intensified case finding, isoniazid preventive therapy and infection control).

Further, a study conducted in a rural region of north–eastern South Africa shows that the value of conducting high resolution spatial analyses in order to understand how local micro–epidemics contribute to changes seen over a wider area (Mee et al., 2014). Comparing two periods 2007 and 2010, there was a 30% decrease in age and gender standardised adult HIV–related and TB (HIV/TB) mortality with no change in mortality due to other causes. There was considerable spatial heterogeneity in the mortality patterns. Areas separated by 2 to 4 km, with very different epidemic trajectories, were identified. There was evidence that the impact of ART in reducing HIV/TB mortality was greatest in communities with higher mortality rates in the earlier period (Mee et al., 2014).

In another study, Zulu, Kalipeni, & Johannes (2014) demonstrated that spatial analysis enhanced understanding of local spatio-temporal variation in HIV prevalence, possible underlying factors, and potential for differentiated spatial targeting of interventions. Their findings suggest that intervention strategies should also emphasize improved access to health/HIV services, basic education, and syphilis management, particularly in rural hotspot districts, as further research is done on drivers at finer scale. Further, the analysis revealed wide spatial variation in HIV prevalence at regional, urban/rural, district and sub-district levels. However, prevalence was spatially levelling out within and across 'sub-epidemics' while declining significantly after 1999. Prevalence exhibited statistically significant spatial dependence nationally following initial (1995-1999) localized, patchy low/high patterns as the epidemic spread rapidly. The epidemiologic and spatialtemporal analysis of TB epidemic and TB/HIV co-infection epidemic demonstrates a potential connection between TB and HIV in Urumqi. Demographic, temporal, geographic factors are the reasons of causing TB and TB/HIV co-infection epidemic (Wei, et al., 2016).

In fact, to allow measuring the evolution of risk in time across regions, spatio-temporal models are used. This is appropriate when data is collected over space and time (GomezBarroso, Rodriguez-Valin, Ramis, & Cano, 2013). These approaches do not detect the cause of disease but rather provides an insight of the areas with high incidences which can be investigated further for causal relationship (Lawson, 2013). In addition, the method can identify the source of disease occurrence with respect to demographic factors, time and space. Spatial-temporal methods are useful in monitoring of disease status in a community as they may provide information on changing disease pattern over time. We apply spatio-temporal models in the survival analysis of TB/HIV co-infected patients in Erongo, Namibia.

METHODS

Study Context

Namibia has a three-tier health delivery system consisting of primary (health care centres), secondary (district level) and tertiary (referral) levels of care. Ownership of the facilities is largely by the public sector (Ministry of Health and Social Services), although reasonable number is also managed by non-governmental organizations, such as NAPPA, WBCG and the private sector. Erongo Region in particular has 24 public health facility, of which 4 are district hospital, 2 health centres and 18 clinics. There are 16 ART sites and four (4) TB clinics.

Erongo Region, situated at the coast of Namibia, is considered as one of the hotspot town of HIV epidemic in Namibia. This is necessitated by the fact that Erongo Region is home to Walvisbay which is a harbour town proving port facility for the import and export of cargo to the rest of Namibia, and neighbouring land-locked countries. According to the Surveillance Report of the 2016 National HIV Sentinel Survey (MoHSS, 2016c) the region had a HIV prevalence rate of 15.2% and that make it one of the highest affected by the HIV/AIDS epidemic in the country. In 2015, the region reported third highest TB cases in Namibia only behind Khomas and Ohangwena region. This makes it one of the hotspot of the HIV epidemic and other opportunistic diseases in Namibia. Erongo Region started providing TB and Antiretroviral Therapy (ART) treatment to its people in 2003, but there have never been and in-depth analysis of the trends in survival of people on TB and ART management, particularly in relation to localities within the region. The U.S President's Emergency Plan for AIDS Relief (PEPFAR) which has supported the region in the provision of general medical support has for the purpose of their project prepared reports that mostly focuses on adherence of patients on treatment. Further studies are therefore required to understand the factors associated with survival and clustering based on individual level locations and investigation of cases and mortality of TB and HIV co-infections.

Study Design

A retrospective cohort study was adopted for the study. The target population of the study included all the patients on co-therapy TB and ART management who attended the health facilities between 1st January 2003 and 31st December 2017. However, the study only used all the available TB and HIV co-infected patient records on the Electronic Patient Monitoring System (ePMS) database throughout all the facilities in Erongo Region.

The data were extracted from the ePMS database at MoHSS Erongo Regional Office into Microsoft Excel 2016. TB and HIV data are captured from a paper-based form, by the clinician conducting diagnosis during the visit of the patient which results into enrolment into the TB and HIV programme. The data forms are then entered into the administrative database in the health management information system section where the database is kept. Considering that subjects who started the treatment outside Erongo Region might have started the treatment under totally different conditions, data relating to patients who were initiated outside Erongo Region (the transfer in) were excluded from the analysis.

Variables

The outcome variable considered for the study was the survival outcome which was time-to-death measured in years from the time of co-infection to death or censored. Time-to-death was censored for those patients who were lost-to-follow, transferred to another health facility and did not died at 31st of December 2017 (at end of the study). The independent covariates considered for the modelling are listed in Table 14.1 below.

Table 14.1. List of covariates used in the study

Variable name	Values of the variable	Type
Age	Years (at baseline or enrollment)	Continuous
Sex	Female, male	Categorical
CD4	CD4 counts	Continuous
Still on care	Alive, dead, default, stopped or transferred out	Categorical
Weight	Weight (at baseline or enrollment)	Continuous
Status of patient	Alive, dead	Categorical
Marital status	Single, married, divorced and widowed	Categorical
Function status	Ambulatory, working or bedridden	Categorical
WHO clinical stage	Stage I, II, II, IV	Categorical
Facility level	Clinic, health centre and hospital	Categorical

Notice that WHO Clinical Stage which is classified into four; I, II, III and IV; where Stage I indicates asymptomatic disease, Stage II indicates mild disease, Stage III indicates advanced disease and Stage IV indicates severe disease. Hence disease severity increases from Stage I to Stage IV. Functional status of the patients is also categorical covariate with three categories: Working, ambulatory and bedridden. Working patients are those patients who can able to work day to day while ambulatory patients are those patients who can able to work some time but bedridden patients cannot able to work due to the infectious disease.

Measures of Spatial Association

For this section the study assume that the data has a spatial process with a mean, say $\mu(\mathbf{s}) = E(Y(\mathbf{s}))$, associated with it and that the variance of $Y(\mathbf{s})$ exists for all $\mathbf{s} \in D$. According to Banerjee, Carlin, & Gelfand (2014), the process $Y(\mathbf{s})$ is said to be Gaussian if, for any $n \geq 1$ and any set of sites $\{\mathbf{s}_1, \ldots, \mathbf{s}_n\}$, $\mathbf{Y} = (Y(\mathbf{s}_1), \ldots, Y(\mathbf{s}_n))^T$ has a multivariate normal distribution. The process is said to be strictly stationary (sometimes strong stationarity) if, for any given $n \geq 1$, any set of n sites $\{\mathbf{s}_1, \ldots, \mathbf{s}_n\}$ and any $h \in R$, the distribution of $(Y(\mathbf{s}_1), \ldots, Y(\mathbf{s}_n))$ is the same as that of $(Y(\mathbf{s}_1 + \mathbf{h}), \ldots, Y(\mathbf{s}_n + \mathbf{h}))$. Here D is envisioned as R^r as well.

Two standard statistics that are used to measure strength of spatial association among areal units are Moran's I and Geary's C (Banerjee, Carlin, & Gelfand, 2014). This study will make use of the Moran's I to measure for spatial association among the health district. These are spatial analogues of statistics for measuring association in time series, the lagged autocorrelation coefficient and the Durbin-Watson statistic, respectively. They can also be seen to be areal unit analogues of the empirical estimates for the correlation function and the variogram, respectively. For point-referenced data, the empirical covariance function (Banerjee, Carlin, & Gelfand, 2014):

$$\hat{C}(t_k) = \frac{1}{N_k} \sum_{(s_i, s_j) \in N(t_k)} (Y(s_t) - \bar{Y})(Y(s_j - \bar{Y}) \tag{1}$$

where $N(t_k) = \{(s_i, s_j) : \| s_i - s_j \| \in I_k\} fork = 1, \ldots, K, I_k$indexes the kth bin, and there are $N(t_k)$ pairs of points falling in this bin. Equation (2) is a spatial generalization of a lagged autocorrelation in time series analysis and semivariogram:

$$\bar{\gamma}(d) = \frac{1}{2N(d)} \sum_{(s_i, s_j) \in N(d)} [(Y(s_i) - Y(s_j -)]^2 \tag{2}$$

where $N(d)$ is the set of pairs of points such that $\|s_i - s_j\| = d$,and $|N(d)|$ is the number of pairs in this set. This two respectively, provide customary nonparametric estimates of these measures of association.

Moran's *I* takes the form:

$$I = \frac{n\sum_i \sum_j w_i (Y_i - \bar{Y})(Y_j - \bar{Y})}{(\sum_{i \neq j} w_{ij})(\sum_i (Y_i - \bar{Y})^2} \tag{3}$$

I is not strictly supported on the interval $[-1,\ 1]$. It is evidently a ratio of quadratic forms in **Y**, which provides the idea for obtaining approximate first and second moments through the delta method (Banerjee et al., 2014). Moran shows under the null model where the Y_i are i.i.d., *I* is asymptotically normally distributed with mean $-1/(n-1)$ and a rather unattractive variance of the form:

$$Var(I) = \frac{n^2(n-1)S_1 - n(n-1)S_2 - 2S_0^2}{(n+1)(n-1)^2 S_0^2} \tag{4}$$

In (4), $S_0 = \sum_{i \neq j} w_{ij}$, $S_1 = \frac{1}{2}\sum_{i \neq j}(w_{ij} + w_{ji})^2$ and $S_2 = \sum_k (\sum_j w_{kj} + \sum_i w_{kj})^2$.

This study recommends the use of Moran's *I* as an exploratory measure of spatial association, rather than as a "test of spatial significance".

Modelling Spatial Survival

In survival analysis the time to endpoint (*T*) is the random variable. Following Banerjee, Carlin, & Gelfand (2014), we assume that a sample of individuals have associated endpoint times. Denote this sample of times as $\{t_i\}, i = 1,2, \dots, m$. For now, assume these are all observed exactly. In addition to an endpoint time a geo-reference is also available. The geo-reference could be an address location, in which case it is denoted as s_i, or a contextual spatial effect, denoted as w_i. The contextual effect is simply a factor that may have spatial correlation so that areas close together have similar risk. In addition to these ingredients, each observation unit can have covariates associated and, for the *i*th unit/person these are denoted by the vector $\mathbf{x}_i$. These covariates could be individual and/or contextual/ecological.

To illustrate, let t_{ij} be the time to death or censoring for subject j ($j = 1, \dots, n_i$) in Enumeration Area (EA) ($i = 1, \dots, I$). Let $\mathbf{x}_{ij}$ be a vector of individual-specific covariates. The usual assumption of proportional hazards $h(t_{ij}; \mathbf{x}_{ij})$ enables models of the form:

$$h(t_{ij}; x_{ij}) = h_0(t_{ij}) exp(\beta^T x_{ij}) \tag{5}$$

where h_0 is the baseline hazard, which is affected only multiplicatively by the exponential term involving the covariates. In the frailty setting, model above is extended to:

$$h(t_{ij}; x_{ij}) = h_0(t_{ij})\omega_i exp(\beta^T x_{ij})$$
$$= h_0(t_{ij})\beta^T x_{ij} + W_i \tag{6}$$

where $W_i \equiv \log \omega_i$ is the stratum-specific frailty term, designed to capture differences among the strata. Typically, a simple i.i.d. specification for the W_i is assumed.

$$W_i \overset{iid}{} N(0, \sigma^2) \tag{7}$$

With the advent of Markov Chain Monte Carlo (MCMC) computational methods, the Bayesian approach to fitting hierarchical frailty models such as these has become increasingly popular (Banerjee et al., 2014). Perhaps the simplest approach is to assume a parametric form for the baseline hazard h_0. While a variety of choices (gamma, lognormal, etc.) have been explored, we will adopt the Weibull, which seems to represent a good tradeoff between simplicity and flexibility. This then produces,

$$h(t_{ij}|x_{ij}) = \rho t_{ij}^{\rho-1} exp(\beta^T x_{ij} + W_i) \tag{8}$$

Now, placing prior distributions on ρ, β, and σ^2 completes the Bayesian model specification. Banerjee, Carlin, & Gelfand (2014) proposed a relaxation of the Weibull model to allow a semi-parametric formulation whereby, with the patients $j, (j = 1, \ldots, n_i)$ in the *i* EAs,

$$h(t_{ij}|x_{ij}) = h_{0i}(t_{ij}) exp\{x'_{(ij)}\beta + w_i\} \tag{9}$$

where h_{0i} denotes the county-specific baseline hazard. For an individual with censoring indicator Υ_{ij} (0 if alive, and 1 if dead) the likelihood contribution is then:

$$h(t_{ij}|x_{ij})^{\gamma_{ij}} exp\left\{-H_{0i}(t_{ij}) exp\{x'_{(ij)}\beta + w_i\}\right\} \tag{10}$$

where $-H_{0i}(t_{ij}) = \int_o^{t_{ij}} h_{0i}(u)\, du$ is a county-specific cumulative baseline hazard, and the covariates are assumed to be not time-dependent. The baseline hazard appears in this likelihood and so must be estimated. Different approaches have been proposed for estimation of this baseline. One approach assumes a gamma process which is a function of a parametric cumulative hazard (Collett, 2015). Another is the use of beta mixtures (Banerjee et al. 2014). A related spatial model was developed by Bastos and Gamerman (2006) whereby they allowed time-dependent covariates which are fixed within small time periods and a CAR spatial frailty. They assume no separate baseline risk however.

The Cox proportional hazards model has been applied in a spatial context by Jerrett, et al. (2013). In the context of population and mortality survival, the authors posited that the partial likelihood could be used without recourse to the estimation of the baseline. For ordered uncensored times $\{t_{(1),\dots\dots\dots,}t_{(m)}\}$, then the partial likelihood is given by:

$$\prod_i [\exp\{\mathbf{x}'_{(i)}\beta\} / \sum_{j\in R_i} \exp\{\mathbf{x}'_{(j)}\beta\}] \qquad (11)$$

where R_i is the set of those individuals at risk just before the ith event time. This can be used for the estimation of regression parameters. This is less parametric than models that include baseline components. Note that spatial effects can be included again as contextual effects by extending the specification of the intensity term $exp\{x'_{(j)}\beta\}$ to include the random effects:

$$exp\{x'_{(j)}\beta\} + w_i + v_i\} \qquad (12)$$

where w_i and v_i are random contextual effects which could be at an aggregate level such as EAs, health district and constituency. In addition, these effects could also be purely individual (in the sense of frailty rather than context).

Modelling Spatial Temporal Survival Data

In this section we use semiparametric (Cox) hierarchical Bayesian frailty model in Banerjee et al. (2014) to capture for spatiotemporal heterogeneity in survival data. We then use these models to describe the pattern of death in Erongo Region while accounting for important covariates, spatially correlated differences in the hazards among the Enumeration Areas, and possible space-time interactions.

We apply the framework in the preceding section to incorporate temporal dependence. Here we have t_{ijk} as the response (time to death) for the jth subject residing in the ith district who died in the kth year, while the individual-specific vector of covariates is now denoted by $\mathbf{x}_{ijk}$, for $i = 1, 2, \dots, I$, $k = 1, \dots, K$, and $j = 1, 2, \dots, n_{ik}$. We note that "time" is now being used in two ways. The measurement or response is a survival time, but these responses are themselves observed at different areal units and different times (years). Furthermore, the spatial random effects W_i in the preceding section are now modified to W_{ik}, to represent spatiotemporal frailties corresponding to the ith district for the kth diagnosis year. Our spatial frailty specification in (5) now becomes:

$$h(t_{ijk}; x_{ijk}) = h_{0i}(t_{ijk}) exp(\beta^T x_{ijk} + W_{ik}) \qquad (13)$$

Our CAR prior would now have conditional representation

$$[\, W_{ik} \,|\, W_{(i' \neq i)k}] \sim N\left(\overline{W}_{ik}, 1/(\lambda_k m_i)\right).$$

Conditional autoregressive (CAR) are network-based models (also known as graphical models) specifically designed to model spatially autocorrelated data based on neighborhood relationships (Ver Hoef, Peterson, Hooten, Hanks, & Fortin, 2018). Ver Hoef et al. (2018) further explain geostatistics as a set of models and tools developed for statistical analysis of continuous data. These data can be measured at any location in space, but they are available in a limited number of sampled points. Applying geostatistical model would cause a challenge as most of the data are count data, hence the CAR model are preferred over geostatistical model.

Assuming that the λ_k are identically distributed from a common hyperprior, we can account for temporal correlation in the frailties. Since this is an appropriate MCMC application, we select a gamma prior. A flat prior for β is typically chosen, this still admits a proper posterior distribution. Then we update $\mathbf{W}_k$ and β by using the Adaptive rejection Metropolis-Hastings sampling in a hybrid sampler. We remark that it would certainly be possible to include both spatial and nonspatial frailties. This would mean supplementing our spatial frailties W_{ik} with a collection of nonspatial frailties, say, $V_{ik} \overset{iid}{} N(0, 1/\tau_k)$. We summarize our full hierarchical model as follows:

$$L(\beta, W; t, x, \gamma) \propto \prod_{k=1}^{K} \prod_{i=1}^{I} \prod_{j=1}^{n_{ik}} \{h_{0i}(t_{ijk}; x_{ijk})\}^{\gamma_{ijk}}$$

$$\times\, exp\{(-H_{0i}) \exp(\beta^T x_{ijk} + W_{ik} + V_{ik})\}, \quad (14)$$

where $p(W_k|\lambda_k) \sim CAR(\lambda_k)$

$p(V_k|\tau_k) \sim N_I(0, \tau_k I)$

$\lambda_k \sim Gamma(a, b),$

$\tau_k \sim Gamma(c, d)$

for $k = 1,2, \ldots, K.$

In the sequel we adopt the beta mixture approach to model the baseline hazard functions $H_{0i}(t_{ijk})$ nonparametrically (Banerjee et al., 2014).

RESULTS

Spatial Autocorrelation

Spatial autocorrelation plays an important role in geographical analysis, Global Moran's *I* is the most common measures of the spatial autocorrelation which determine the overall strength of spatial dependence (Ntirampeba, Kazembe, & Neema, 2017). This statistic is a useful measure of the overall clustering. The global Moran's *I* statistic was 0.173 (p-value < 0.001) with a variance of 0.0008. Thus, the positive significant Moran's *I* value indicates that values in neighbouring Enumeration Areas tend to cluster. To detect local spatial patterns, we have used local Moran's *I*. It enhances to identify clusters, which are observations with very similar neighbours and hotspots, which are characterised by observations with very different neighbours. Figure 14.1 shows the map of local Moran's *I* statistics. From this figure, it can be noted that EAs around Karibib and Usakos districts have elevated positive Moran's *I* values. These results have inspired the spatial analysis undertaken in the subsequent sections in the study.

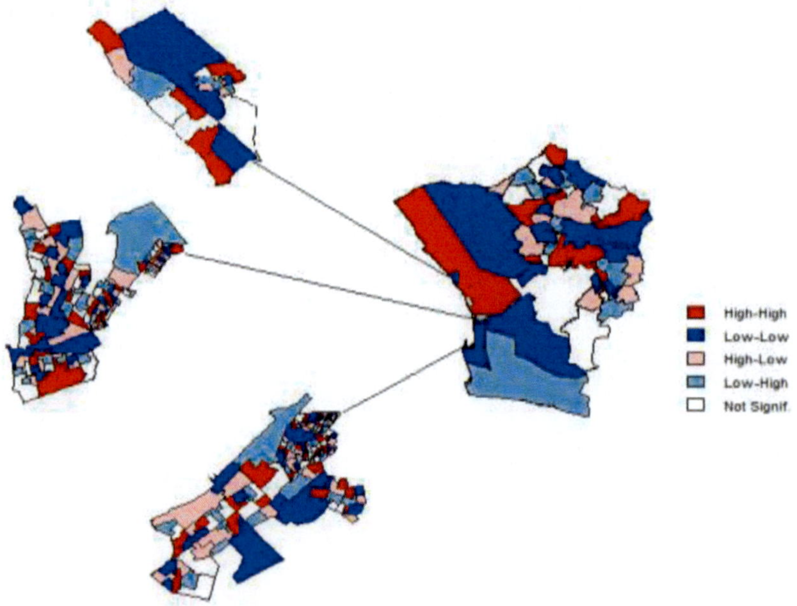

Figure 14.1. LISA Clustered map for TB and HIV death cases.

Frailty and Spatial Effects

Models referred to in this section are those under spatial survival modelling under section above. The study implemented a simple nonhierarchical ("no-frailty") model, spatial ("frailty") model (5) and random effect model with spatially-correlated term (w_i) added (6), and finally the last model is a convolution model with the uncorrelated random effect (v_i) and w_i added to the basic model (12). Table 14.2 below compares our four models in terms of the criteria discussed in sections below.

The models fitted were as follows: The first line 1 displays the overall goodness-of-fit results for the basic model with no spatial random effects. Under this model, the DIC was found to be 3394.61, with effective number of parameter pD = 15.19. The results of fitting these different models are found on Table 14.2. It is interesting to note that all models that included spatially referenced random effects (models 1, 2, 3) have a considerably lower DIC than the basic model. The best fitting model by the DIC criterion is model 2 with the uncorrelated random effect that has a DIC = 3269.11 and effective number of parameter pD = 43.25. Although the convolution model yields a slightly higher DIC, it has lower DIC and pD than a model with only the correlated spatial effect.

Table 14.2. Comparison of four spatial models for seizure data: basic model and models 1 through 3

Model	$\overline{D}$	pD	DIC
Basic model (Null)	3294.61	15.19	3324.99
Spatial (1)	3152.38	71.80	3295.98
Random (2)	3184.60	43.25	3269.11
Spatial + Random (3)	3169.37	50.20	3269.76

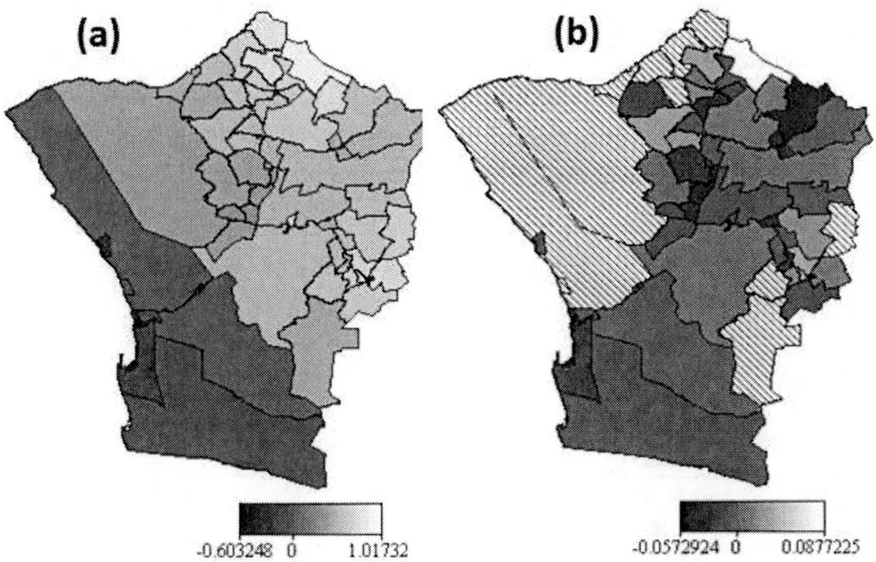

Figure 14.2. (a) Posterior mean of the structured spatial effect for each area and (b) posterior mean of the unstructured spatial effect. Areas with no or insufficient data are marked with diagonal solid lines.

It is reassuring that the uncorrelated noise model has best fit as we did not present any overt spatial structure in the data. However, the fact that any spatially-referenced model (whether correlated or not) leads to much reduced DIC is an important consideration. The reason for the closeness of models 1 and 2 is largely because the spatially-correlated term (w_i) will often mimic small scale variation found. A further advantage of the graphs is, that they allow to visualise the estimation results for the uncorrelated spatial effects. Figure 14.2 displays the results of the structured (a) and unstructured posterior mean spatial effect (b). The map further illustrated that the structured model and unstructured have posterior mean ranging from -0.60 to 1.02 and -0.06 to 0.09 respectively. Enumeration Areas with no or insufficient data are marked with diagonal solid lines.

Spatial-Temporal Patterns of Mortality of TB and HIV

Models referred to in this section are those under spatial temporal survival modelling under section above. The first three models are simple separable models with no spatial-temporal interaction: a model with a random spatial and temporal term and an autoregressive temporal effect (model 1); a model with random spatial terms only (model 2); and a model with spatial and uncorrelated random effect (model 3). The last three have a temporal interaction term. Model 4 has an interaction of temporal and spatial effect, and the last model involves temporal, spatial and also added an uncorrelated random effect. Table 14.3 displays the DIC results for models 1 through 6. It is clear that models 1 and 2 while parsimonious are far from the best models. It can be also observed clearly that model 6 that has an interaction of temporal and spatial component, yielding the lowest DIC of 3271.51 and effective number of parameter *pD* of 53.76. Of course these results depend on prior specifications and in any particular applications, sensitivity to prior specification should be examined (Lawson, 2018).

Table 14.3. Space-time models for TB and HIV related data set

Model	$\overline{D}$	pD	DIC
Model 1	3673.59	45.71	3765.02
Model 2	3182.61	44.67	3271.94
Model 3	3673.59	45.71	3765.02
Model 4	3671.60	46.69	3764.98
Model 5	3182.61	44.67	3271.94
Model 6	3163.98	53.76	3271.50

Table 14.4. Posterior summary statistics: fixed effects for model 6

Covariates	Posterior Mean	Std. Dev.	2.5%	50%	97.5%
Intercept	-4.52	0.91	-6.35	-4.52	-2.85
Sex: male	0.10	0.11	-0.11	0.10	0.32
Age category	0.42	0.11	0.21	0.42	0.62
CD4 category	-0.47	0.13	-0.74	-0.47	-0.23
Marital status	0.25	0.14	-0.01	0.25	0.53
Facility level	0.33	0.13	0.08	0.33	0.60
District	0.05	0.11	-0.18	0.05	0.26
WHO stage: stage 4	0.15	0.05	0.06	0.15	0.25
Function status: working	-0.74	0.07	-0.87	-0.74	-0.61

For the model with lowest DIC, various posterior summaries are available. Table 14.4 displays the summary of posterior mean, standard deviation and the three posterior quantiles of q2.5%, q50% and q97.5% for the main effect of the model. All of the predictors were considered significant at 0.05 level. Higher age at diagnosis also increases the hazard (Coeff: 0.42, 95% CI: 0.21, 0.62). Mortality was also associated with CD4 counts. Higher CD4 counts was likely to be associated with low mortality (Coeff: -0.47, 95% CI: -0.74, -0.23). We aso observed that mortality was associated with WHO clinical stages, with stage 4 related to lower risk of mortality (Coeff: 0.15, 95% CI: 0.06,0.25). With regards to functional status, we observed that working patients were associated with reduced mortality compared to those bedridden (Coeff: -0.74, 95% CI: -0.87,-0.61).

Figure 14.3 present the the posterior mean of the structured spatial temporal effect for the lowest DIC model fitted for the data (model 6). The maps used colours ranged from light grey to dark grey with the extreme negative random effects corresponding to extreme dark grey and the extreme positive random effects corresponding to extreme light grey. Three different numbers were used to distinguish significant observed random effects. dark grey denoted by (-1) indicated significant negative random effects, (0) indicated non-significant random effects, and light grey denoted by (1) represented significant positive random effects.

The posterior mean were the highest in 2013, ranging from -1.6133 to 2.8277, while the lowest was in year 2011 with mean values between -1303 and 0.0716, followed by the year 2008 ranging between -0.385002 and 0.409772. Overall, the risk surfaced changed little between 2007 and 2017. Generally, the population at highest risk was in the northern-eastern part of the country (upper right corner of the maps), while low risk was registered in the western part of the country (left corner of the maps).

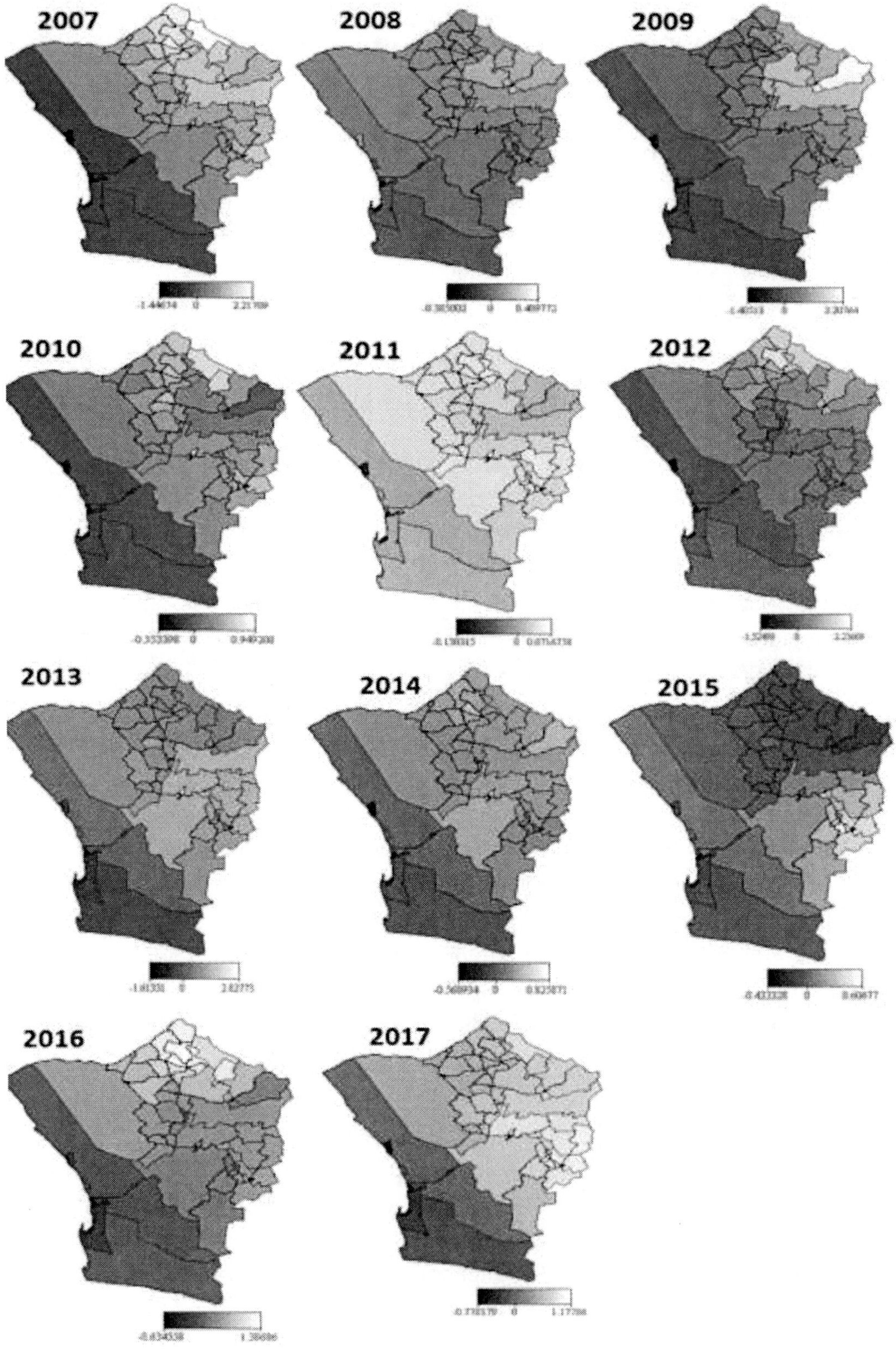

Figure 14.3. Posterior means of the temporal variability in risk between 2007 and 2017.

DISCUSSIONS

The study shows that areas around Swakopmund, Walvisbay, Usakos, Karibib and Otjimbingue are a hotspot, using LISA, and this has been shown on Figure 1. The positive significant Moran's I value indicates that values in neighbouring Enumeration Areas tend to cluster and neighbouring EAs have the same characteristics. Neverthless, these have to be interpreted with caution, county-level direct estimates of less common mortality outcomes (like TB) are often highly unstable. Many prior studies on county-level variations in less common causes of mortality outcomes have relied on estimates aggregated over time or larger geographic areas (Khana, Lauren, Holly, & Margaret, 2018). However, this type of aggregation precludes the examination of detailed temporal and spatial trends. To overcome these limitations, this study used hierarchical Bayesian methods to generate robust model-based estimates of regional-level to examine the spatial and temporal variations across a span of 14 years.

We found spatial and spatial-temporal clusters and variations in the distribution of mortality of TB and HIV in the western and east districts of the study area. These clusters were stable over the years with the exception of a few location differences. This could be explained by a high transmission over many years due to the existence of disproportionate high-risk factors, and a varying program performance.

In our finding, most cluster locations were identified in urban areas, semi-urban areas with a high population density, as well as neighbouring areas close to towns and areas near road networks, which connect major towns. Various studies have reported that poor socioeconomic conditions such as social inequality, low income, poverty, poor housing conditions, overcrowding and social unrest could all be risk factors for the high burden and variations of disease occurrence (Corbett, et al., 2009; Couceiro, Santana, & Nunes, 2011; Gomez-Barroso, Rodriguez-Valin, Ramis, & Cano, 2013; Maciel, et al., 2010; Oren, Koepsell, Leroux, & Mayer, 2012; Wong, Yadav, Nishikiori, & Eang, 2013). In addition, patient care factors (Randremanana, Richard, Rakotomanana, Sabatier, & Bicout, 2010) and poor access to health care and TB and HIV control services could also contribute to a high rate of the disease (Ng, Wen, Wang, & Fang, 2012), since infectious cases may remain undiagnosed and may not acquire treatment, which could consequently contribute to the transmission dynamics of the disease.

Furthermore, most urban towns where the clusters identified were the capitals of the districts, had market places, had public transportation routes and were the hub for different socio-economic activities. The better access to road and movement using public transportation in a crowded and poorly ventilated environment may assist in facilitating contact with infectious cases, which could favour a transmission of the disease in the locations where clusters were detected (Feske, Teeter, Musser, & Graviss, 2011; Edelson & Phypers, 2011) . Therefore, the inclusion of these factors in cluster analyses in the future may help to improve the understanding of their effect on the clusters of the diseases in the study area.

Improving TB and HIV control efforts could help reduce the transmission and change the geographic distribution of TB. In our study, despite different intervention programs aimed at reducing disease transmission and improving case detection over many years, the unusually high rates of the diseases persisted in the same places, with the most likely spatial clusters showing a stable pattern in the preceding years during the study. This could explain, at least in part, that the interventions may not be properly focused on influencing the disease epidemiology or could be due to a low case finding and poor treatment of infectious cases, which may indicate the continued transmission of the two disease.

REFERENCES

Banerjee, S., Bradley, C. P., & Gelfand, A. E. (2014). *Hierarchical Modeling and Analysis for Spatial Data.* Boca Raton: CRC Press.

Bastos, L. S., & Gamerman, D. (2006). Dynamic survival models with spatial frailty. *Lifetime data analysis, 12(4)*, 441-460.

CDC. (2017, May 30). *Centers for Disease Control and Prevention*. Retrieved from https://www.cdc.gov: https://www.cdc.gov/hiv/basics/livingwithhiv/opportunisticinfections.html.

Collett, D. (2015). *Modelling survival data in medical research.* New York: Chapman and Hall/CRC.

Corbett, E. L., Bandason, T., Cheung, Y. B., Makamure, B., Dauya, E., & Munyati, S. S. (2009). Prevalent infectious tuberculosis tuberculosis in Harare, Zimbabwe: burden, risk factors and implications for control. *The International Journal of Tuberculosis and Lung Disease: The Official Journal of the International Union against Tuberculosis and Lung Disease, 13(10)*, 1231–1237.

Couceiro, L., Santana, P., & Nunes, C. (2011). Pulmonary tuberculosis and risk factors in Portugal: a spatial analysis. *The International Journal of Tuberculosis and Lung Disease: The Official Journal of the International Union Against Tuberculosis and Lung Disease, 15(11)*, 1445–1454.

Edelson, P. J., & Phypers, M. (2011). TB transmission on public transportation: a review of published studies and recommendations for contact tracing. *Travel medicine and infectious disease*, 9(1): 27–31.

Feske, M. L., Teeter, L. D., Musser, J. M., & Graviss, E. A. (2011). Giving TB wheels: Public transportation as a risk factor for tuberculosis transmission, *Tuberculosis (Edinburgh, Scotland), 91(1)*, 16–23.

Getahun, H., Gunneberg, C., Granich, R., & Nunn, P. (2010). HIV infection—associated tuberculosis: the epidemiology and the response. *Clinical Infectious Diseases, 50(Supplement_3)*, S201-S207.

Gomez-Barroso, D., Rodriguez-Valin, E., Ramis, R., & Cano, R. (2013). Spatio-temporal analysis of tuberculosis in Spain, 2008–2010. *The International Journal of Tuberculosis and Lung Disease: The Official Journal of The International Union against Tuberculosis and Lung Disease, 17(6)*, 745–751.

Grover, G., Swain, P. K., & Ravi, V. (2014). A Competing Risk Approach with Censoring to Estimate the Probability of Death of HIV/AIDS Patients on Antiretroviral Therapy in the Presence of Covariates. *Statistics Research Letters (SRL), 3*, 7-16.

Khana, D., Lauren, R. M., Holly, H., & Margaret, W. (2018). A Bayesian Spatial and Temporal Modeling Approach To Mapping Geographic Variation in Mortality Rates for Subnation Areas With R-INLA. *HHS Public Access, 16(1)*, 147–182.

Lawson, A. B. (2013). *Bayesian disease mapping: hierarchical modeling in spatial epidemiology.* New York: CRC Press.

Lawson, A. B. (2018). *Bayesian Disease: Hierarchical Modeling in Spatial Epidemiology.* Boca Raton: CRC Press.

Maciel, E. L., Pan, W., Dietze, R., Peres, R. L., Vinhas, S. A., & Ribeiro, F. K. (2010). Spatial patterns of pulmonary tuberculosis incidence and their relationship to socioeconomic status in Vitoria, Brazil. *The International Journal of Tuberculosis and Lung Disease: The Official Journal of the International Union against Tuberculosis and Lung Disease, 14(11)*, 1395–1402.

Mee, P., Collinson, M. A., Madhavan, S., Root, E. D., Tollman, S. M., Byass, P., & Kahn, K. (2014). Evidence for localised HIV related micro–epidemics associated with the decentralised provision of antiretroviral treatment in rural South Africa: a spatio– temporal analysis of changing mortality patterns (2007–2010). *Journal of Global Health, 4(1)*, 010403.

Menges, W. (2018, Feruary 12). *The Namibian Newspaper.* Retrieved from The Namibian Newspaper Website: https://www.namibian.com.na/174350 /archive-read/GlobalFund-cuts-deepen-Namibias-woes.

Miller, H. J. (2004). Tobler's First Law and Spatial Analysis. *Annals of the Association of American Geographers*, 284-289.

MoHSS. (2017). *National Staratergic Fframework for HIV and AIDS Response in Namibia.* Windhoek: MoHSS.

MoHSS. (2016a). *National Tuberculosis and Leprosy Programme.* Windhoek: MoHSS.

MoHSS. (2017b). *Third Medium Term Strategic Plan for Tuberculosis and Leprosy 2017/18 – 2021/22.* Windhoek: Ministry of Health and Social Services.

MoHSS. (2016c). *Surveillance Report of the 2016 National HIV Sentinel Survey.* Windhoek: MoHSS.

Ng, I. C., Wen, T. H., Wang, J. Y., & Fang, C. T. (2012). Spatial dependency of tuberculosis incidence in Taiwan. *PloS One, 7(11)*, e50740. doi:10.1371/journal.pone.0050740.

Ntirampeba, D., Kazembe, L. N., & Neema, I. (2017). Modelling spatial patterns of misaligned disease data: An application on measles incidence in Namibia. *Clinical Epidemiology and Global Health, 5(4)*, 190-195.

Oren, E., Koepsell, T., Leroux, B. G., & Mayer, J. (2012). Area-based socio-economic disadvantage and tuberculosis incidence. *The International Journal of Tuberculosis and Lung Disease: The Official Journal of The International Union against Tuberculosis and Lung Disease, 16(7)*, 880–885.

Randremanana, R. V., Richard, V., Rakotomanana, F., Sabatier, P., & Bicout, D. J. (2010). Bayesian mapping of pulmonary tuberculosis in Antananarivo, Madagascar. *BMC infectious diseases*, 10-21.

UNAIDS. (2017, November 3). *UNAIDS.* Retrieved from http://www.unaids.org: http://www.unaids.org/en/topic/tuberculosis

Uthman, O. A., Yahaya, I., Ashfaq, K., & Uthman, M. B. (2009). A trend analysis and sub-regional distribution in number of people living with HIV and dying with TB in Africa, 1991 to 2006. *International Journal of Health Geographics, 8(1)*, 65.

Ver Hoef, J. M., Peterson, E. E., Hooten, M. B., Hanks, E. M., & Fortin, M. J. (2018). Spatial autoregressive models for statistical inference from ecological data. *Ecological Monographs, 88(1)*, 36-59.

Wei, W., Wei-Sheng, Z., Ahan, A., Ci, Y., Wei-Wen, Z., & Ming-Qin, C. (2016). The characteristics of TB epidemic and TB/HIV co-infection epidemic: a 2007–2013 retrospective study in Urumqi, Xinjiang Province, China. *PloS One, 11(10)*, e0164947.

WHO. (2018, January 30). *World Health Organization*. Retrieved from World Health Organization: http://www.who.int/mediacentre/factsheets/fs104/en/

WHO. (2016a). *Consolidated guidelines on the use of antiretroviral drugs for treating and preventing HIV infection: recommendations for a public health approach.* Geneva: World Health Organization.

WHO. (2018b, January 30). *World Health Organization*. Retrieved from World Health Organization Web site: http://www.who.int/mediacentre/factsheets /fs104/en/

Wong, M. K., Yadav, R. P., Nishikiori, N., & Eang, M. T. (2013). The association between household poverty rates and tuberculosis case notification rates in Cambodia, 2010, 4(1). *Western Pacific surveillance and response journal: WPSAR*, 25–33.

Zulu, L. C., Kalipeni, E., & Johannes, E. (2014). Analyzing spatial clustering and the spatiotemporal nature and trends of HIV/AIDS prevalence using GIS: the case of Malawi. *BMC infectious diseases, 14(1)*, 285.

In: Statistical Modelling …
Editors: N-B. Kandala and L. Kazembe

ISBN: 978-1-53615-981-3

Chapter 15

ATTRITION OF WOMEN INITIATING ANTIRETROVIRAL THERAPY (ART) UNDER OPTION B+: COX PROPORTIONAL HAZARDS, COMPETING RISKS AND MULTISTATE SURVIVAL MODELS

Andrew Mganga and Lawrence N. Kazembe*
Mathematical Sciencies Department, University of Malawi, Zomba, Malawi
Department of Statistics and Population Studies, University of Namibia, Windhoek, Namibia

ABSTRACT

Prevention of Mother to Child Transmission (PMTCT) remains critical to decreasing the number of exposed children infected with HIV. Women need to remain on antiretroviral therapy (ART) in order for PMTCT to be effective in preventing HIV transmission. In 2011, the Malawi programme changed the clinical guidelines for the provision of PMTCT to Option B+. However, Option B+ intervention was implemented without any trials to assess effectiveness. To-date, few studies have evaluated outcomes of women on Option B+. In order to inform effective and efficient implementation of the interventions, this study explored facility level and individual level factors that affect attrition among women on Option B+ in Malawi. Secondary data from PMTCT Retention of Option B+ Evaluation (PROBE) study, collected by Management Sciences for Health (MSH), was used. Three approaches were used to analyse attrition from ART: Cox proportional hazards model, competing risks model and multi-state models. The findings suggest that management authority, facility level, status at ART initiation, age group, ART education and clinical condition were significantly associated with attrition from ART. Nevertheless, the association of the predictors and the outcome varied when

* Corresponding Author's Email: lkazembe@yahoo.com.

stratified by ART registration, ART treatment follow-up, and follow-up outcomes. In conclusion, ART attrition was high, and interventions should be targeted at improving clinical education, those on those who start late, or those at an early age.

Keywords: PMTCT, ART attrition, Option B+, Cox proportional hazard model, competing risk, multi-state models, Malawi

INTRODUCTION

Prevention of Mother-to-Child Transmission (PMTCT) remains critical to decreasing the number of exposed children infected with HIV. PMTCT interventions considerably reduce vertical transmission of HIV through the provision of Antiretroviral Therapy (ART) to infected mothers and Antiretroviral (ARV) prophylaxis to infants born to infected mothers (Schouten et al, 2011, 2013). The World Health Organisation (WHO) noted that in the absence of any intervention a substantial proportion of children born to women living with HIV acquired the virus during pregnancy, labour, delivery and breast-feeding. The risk of transmission without intervention is 15-30% in non-breastfeeding populations and 20-45% in breastfeeding populations. However, the transmission rate can be reduced to below 5% with effective interventions during the periods of pregnancy, labour, delivery and breastfeeding (WHO, 2010, 2013). The WHO currently recommends three PMTCT approaches (WHO, 2010, 2013):

- Option A: recommends maternal AZT for mothers with low CD4 count and ARV prophylaxis for infants.
- Option B: recommends maternal triple ARV prophylaxis for low CD4 count women through breastfeeding.
- Option B plus (Option B+): recommends triple ARVs starting as soon as a mother is diagnosed regardless of CD4 count, and continues this treatment for life.

In 2011, the Malawi programme changed the clinical guidelines for the provision of PMTCT to Option B+. Specifically, all women attending antenatal clinic are offered HIV Testing and Counselling (HTC) services in Option B+. All consenting HIV positive women are then put on lifelong ART. Furthermore, initiation on ART treatment continues at maternity and during postnatal period (for those are classified as having initiated treatment during breastfeeding). For infants born to infected mothers, Neverapine (NVP) prophylaxis/preventive is provided to all babies born to Human Immunodeficiency Virus (HIV) infected mothers. This is aimed at slowing disease progression, increase survival and reduce HIV transmission to the baby.

The main outcomes of interest are retention of mother on ART, retention in exposed children follow-up and HIV free survival of the new born baby at 24 months of age or more precisely a confirmed HIV negative result at least 6 weeks after weaning. Children born to infected mothers are tested for HIV at 6 weeks, 12 months and 24 months (Ministry of Health, 2011).

There are several possible challenges that may affect the realisation of Option B+ objectives. Ideally, women are supposed to remain in care for life i.e., retention-in-care. Retention-in-care is commonly used as a proxy for treatment adherence and viral suppression. It is assumed that patients who are retained-in-care receive appropriate care at relevant time points, whereas those Lost-to-Follow-Up (LTFU) have stopped taking their treatment (Rollins et al., 2014). Furthermore, women who are retained-in-care are assumed to be virally suppressed and therefore at a low risk of transmitting HIV. PMTCT interventions are therefore considered effective if women are virally suppressed, and at low risk of transmission. Some women, however, default from treatment, transfer to another health facility, die, and/or stop treatment. These are considered to be adverse outcomes.

In Malawi, women have been initiating ART under Option B+ since July 2011. The fourth quarter 2014 Ministry of Health (MoH) integrated HIV programme report indicated that retention on Option B+ was 78%, 72% and 68% retained on ART at 6, 12 and 24 months after initiation, respectively (Ministry of Health, 2014). These retention rates were among women who had not transferred out as of end June 2014. However, limited evaluation has been done to determine the outcomes of women initiated under Option B+. Furthermore, Option B+ was implemented without any trials to assess effectiveness as a PMTCT intervention (Corlten et al, 2014). To-date, few studies have evaluated outcomes of women on Option B+. Therefore, knowledge on effectiveness of Option B+ is limited. There is a need to study the outcomes and characteristics of women who initiated ART on Option B+. In a viewpoint, Schouten et al., agreed that there is indeed a strong case for expanding research on Option B+, but not for impeding countries that implement Option B+ on the basis of available evidence and programmatic experience (Schouten et al., 2013). Outcomes of the intervention strategy are therefore not comprehensively understood. Subsequently, effectiveness of the intervention is not fully understood.

Retention of Option B+ women data is an example of censored survival data that is routinely collected. Censored survival data is common in health, therefore survival analysis methods are used in the analysis of such data. A widely used survival analysis method is Cox proportion hazards model. Competing risks and multi-state survival models are extensions of the basic survival models to contexts with multiple events of interest and across a continuum. However, there are other approaches to modelling survival data that have been put forward in literature, but the above three remain popular.

This study applied the three approaches, i.e., Cox proportional hazard, competing risk and multi-state models to analyse attrition on ART in Malawi.

Study Methodology

PROBE Study: Study Description, Sampling and Sample Size

This study used secondary data from PMTCT Retention of Option B+ Evaluation (PROBE) study collected by Management Sciences for Health (MSH), in Malawi, in 2013. The goal of the PROBE study was to evaluate the extent to which HIV-infected pregnant women initiating ART under Option B+ were retained-in-care, both during pregnancy and beyond. The results of the study were to inform additional analysis and project design for future research. The results would further assist in examining adherence to other care and support strategies among HIV positive pregnant women, including tuberculosis prophylaxis, malaria prevention, and sexually transmitted infection (STI) diagnosis and treatment. Data for PROBE were consolidated from routinely collected information from clinic records. The outcome measure obtained from selected clinic registries and other sources was documentation of repeat clinic visits for medication renewals during pregnancy and following delivery. Data collected included information on: initial presentation at antenatal clinic, date of HIV diagnosis, date of ART initiation under Option B+, other clinical prophylaxis and tracking additional presentation at clinics to refill prescriptions for ART according to the prescribed ART regimen.

In brief, PROBE study focused on analysing attrition of mothers from ART care and treatment in relation to gestational age/age of infant. A stratified multi-stage sampling was used. To ensure that the sample was representative, the country was stratified into 4 geographical areas: (1) rural health facilities from north & central; (2) rural districts of the southern region; (3) Blantyre urban; and (4) Lilongwe urban. The geographical areas were divided based on the epidemiological characteristics of the area (HIV prevalence and rural & urban facilities).

In the second stage of sampling, facilities were sampled from each of these strata through systematic random sampling. The sample frame was limited to facilities that had more than 100 exposed infants ever enrolled. These were 76% of the 600 total ART facilities in the country. To achieve a total sample of 30 facilities, every 15th facility was selected from the sample frame. Finally, the study subjects were selected through systematic random sampling. The final sample obtained was 3,000, using the following approach. In facilities with less than 150 pregnant or breastfeeding women started on ART, data was collected from all potential subjects. In facilities with 151 to 300 eligible patients, data was collected from every second potential subject. In clinics with 301 to 450 or more eligible patients, data was collected from every third potential subject.

The inclusion criteria of the study was all women who were initiated on ART while pregnant or breastfeeding in 2012. The study used the following Ministry of Health data collection tools, i.e., health facility source documents, to assemble data: ART patient card/ARV Formulations card for mother; ANC register; ART register; and HIV Care Patient Card for exposed child under 24 months.

Study Variables

Dependent Variable

Attrition from ART treatment was the outcome of interest. In this study, attrition was defined as time (in months) spent on ART, with exits as either a default or a transfer out. Deaths and stops were considered as censored events. Cornell et al. (2014) recommended separately studying events rates among transfer outs, defaulters and patients retained. Studies have shown that the cascade of care differed among patients that remain and transfer (Krentz et al., 2015), thus there was need to study Transfer Out (TO) separately in measuring performance within the cascade of care. Performance analysis of cascade of care without accounting for TO could therefore lead to imprecise information for public health initiatives and program metrics. Joseph et al., (2008), in a study in Mzuzu, Malawi, found that TO had similar treatment outcomes to patients that remain in one facility. The assumption that TO have similar outcomes therefore requires further verification through studies. Nevertheless, the above studies emphases the importance of studying TO separately in attrition studies.

Independent Variables

These were split into individual and health facility variables. At individual level we considered (i) age of the woman at ART registration, with those at 24 years or younger considered adolescents, while women registering at 25 years or above were considered adults; (ii) whether a woman was pregnant or lactating at ART registration; (iii) whether she received patient education at ART registration or not; and (iv) clinical condition at the time of enrollment. Facility characteristics included as covariates in this study were: (i) managing authority (public or CHAM- Christian Hospitals Association of Malawi); (ii) level of service provision (health centre, hospital and referral hospital) and, (iii) location of the facility (urban, semi-urban and rural). All the independent variables in this study were modelled as categorical.

Model Description

Three types of models were considered to analyze the outcome- time (in months) spent on the ART programme before defaulting, transfer out, death or stopping treatment. We outline the three approaches below.

Cox Proportional Hazards Model

In studying attrition, we assume T as time to event or attrition time with t as the actual realization. The probability that an attrition time T is less than or equal to some value t is given by

$$F(t) = \int_0^t f(u)du = \Pr(T \leq t).$$

In our context, $F(t)$ is the cumulative probability that a woman attrite at or before some given time t, thus $F(t) = P$[woman attrition at time $\leq t$]. The instantaneous probability that an event will occur at any given small interval is defined as $F'(t) = f(t)$. The proportion of women surviving to time t or beyond is derived as $S(t) = 1-F(t)$, which is also known as the survivor function. An important approach is to consider the duration analysis through the hazard rate. The hazard rate, which links the the survival and failure functions, is of the form $h(t)=\frac{f(t)}{S(t)}$ or equivalent to $h(t) = -logS'(t)$. The hazard rate, unlike the survivor function, describes the risk or event of "failure" given that the individual has survived (i.e., retained in care) all along up to point t.

In studying attrition, our interest is to answer this question: Given that the woman has survived up (i.e., retained in care) to month t, what is the likelihood she will survive (i.e., will be retained in care) in the subsequent months? Further to this, one is interested in how the hazard rate varies with respect to some covariates, e.g., age, facility characteristics, etc.

A standard approach is to analyze the duration time as a regression using the Cox proportional hazard function. This is defined by:

$$h_i(t|x) = h_0(t)e^{\beta x_i}$$

where $i = 1,2,\dots\dots\dots,n$ are individuals in the study, such that $h_0(t)$ is the baseline hazard function. The baseline hazard characterises how the hazard function changes as a function of time, while $e^{\beta x_i}$ characterises hazard function of subject covariates (Hosmer and Lemeshow, 1999). The hazard function is equal to $h_0(t)$ when all the variables included in the model are zero.

The ratio of hazard functions of two subjects with dichotomous covariate values denoted as x_1 and x_0 is given by:

$$HR\ (t, x_1, x_0) = e^{\beta(x_1 - x_0)}$$

where HR is the hazard ration in the Cox proportional hazards model. When $x_1 = 1$ and $x_0 = 0$, the Cox model is defined by the following:

$$HR\ (t, x_1, x_0) = e^{\beta}\ .$$

Competing Risks Survival Model

Competing risk often occurs when the main interest in estimating the time until a first event [i.e., defaulting] can compete with other type of the events[e.g., transfer out, death and complete stopping]. Subjects are at risk of several different types of events during the follow-up period. Unlike censoring, which merely obstructs one from viewing the event, a competing event prevents the event of interest from occurring altogether, and analysis has to be adjusted accordingly. In Cox regression, the focus is on the survivor function or the hazard, which models the probability of surviving beyond a given time. In competing-risks regression, instead, the focus is on the cumulative incidence function, which indicates the probability of the event of interest happening before a given time. In other words, in the standard Cox regression, the competing risks are treated as censored observations.

Put differently, the hazard function is specific to a particular event in competing risks (Kleinbaum and Klein, 2012). The hazard is therefore known as cause-specific hazard. The cause-specific hazard is defined by:

$$h_{ij}(t, x_i) = \lim_{\Delta t \to 0} \frac{\Pr(t < T \le t + \Delta t, J = j\ | T \ge x_i)}{\Delta t}$$

which is the instantaneous risk of event of cause j. In other words, it is the conditional probability that a subject with covariates x has an event in the interval $[t, t + dt)$ from the $J - th$ cause, given that the subject had not experienced any other just before time t. Dividing by dt and then taking the limit as $dt \to 0$ converts to probability. The overall survivorship is therefore defined by:

$$S(t, x_i) = e^{-H(t, x_i)}$$

where $-H(t, x_i)$ is the cumulative risk obtained by integrating the overall hazard.

$$-H(t, x_i) = \int_0^t h(u, x_i)du$$

Direct modelling of the effect of covariates on the cumulative incidence function (CIF) for competing risk data has been proposed by Fine and Gray (1999), among others. The Fine and Gray model assumes sub-distributions hazard for the CIF. The subdistribution hazard for cause j is written as

$$h_{ij}(t, x_i) = \lim_{\Delta t \to 0} \frac{\Pr\big(t < T \leq t + \Delta t, J = j \,|T \geq t \cup (T \leq t \cap J \neq j)\big)}{\Delta t}$$

or

$$h_{ij}(t, x_i) = -\frac{d}{dt}\log(1 - I_r(t))$$

where $I_r(t) = \Pr(T \leq t, J = j)$ is the CIF for cause j $(j = 1,2, \ldots k)$. Fine and Gray adopted a semi-parametric proportional hazards model for the sudistributional hazard of cause j for a woman with covariates X, given by

$$h_{ij}(t, x_i) = h_{oj}(t)e^{\beta_r x_{ri}}$$

where $h_{oj}(t)$ is the baseline subdistribution hazard of cause j, and β_r are vector of regression coefficients corresponding to covariates x_{ri}.

Multi-State Survival Model

In multi-state survival models, the main interest is the occurrence of different types of events and the subsequent events. The subsequent events are usually of several different types. This is in contrast to competing risks where the main interest is the first event and event type. Multi-state models are therefore a generalisation of basic survival data analysis to a case where several competing events occur successively over time. The occurrence of successive events constitutes the transitions from an initial state to a final state.

The multi-state hazard function is specific to a particular transition from one competing risk set to another nested competing risks set. This transition is also referred to as transition from one state to another. The multi-state transition hazard is therefore given by:

$$h_{ij}(t) = P\left(X_{(t+dt)}\,\middle|X_t = i\right)$$

where $i = 0, 1, 2, \ldots\ldots I; j = 0, 1, 2, \ldots\ldots J$ and $i \neq j$. It is the instantaneous risk of transitioning. In other words, it is the probability of making $I \rightarrow j$ transition in the interval dt.

Data Analysis

Analysis proceeded as follows. First, descriptive and exploratory analysis using tabular summaries and Kaplain-Meier (KM) plots were generated. Log-rank test was used to test equality of variable strata and KM methods were used to describe the survival time distribution. This provided insight into the shape of the survival function for each group and gave an idea of whether or not the groups were proportional, that is, if the survival functions were approximately parallel. To explore whether or not to include the variable in the final model, test of equality across strata was done. Log-rank test of equality across strata was used for categorical variables. A predictor was included in the model if the test had a p-value of 0.2 or less. In variable selection, a less stringent significance level is recommended in order to avoid rejecting potentially important variables (Machin and Cheung, 2006). All tests in this study were done at 95% significance level. Second, various models using Cox regression, competing risk and multi-state models were fitted. Tests for validity of the proportional hazard assumptions were carried out, as well as, model fit assessment using Cox-Snell residuals were conducted.

In fitting the Cox proportion hazards model, the event of interest was default from treatment. The Ministry of Health considered patients defaulted if they had not returned to the clinic and were not known to have transferred-out, stopped or died (Tweya et al., 2013; Tenthani et al., 2014). The date of entry into follow-up in Cox model was the date of ART registration. This study was censored on 1st October 2013. Subjects were also censored if they had died, stopped treatment or transferred-out of the facility.

In multivariate Cox proportional hazards model, predictors that had a p-value of 0.2 or less in univariate analysis were first included in the model. The p-values from the Wald test were then used to check if the coefficients of the fitted model were different from zero (Vittinghoff et al., 2005). In the initial multivariate model, all the coefficient were significantly different from zero. Finally, all possible interactions of the predictors were then considered. The p-value from the Wald test of the interaction terms was used to select significant interactions. The two factor interactions of pregnancy and ART education, age group & manage and age group & pregnancy were significant. A model with all the interaction terms was then compared with the nested models using the goodness (BIC) test. Lawl (2003) discusses model selection criteria. They noted that parsimony is one of the basic characteristics in model selection. Specifically, there is need to balance between complexity of the model and easiness of model interpretation. BIC takes account of not only the model complexity, but also of the sample size. This is unlike the other penalised criterion. The main effects model that had the lowest BIC is considered the most parsimonious model.

With regards to competing risk models, four outcomes (default, transfer-out, dead, and stopped), from the date they initiate or registered ART, were considered as possible competing events. This study focused on default and transfer out as the competing events. Subjects were also censored if they had died or stopped. Transfer-outs were censored in modelling the sub-hazard of default, while default was censored in modelling the sub-hazards of transfer-outs.

For the multi-state models, two nested competing risks constituted the multi-state survival model in this study. Figure 15.1 presents a diagram showing the states and transitions. The first competing risks set, in the multi-state was modelled from the date the woman initiated ART. The three possible risks were:

1. Failed registration.
2. on ART
3. Transferred out

In this study, *failed registration* refers to patients who missed their first follow up visit after the first visit and did not return to care for at least 60 days. This is similar to the definition used by Tenthani and colleagues in their study on Option B+ in Malawi (Tenthani et al., 2014). This was also noted by Rollins and colleagues in their article on defining retention-in-care. They noted that women who did not have a first follow-up visit were never in care, because including these women in definition retention-in-care distorts the understanding of retention-in-care (Rollins et al., 2014). Transition probabilities were estimated between *failed registration* and *on ART*, while *transfer out* were censored events.

The second competing risk set was nested within the state of *on ART* (Figure 15.1). The *on ART* state therefore is a transient state. *Failed registration* was an absorbing state. Below are the states that can be reached from the state of *on ART*:

1. Default
2. Transfer Out
3. Dead
4. Stop

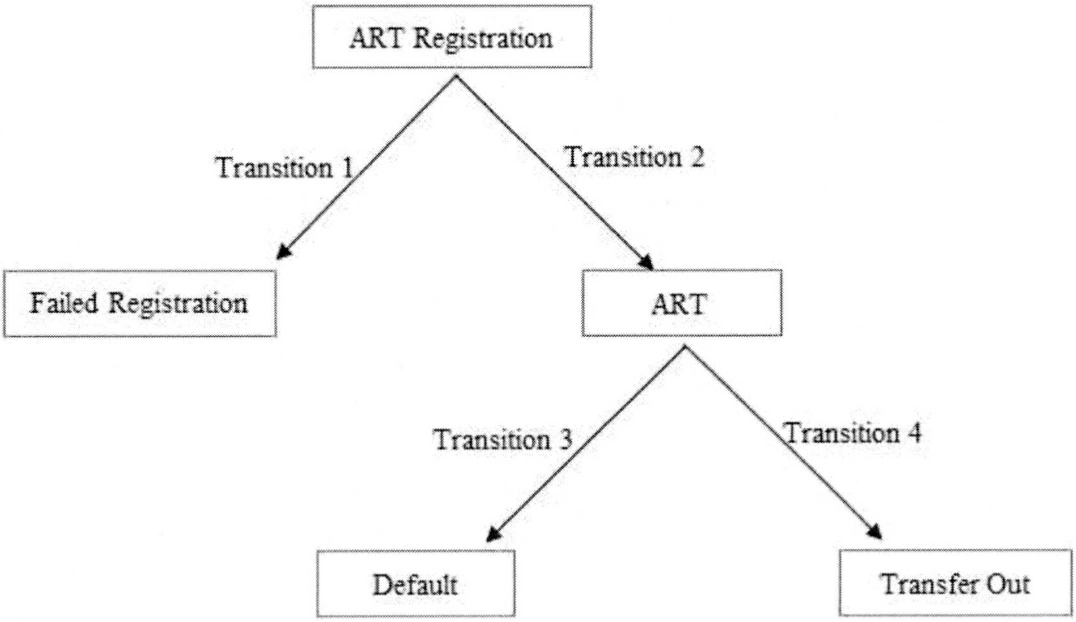

Figure 15.1. Scheme for multi-state models.

Transition probabilities were estimated for *defaults* and *transfers-outs*, while *stops* and *deaths* were censored events. *Default* and *transfer-out* were absorbing.

Model Fitting and Assessment

Model fitting and data analysis were done in STATA (version 13; Stata Corporation, College station, Texas, USA) and R statistical software. Cox proportion hazards and competing risks survival model were fitted in STATA, whereas the multi-state model was fitted in R using mstate package.

The model fit was assessed by using the Cox-Snell residuals. These residuals were used to assess the model through evaluating whether the residuals were indeed a plausible sample from a unit exponential distribution. This is based on the fact that if a random variable T has an exponential distribution with unit mean, the survival function of T is e^{-t}. A plot of the cumulative hazard function $H(t) = -logS(t)$ against t will have a straight line through the origin with unit slope. Then one can graph the Nelson-Aalen cumulative hazard function and the Cox-Snell variable to compare the hazard function to the diagonal line. If the hazard function follows the 45 degree line then we know that it approximately has an exponential distribution with a hazard rate of one and that the model fits the data well.

Results and Discussion

Descriptive Summaries

PROBE study collected data on 2,979 Option B+ women. These women had 20,960 observations. A total of 2,739 women with 17,769 observations had complete information on all the studied variables and hence their data was analysed in this study. The incomplete data was dropped based on the assumption that the data was missing at random, and that these could produce similar results if analysed separately. The 2,739 patients accumulated 2,033 person-years. At the end of follow-up, 410 had defaulted, 39 transferred out, 14 died and 4 stopped ART.

Table 15.1 presents results from the log-rank tests, trend test and frequencies of the independent variables. From that the Table, most of the studied women were from urban areas (60%), were enrolled in MoH (public) facilities (73%), had accessed services in health centres (72%), were pregnant (75%), were adults (77%), had ART patient education (69%), and finally were asymptomatic (94%). The corresponding log-rank test of equality across strata shows that there were no differences in defaulting across the strata of location. Location with a p-value of 0.531 was excluded in the fitted models. The default rates among women registered in rural, semi-urban and urban facilities were similar. Managing authority was significant in the log-rank test of equality across strata. There were significant differences in default rates between women accessing ART through MoH facilities and those accessing ART through CHAM facilities. Managing authority with a p-value of 0.008 was therefore included in the model.

Similarly, facility level of service provision was significant in the log-rank test of equality across strata. Women accessing ART through health centres and hospitals had different rates of defaulting. Facility level of accessing ART with a p-value of 0.014 was therefore included in the model.

Status at ART registration was significant in the log-rank test of equality across strata. Among women initiated on ART while lactating or pregnant, there were differences in defaulting from ART treatment. Status at initiation with a p-value of < 0.001 was therefore included in the model.

Age group was significant in the log-rank test of equality across strata. There were differences in defaulting from ART treatment among young women relative to older women. Age group with a p-value of 0.038 was therefore included in the model. Patient ART education was significant in the log-rank test of equality across strata, with different defaulting rates among women who had ART education and those who did not had patient ART education. ART education with a p-value of < 0.001 was therefore included in the model. WHO HIV clinical condition was also significant with respect to the log-rank test of equality across strata. There were therefore significant differences among women across the strata of clinical condition. Clinical condition with a p-value of 0.031 was therefore included in the model.

Table 15.1. Basic characteristics, log rank test and trend test

Variable	Category	n (%)	Log-rank p-value	Trend test p-value
Location	Urban	1,631 (60)	0.531	0.297
	Semi-urban	566 (21)		
	Rural	542 (19)		
Management	CHAM	731 (27)	0.008	
	MoH	2,008 (73)		
Facility level	Health centre	1,984 (72)	0.055	
	Hospital	755 (28)		
Status at registration	Lactating	672 (25)	<0.001	
	Pregnant	2,067 (75)		
Age-group	Adolescent (<=24 years)	642 (23)	0.038	
	Adult (>=25 years)	2,093 (77)		
ART education	No	837 (31)	<0.001	
	Yes	1,902 (69)		
Clinical condition	Asymptomatic	2,565 (94)	0.064	
	Symptomatic	174 (6)		

Note: CHAM= Christian Hospitals Association of Malawi; MoH=Ministry of Health.

Figure 15.2 presents a Kaplan Meier curve describing survival time distribution of the women. Overall, the Kaplan Meier graph indicates that there were very few defaulters and the survival rate remained at almost 100% until about 3 months on ART. There was a steep drop in survival probability soon after the third month until around the sixth month. The survival probability then stabilises at around 75%.

This is likely a result of the definition of defaulter in the clinical guidelines. According to the guidelines (Ministry o Health, 2011), an ART patient was considered a defaulter at two months after the patient run outs of ARVs. On the other hand, newly registered ART patients were given one-month worth of drugs in order to necessitate clinical reviews. This therefore means that most cases can only be considered as defaulters at least 3 months after registering. The women indicated as defaulters at third month might never have started ART or they stopped ART during the first month, however, considering the guidelines and the monthly drug dispersion at ART registration, they can only be considered defaulters after 3 months. The KM curves for the various predictors are presented in Figure A.15.1 in Appendix 1.

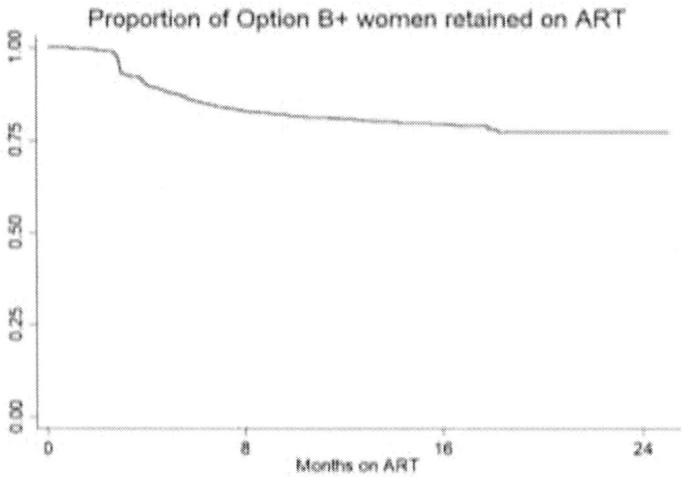

Fgure 15.2. Kaplain-Meier curves.

Cox Proportional Hazards Model

Table 15.2 presents results from the final Cox model. All the covariates included in the model, with the exceptional of facility level, were significantly associated with defaulting from ART treatment. Women in MoH facilities were at a 30% lower risk defaulting relative to women in CHAM facilities ($HR = 0.70; 95\%\ CI = 0.56, 0.87$). The hazard ratio of women in hospitals relative to women in health centres was not significantly different.

Women who initiated ART while pregnant had 64% higher risk of defaulting relative to women who initiated while lactating. The hazard ratio among pregnant relative to breastfeeding women was $HR = 1.64; 95\%\ CI = 1.29, 2.07$. Adult women were 21% less likely to default compared to adolescent women with a hazard ratio of $HR = 0.79; 95\%\ CI = 0.62, 0.99$. Similarly, women who had ART patient education had much lower rate of default relative to women who did not have patient education at registration. Women with ART education have a 40% lower risk of default with a hazard ratio of $HR = 0.60; 95\%\ CI = 0.47, 0.77$. Furthermore, symptomatic women at ART registering had a higher default rate relative to asymptomatic women. Symptomatic had a hazard ratio of $HR = 1.61, 95\%\ CI = 1.05, 2.48$.

Management and facility level are facility characteristic in predicting ART treatment default. These results suggest that there were important variations in provided services by management authority. The result confirm recent study which Malawi Service Provision Assessment (MSPA) reported a higher proportion of MoH compared to CHAM facilities with at least one trained ART provider (Ministry of Health and ICF International, 2015). These findings are therefore consistent with the MSPA findings. The same 2015 MSPA assessment reported a higher proportion of hospitals compared to health centres with at least one trained ART provider. However, hospitals are most likely to have a high burden of patients and further distant from most of the catchment population.

Table 15.2. Estimates for the Cox proportional hazards model

Predictor	Obserced HR	Bootstrap Std. Error	Z	P-Value	Normal based 95% CI
Management					
CHAM	Ref				
MoH	0.70	0.08	-3.27	0.001	0.56, 0.87
Facility Level					
Health Centre	Ref				
Hospital	1.32	0.20	1.88	0.060	0.99, 1.77
Status at Registration					
Lactating	Ref				
Pregnant	1.64	0.20	4.09	<0.001	1.29, 2.07
Age group					
Adolescent	Ref				
Adult	0.79	0.09	-2.01	0.045	0.62, 0.99
ART Education					
No	Ref				
Yes	0.60	0.07	-4.13	<0.001	0.47, 0.77
Clinical condition					
Asymptomatic	Ref				
Symptomatic	1.61	0.35	2.17	0.030	1.05, 2.48

Note: Ref is the reference category

The higher default rate among pregnant compared to breastfeeding women suggest that pregnant women are relatively unprepared for ART. This could be due to poor counselling of pregnant women in the antenatal clinic leading to low spousal and family disclosure. Adult women expectedly had better retention rates. These women might have better spousal and family support, making it easier for them to remain on treatment. Secondly, older women are comparably likely to remain on treatment solely for PMTCT. Specifically, the desire to protect the child from infection might be higher among adult women.

These finding also stresses the importance of ART education. Education prepares the women in mitigating challenges faced when on ART treatment and emphases the importance of remaining on treatment. It is therefore not surprising that women with ART education had better retention rates.

The findings above suggest that symptomatic women have a higher default rate. These women are clinically unwell and therefore anticipated to have better retention rates. In this study, we found 78% $(n = 135)$ of the symptomatic women in this study had mild symptoms in WHO clinical stage 2, therefore, were clinically unwell but not feeling ill. The women therefore were on ART primarily for PMTCT but not for their own health. This is even though they were symptomatic.

Test of PH Assumptions

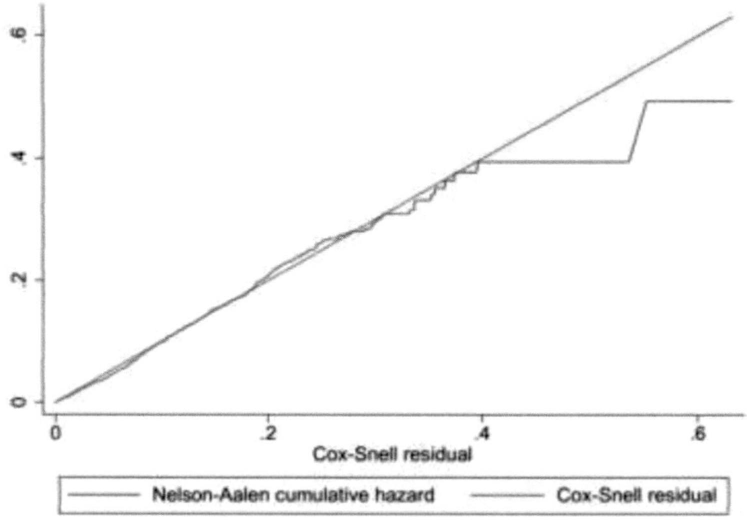

Figure 15.3. Cox-Snell residuals.

Cox proportional hazard assumptions were then assessed. Cox model assumptions are that all the covariates included in the model are proportional and time independent. Hosmer and Lemeshow (1999) recommended using a two-step procedure for assessing the proportional hazards assumptions. First, add the interaction term of time and covariates to the model, then assess the significance of the interaction terms using partial likelihood ratio test, score test or Wald test. Second, plot the scaled and smoothed scaled residuals obtained from model without the interaction terms. The assumptions are met if the results of the two step procedure are consistent (Hosmer and Lemeshow, 1999). In the second step of the two-step assessment process of proportionality assumption, scaled Schoenfeld residuals were plotted. A plot of scaled Schoenfeld residuals should have a slope of zero if the proportional hazards assumption was met. In summary, only facility level and ART education do not have a zero slope over time. Thus, results for facility level, clinical conditions and ART education were inconsistent over the two-steps of the assessment process. Therefore, all the covariates included in the model met the proportional hazards assumption (see Figure A.15.2, Appendix 1). Further the model fit was assessed using Cox-Snell approach (Figure 15.3). Based on Figure 15.3, we can conclude that the final model fits the data well.

Competing Risks Model

Cumulative incidence are used to describe the survival experience when there are competing events. Kaplan-Meier estimate ignores events of all types other than the one of interest and the probability is calculated whenever the event of interest occurs. The CIF estimator for a particular event depends not only on the number of individuals who have experienced the event of interest, but also on the number of individuals who have not experienced any other type of event. The CIF represents the probability that an individual will experience an event of the particular type given that the individual has not experienced any event.

In the fitted competing risks model, the main event of interest were default from treatment and transfer out. Figure 15.4 present a plot of the cumulative incident of default and transfer out.

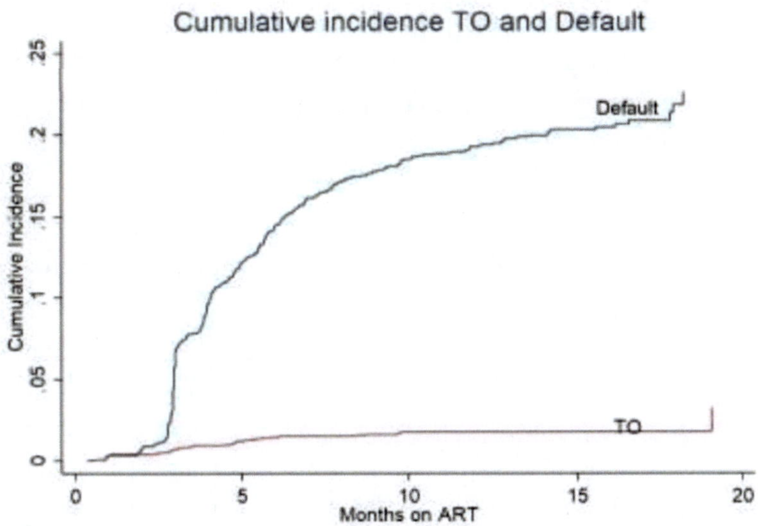

Figure 15.4. Cumulative incidence functions for default and transfer-out (TO).

The estimated probability that an individual experienced the event of default was higher over the whole follow-up period. There were a few incidences of both events in the first 3 months of ART treatment. The chance of a woman experiencing either event was therefore very small. The probability of defaulting increased steeply from 3 months, then the increase was slower from around 10 months. The probability of transfer-outs, however, remained relatively low over the entire follow-up period. This is very likely because of the few incidences of transfer out over the entire follow-up period.

Competing Risk Model Test of PH Assumptions

Having fitted a competing risks model, the proportional hazard assumptions were then assessed. Similar to Cox proportion hazards model, proportionality was checked by including time-varying covariates in the model. If a time-dependent covariate was significant, it was an indication of a violation of the proportionality assumption for that specific predictor.

In the competing risk model of default, the log-time interactions with management, status at registration, age group and ART education were all non-significant. The interactions of log-time with facility level and clinical condition were significant. This indicates that facility level and clinical condition violated the Cox model assumption of proportional hazards. The Cox proportional hazards assumptions, therefore, were valid only in 4 of the 6 covariates included in the model. Table 15.3 presents summaries of the models.

In TO model of the competing risks model, all the log-time interaction terms of management, facility level, status at registration, age group, ART education and clinical condition were all independently non-significant. The Cox proportional hazards assumptions therefore were valid for all the covariates included in the model (Table 15.3).

Table 15.3. Competing risk model assessment of proportional hazard assumptions

Predictor	**Defaulting**			**Transfer-out**		
	SHR	**p-value**	**95% CI**	**SHR**	**p-value**	**95% CI**
Time-dependent						
Management	0.94	0.745	0.63,1.40	1.70	0.284	0.65,4.46
Facility level	0.60	0.031	0.38,0.95	1.00	0.996	0.40,2.48
Status at registration	0.85	0.508	0.52,1.38	1.18	0.693	0.53,2.62
Age-group	0.92	0.692	0.60,1.40	1.38	0.469	0.57,3.33
ART education	1.18	0.432	0.78,1.78	1.52	0.418	0.55,4.41
Clinical conditions	0.36	0.013	0.16,0.80	0.55	0.385	0.14,2.24

*SHR=Sub-hazard ratio.

Stratified Competing Risks Model

Stratified competing risks model was fitted in order to adjust for violation of Cox model assumptions. Therefore, separate models were fitted stratified by the two variables (facility level and clinical conditions) that violated the Cox model assumptions. Stratifying by clinical condition produced inestimable results because of imbalanced data (since 174 (6%) women were symptomatic and 2,565 (94%) were asymptomatic). Hence, results from stratifying by clinical condition have not been presented and subsequently have not been included interpretation of results. On the other hand, facility level had large enough sample to produce estimates, and subsequently a competing risks model with the sub-hazard of default was fitted, stratified on facility level. Table 15.4 presents the stratified competing risks model.

All sub-hazards for covariates included in the stratified competing risks for hospital were not significant. Nevertheless, the sub-hazard model of default was 3% higher in MoH relative to CHAM facilities ($SHR = 1.03; 95\%\ CI = 0.56, 1.91$). Pregnant women also had a 3% higher sub-hazard of default compared to lactating women ($SHR = 1.03; 95\%\ CI = 0.54, 1.96$). Adults relative to adolescents had a 26% higher risk of default ($SHR = 1.26; 95\%\ CI = 0.63, 2.49$). Similarly, estimates of management, status at registration and age group showed a higher risk, though not significant. Finally, symptomatic women had lower rates of default although not significant ($SHR = 0.83; 95\%\ CI = 0.14, 4.92$).

Table 15.4. Estimates for the Strafied competing risks model

Predictor	Hospital			Health Centre		
	Observed SHR	P-Value	Norm. based 95% CI	Observed SHR	P-Value	Norm. based 95% CI
Management						
CHAM	Ref					
MoH	1.03	0.924	0.56, 1.91	0.66	<0.001	0.53, 0.82
Facility Level						
Health Centre	Ref					
Hospital	1.03	0.924	0.54, 1.96	1.74	<0.001	1.33, 2.27
Age group						
Adolescent	Ref					
Adult	1.26	0.512	0.63, 2.49	0.74	0.016	0.58, 0.95
ART Education						
No	Ref					
Yes	1.07	0.968	0.43, 2.68	0.55	<0.001	0.43, 0.71
Clinical condition						
Asymptomatic	Ref					
Symptomatic	0.83	0.979	0.14, 4.92	1.60	0.008	1.13, 2.26

Note: Ref is the reference category

In the stratified competing risks model for health centre, all the sub-hazards for co-variates included were significant. Women in MoH facilities had a 34% lower sub-hazard relative to women in CHAM facilities ($SHR = 0.66; 95\%\ CI = 0.53, 0.82$). The sub- hazard of default among pregnant women in health centres was 74% higher ($HR = 1.74; 95\%\ CI = 1.33, 2.27$). The sub-hazard for adults was lower relative to adolescents ($SHR = 0.74; 95\%\ CI = 0.58, 0.95$). Women with patient education were at 45% lower risk of default in health centres ($HR = 0.55; 95\%\ CI = 0.43, 0.71$).

Symptomatic women in health centres were 60% more likely to default ($HR = 1.60; 95\%\ CI = 1.13, 2.26$).

These results suggest that management, status at registration, age group, ART education and clinical condition were important factors in predicting default rate in health centres but not hospitals. However, overall the estimates of hospital level had wider confidence intervals relative to estimates of health centre.

Multi-State Models

Table 15.5. Regression model for all women who started in the registration state to failed initial or ART

Multi-Sate Model from Registration to Failed Initiation or ART

Predictor	Failed Registration			ART		
	Observed SHR	P-Value	Norm. based 95% CI	Observed SHR	P-Value	Norm. based 95% CI
Management						
CHAM	Ref					
MoH	1.73	0.93	0.95, 1.99	1.04	0.383	0.95, 1.14
Facility Level						
Health Centre	Ref					
Hospital	10.75	<0.001	8.45, 13.67	0.35	<0.001	0.30, 0.41
Status at Registration						
Lactating	Ref					
Pregnant	1.10	0.387	0.89, 1.37	0.98	0.638	0.98, 1.19
Age group						
Adolescent	Ref					
Adult	1.03	0.695	0.90, 1.17	1.08	0.114	0.98, 1.19
ART Education						
No	Ref					
Yes	0.09	<0.001	0.07, 0.12	2.82	<0.001	2.43, 3.26
Clinical condition						
Asymptomatic	Ref					
Symptomatic	1.13	0.707	0.59, 2.19	0.81	0.005	0.70, 0.94

Note: Ref is the reference category

A multi-state model was fitted to further study whether the effect of covariates changed over time. Transition-specific covariates were used to estimate the covariate effects across states. A stratified baseline hazards models was fitted. In total, the model had 4 transitions (See Figure 15.1). Results for the multi-state models are shown in Tables 15.5 and 15.6.

ART Registration State

Table 15.6. All women from starting on ART to default or transfer out

Multi-Sate Model from ART to Default or Transfer Out

Predictor	Default			TO		
	Observed SHR	P-Value	Norm. based 95% CI	Observed SHR	P-Value	Norm. based 95% CI
Management						
CHAM	Ref					
MoH	0.76	0.134	0.53, 1.09	2.89	0.055	0.98, 8.54
Facility Level						
Health Centre	Ref					
Hospital	0.96	0.894	0.55, 1.69	1.69	0.292	0.64, 4.50
Status at Registration						
Lactating	Ref					
Pregnant	1.40	0.096	0.94, 2.09	0.49	0.091	0.22, 1.12
Age group						
Adolescent	Ref					
Adult	0.83	0.369	0.55, 1.24	0.79	0.591	0.33, 1.87
ART Education						
No	Ref					
Yes	0.75	0.236	0.47, 1.20	1.86	0.406	0.43, 7.99
Clinical condition						
Asymptomatic	Ref					
Symptomatic	1.39	0.323	0.72, 2.69	0.45	0.445	0.06, 3.45

Note: Ref is the reference category

The rates of transition to failed registration state were significantly different across categories of facility level and ART education at registration (Table 15.5). The rates were not significantly different across management, status at registration, age group and clinical condition. The risk of transition was 37% higher among the women in MoH facilities compared to women in CHAM facilities ($HR = 1.37; 95\%\ CI = 0.95, 1.99$), however, these were not significant. On the other hand, women in hospitals had significantly higher transition rate of more than 10 times than women in health centres ($HR = 10.75$, 95% CI of $8.45, 13.67$). Failed registrations among pregnant women, relative to lactating women, were 10% higher, although not significant, with $HR = 1.10$ with a 95% CI of $0.89, 1.37$. Adult and adolescent women had a very similar risk of a failed registration. The hazard ratio of adults relative to adolescents was $HR = 1.03$ with a 95% CI of 0.90 to 1.17. ART education reduced the risk of a failed registration by 91% ($HR = 0.09$); $95\%\ CI = 0.07, 0.12$. The risk of a failed registration among symptomatic women was 13% higher compared to asymptomatic women (Table 15.5).

ART State

The rates of transition at state 2 (ART state) were significantly different across categories of facility level, ART education at registration and clinical condition (Table 15.5). The rates of transition were not significantly different across management, status at registration and age group. The risk of making transition to state 2 among women MoH facilities was 4% higher, but not significant, compared to women in CHAM facilities ($HR = 1.04; 95\% \ CI = 0.95, 1.14$). Women in hospitals had a significant lower risk of transitioning to ART relative to health centres. The risk of transitioning to ART in hospitals 65% lower ($HR = 0.35; 95\% \ CI = 0.30, 0.41$). The risk of transition to ART state among pregnant was 2% lower relative to lactating women, again there was not evidence of any difference between the two. The hazard ratio was $HR = 0.98$ with a $95\% \ CI \ of \ 0.90, 1.09$. The transition probability to ART state among adult women relative to adolescent women 8% higher. The hazard ratio of adults relative to adolescents was $HR = 1.08$ with a 95% CI of $0.98, 1.19$. ART education increased the rate of transition to ART by 182% ($HR = 2.82; 95\% \ CI = 2.43, 3.16$). Table 5 further shows that transition rates among symptomatic women was 19% lower ($HR = 0.81; 95\% \ CI = 0.70, 0.94$).

Default State

The rates of transition to state 3 (Default) were all not significantly different across all independent variables (Table 15.6). The risk of transition was 24% lower among the women in MoH facilities compared to women in CHAM facilities ($HR = 0.76; 95\% \ CI = 0.53, 1.09$). Risk of making transition to state 3 was 3% lower among women in hospitals compared to health centres. On the other hand, the risk of default among pregnant women was 40% higher relative to lactating women at $HR = 1.40$ with a 95% CI of $0.94, 2.09$. The risk of default among adults relative to adolescents was lower. The hazard ratio of adults relative to adolescents was $HR = 0.83$ with a 95% CI of $0.55, 1.24$. The risk for transition to default state was lower among women with ART education, with a hazard ratio of $HR = 0.75(95\% \ CI = 0.47, 1.20)$. Defaulting was 39% higher for symptomatic woman relative to asymptomatic women ($HR = 1.39; 95\% \ CI = 0.72, 2.69$).

Transfer-Out State

The rates of transition to state 4 (TO) were also not significantly different across categories of all the covariates (Table 15.6). The risk for transition to transfer out state was 189% higher among the women in MoH facilities compared to women in CHAM facilities with a hazard ratio and confidence interval of $HR = 2.89; 95\% \ CI = 0.98, 8.54$. Women in hospitals had a higher risk of transitioning to transfer out state relative to health centres. The risk of transitioning to transfer out among women in hospitals 69% higher ($HR = 1.69; 95\% \ CI = 0.64, 4.50$). Transfer out among pregnant women was 51% less relative to lactating women at $HR = 0.49$ with a 95% CI of 0.22, 1.12. The risk of transfer out among adults and adolescent was lower, with a hazard ratio of $HR = 0.79$ with a 95% CI of 0.33, 1.87. Transition rates were 86% higher among women who ART education ($HR = 1.86; 95\% \ CI = 0.43, 7.99$). The rates of making transition 4 were 55% lower among symptomatic patients relative to the asymptomatic ones ($HR = 0.45; 95\% \ CI = 0.06, 3.45$).

Results from the fitted multi-state models suggest management authority is not an important factor predicting all the 4 transitions. This is not consistent with findings from both the Cox model and competing risks model presented above. The results of the Cox model and competing risks models summarises the effects of management over the whole follow period while multi-state model splits the period into states.

Facility level was significant in predicting failed registration and ART initiation. This is likely because hospitals have a larger catchment area relative to health centres. Further, the majority of the catchment population was relatively closer to health centres. Women are therefore likely to transfer from hospitals to health centres that are closer to their residence. Furthermore, ART patient sometimes transfer without notifying the originating facility (i.e., silent transfers). This is likely an attempt to reduce the cost of accessing health services. Therefore, some of the failed registration might include silent transfers.

Status at registration and group were both not significant in all the 4 transitions. Similar to facility level, these results are not consistent with findings from both the Cox model and competing risks model.

ART education was significant in predicting failed registration and ART initiation. We found that women with ART education were better informed and subsequently prepared for ART. The challenges that might force women out of treatment are discussed during the education sessions. For example, how to disclose their HIV status to spouse and the whole family, therefore women with ART education were likely to have support from spouse and family members.

Clinical condition was significantly associated with ART initiation. Symptomatic women had a lower rate of ART initiation. Similar to discussion in the Cox model, this is likely because a bigger proportion were mildly symptomatic. These women are clinically symptomatic but they do not feel ill themselves. Therefore, the default patterns among symptomatic women are not reflective of the clinical condition.

Finally, all the predictors were non-significant in predicting the default and TO states. This is likely due to the pattern of attrition that is high from 3 until 6 months. Subsequently, the majority of defaulters are already out of treatment by the time a cohort reaches ART state.

CONCLUSION

The study sought to explore facility level and individual level predictors of attrition among Option B+ women in Malawi. It further described the changes in predictors of attrition from ART treatment in the context of competing events. The study also measured the default rates in the first 21 months of ART among option B+ women. This was to inform effective and efficient implementation of Option B+ intervention. The findings of this study show that attrition was highest from 3 months until 6 months. Further, the study revealed important factors at both facility level and individual level affecting PMTCT programming. In addition, this study has also revealed changes of hazards in the context of competing events and changes in the first 21 months after ART registration.

The findings suggest that management authority, facility level, status at initiation, age group, ART education and clinical condition are significantly associated with attrition from ART. The association of the predictors and attrition, however, varied from ART registration to ART treatment follow-up and follow-up outcomes. On the other hand, location of residence did not have an effect on attrition from ART.

Facilities managed by MoH, being of older age and having received ART education were found to be significantly associated with lower default rate among Option B+ women. Hospitals, pregnancy and symptomatic women were found to be significantly associated with a higher default rate. These results are similar to results found in other studies. Boyles et al. (2011), in their study on general ART cohort, found that pregnancy and younger age were associated with an increased risk of attrition. This bears similarity to our studies of Option B+ women. In another study, Tenthani et al. (2014) found that women who started under Option B+ compared to women who started for their own health, had high attrition rates as well as led to health facilities having a high patient burden. Tenthani and colleagues, however, limited their study to facilities that had electronic medical records. This limitation is inherently biased as only high burden facilities have electronic medical records. Furthermore, follow-up period was limited to the first 6 months after ART registration. This study has gone further to study attrition from randomly selected facilities from across the country, with a follow-up period which was also longer. The results were, however, still similar probably because most of attrition from ART among Option B+ occurred in the first 3 months after registration.

Others studied a different aspect of attrition. Tweya et al. (2013) followed up on patients that had been considered defaulted and examined factors associated with defaulting. Transfers to other facilities and financial constraints were found to be primary causes of ART discontinuation. Du Toit et al. (2014) studied attrition among pre-ART patients, and found that none of the factors were significantly associated with retention in pre-ART. Overall, these factors might have contributed to the attrition among Option B+ women, however, this study focused on the facility level factors.

Various models were used in the reviewed studies. Boyles et al. used multiple Cox proportional hazard regression analysis, adjusted for competing risks. Tenthani et al. used multiple logistic regressions with a random effect for the cohorts and competing risk regression to examine variables associated with attrition. Similarly, this study used competing risks survival model. Across the studies, modelling sub-hazards was probably an appreciation that the other possible events are competing in an ART cohort. Du Toit et al. used logistic regression to explore factors associated with retention in HIV care in the pre-ART group. In logistic regression, odds ratio of an outcome given the predictors was modelled. The studies that used logistic model were primarily interested in the association between attrition and other factors. This study however aimed at modelling rates of change of association between attrition, and other factors, in the first 21 months of ART. Tweya et al. used log-binomial regression to compare categories of tracing outcomes with baseline characteristics of patients considered defaulted. In log-binomial, relative risk of an event is modelled. Finally, both logistic and log-binomial models were used to model the probability of an outcome but not the changes of probability of an outcome over time.

Considering default and transfer out as competing events, the overall conclusion was slightly different. Management, status at registration, age group, ART education and clinical condition were associated with attrition from ART. Considering default as the event of interest, facilities managed by MoH, pregnant women, adult women, women with ART education and clinical condition were still significantly associated with a lower default rate. Hospital was still found to be non-significantly associated with a higher default rate. with regards to transfer out as the event of interest, pregnancy, adult women and symptomatic women were still associated with a lower transfer out rate.

Finally, facilities managed by MoH, hospital and ART education had an increasing non-significant effect transfer out rate. Competing risks models were also used in other studies on attrition (Boyles et al., 2011, Tenthani et al., 2014). Similar results were found in these studies.

There were changes in hazards ratios across the 4 states covering the first 21 months after ART registration. Facilities managed by MoH were found to be non-significantly associated with a higher risk of failure to register, transfer out and a non-significant association with ART initiation and default. Hospitals were significantly associated with a lower risk of failed registration and ART initiation. After ART initiation, hospitals were found to be non-significantly associated with a higher risk of default and transfer out. Referral hospitals were found to be significantly associated with a higher risk of failed registration, ART initiation and default. Status at initiation hazard ratios varied across the 4 states. Pregnancy was associated with a higher risk of failed registration, ART initiation and default. Lower risk of transfer out. The association between pregnancy with failed initiation and default was significant. Association of pregnancy with ART state and transfer out state was non-significant. Adults were associated with lower risk of all outcomes across the whole continuum of ART treatment in the first 21 months. The association however was non-significant across all outcomes. ART education was associated with a significant lower risk of a failed registration and default. A higher risk of ART initiation and transfer out. The association with ART was significant. There were variations in effect of clinical condition across both the 4 states.

Cause-specific proportional hazards and proportional sub-hazards modelling approaches produced similar results in modelling default rate from ART. However, proportional sub-hazards modelling accounts for the possibility that other events can occur. Accounting for competing events presents a more realistic reflection of reality. Therefore, a thorough understanding of the characteristics affecting outcomes can be gained through fitting sub-hazards with transfer out as competing events. Boyles and colleagues considered transfer outs as still in follow-up on ART. Transfer outs were therefore neither censored nor competing events (Boyles et al., 2011).

STUDY LIMITATIONS

Several limitations affected the study. First, the study used secondary data from the PROBE study. PROBE study was not powered sufficiently enough to satisfy the objective of this study. Specifically, the variable clinical condition had insufficient sample of Option B+ women that were symptomatic. This resulted in failure to estimate parameters a competing risks model stratified by clinical condition.

Second, the study used routine hospital data. Routine hospital data is inherently poor particularly in resource limited context like Malawi. For example, routine hospital data is usually incomplete and sometimes inconsistent. It is therefore important to interpret studies that use routine hospital data within the context of the limitations.

Finally, the study would have been stronger if other characteristics were included in the modelling. It was however not possible to include all characteristics that have been found to be significantly associated with attrition from ART in similar studies. Distance from the health facility to patients' place of residence has been found to be an important predictor of attrition. However, routine hospital data does not include distance in Malawi. PROBE study dataset subsequently did not include this variable.

APPENDIX 1

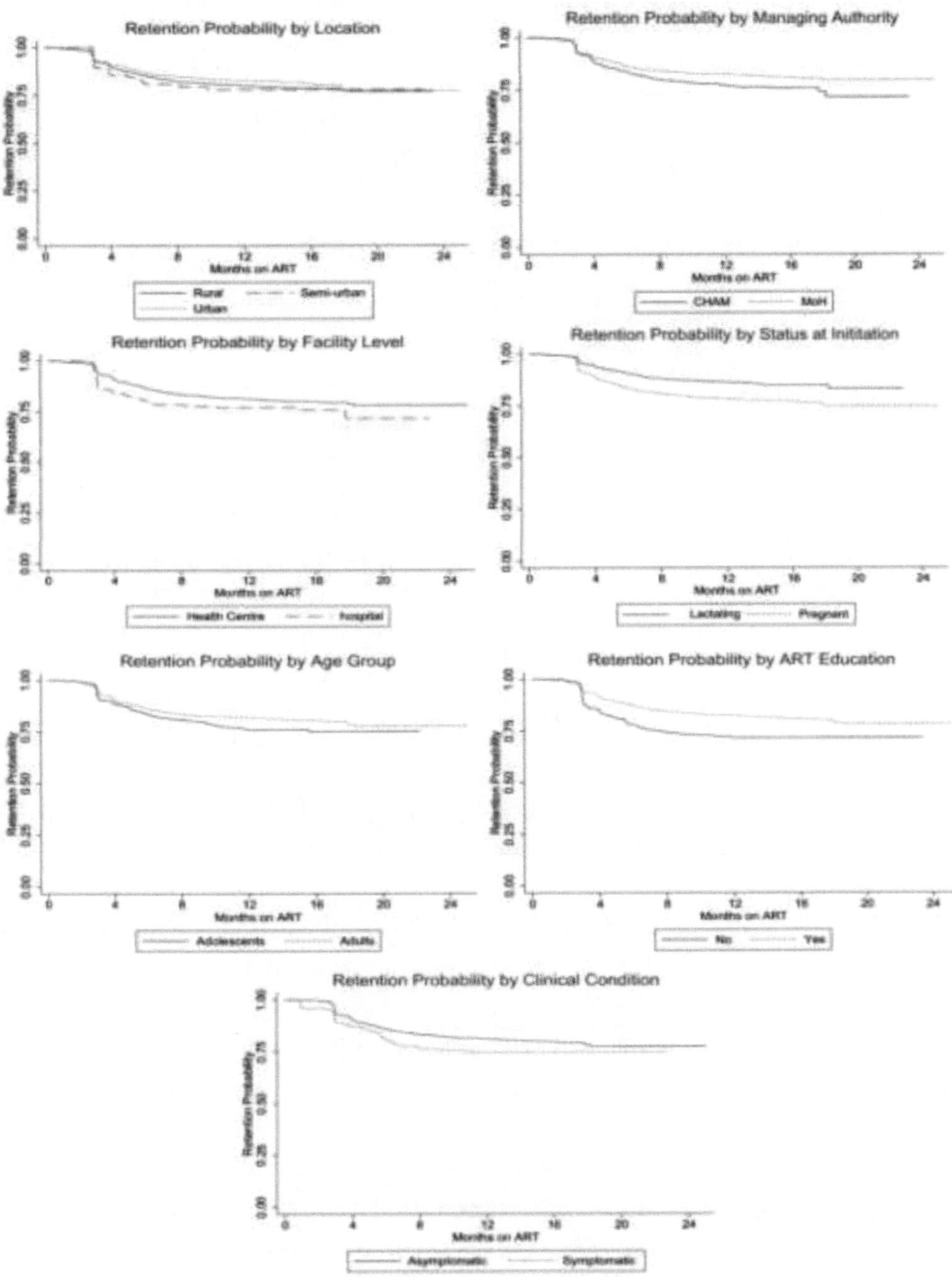

Figure A.15.1. Kaplain-Meier curves.

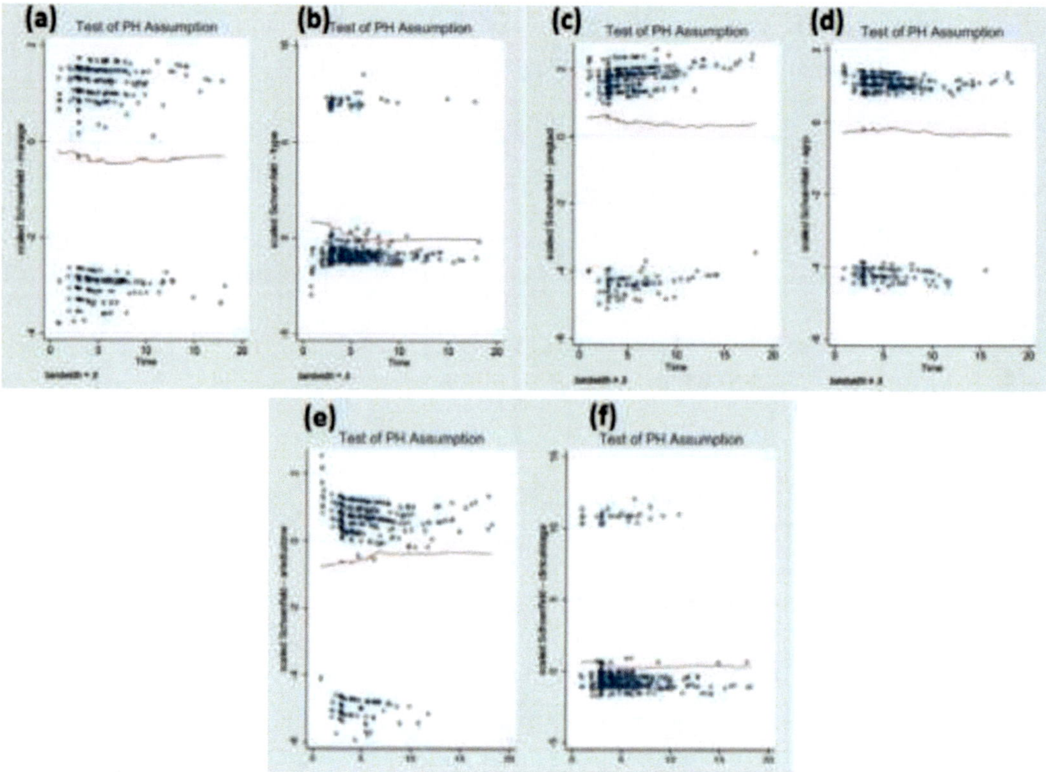

Figure A.15.2. Graph of the scaled Schoenfeld residuals and their lowess smooth for: (a) management authority; (b) facility level; (c) status at registration; (d) age group; (e) ART education; and (f) clinical condition.

References

Beyersmann, J., Allignol, A. S. M. (2012). *Competing Risks and Multistate Models* with R. Springer Science+Business Media, LLC, Spring Street, New York, NY 10013, USA.

Boyles, T. H., Wilkinson, L. S., Leisegang, R., and Maartens, G. (2011). Factors influencing retention in care after starting antiretroviral therapy in a rural South African programme. *PLoS ONE*, 6(5):e19201.

Collett, D. (2003). Modelling Survival Data in Medical Research. Taylor and Francis Inc.

Cornell, M., Lessells, B. A. (2014). Mortality among adults transferred and lost to follow-up from antiretroviral therapy programmes in South Africa: A multicenter cohort study. *Journal of Acquired Immune Deficiency Syndromes*, 67(2):67-75.

Coutsoudis, A., Goga, A., Desmond, C., Barron, P., Black, V., and Coovadia, H. (2013). Is Option B+ the best choice? *The Lancet*, 381(9863):269-271.

du Toit, E., van Schalkwyk, C., Dunbar, R., Jennings, K., Yang, B., Coetzee, D., and Beyers, N. (2014). Missed opportunities for retention in pre-art care in Cape Town, South Africa. *PLoS ONE*, 9(5):e96867.

Fine, J. P., Gray R. J. A proportional hazards model for the sub-distribution of a competing risk. *J. Am Stat Assoc*; 94: 496-509.

Harrell, F. E. J. (2001). *Regression modelling strategies with applications to linear models, logistic regression, and survival analysis.* Springer Verlag New York, Inc.

Hosmer, D., Lemeshow, S. (1999). *Applied Survival Analysis: Regression Modelling of Time to Event Data.* John Wiley and Sons, Inc.

Klein, J. P., Moeschberger, M. L. (2003). *Survival analysis: techniques for censored and truncated data.* Springer-Verlag New York, Inc., 2nd edition.

Kleinbaum, D., Klein, M. (2012). *Survival Analysis: A Self Learning Text.* Springer Science and Business Media.

Krentz, H.B., Worthington, H. G. M. J. (2011). Adverse health effects for individuals who move between HIV care centers. *Journal of Acquired Immune Deficiency Syndromes*, 57(1):51-54.

Krentz, H.B. MacDonald, J. G. M. J. (2015). The impact of transfer patients on the local cascade of HIV care continuum. *Journal of Acquired Immune Deficiency Syndromes,* 68(2):236240.

Kwong-Leung Yu, J. (2008). What happens to patients on antiretroviral therapy who transfer out to another facility? *PLoS ONE*, 3(4).

Lamb, M. R., El-Sadr, W. M., Geng, E., and Nash, D. (2012). Association of adherence support and outreach services with total attrition, loss-to- follow-up, and death among art patients in sub-Saharan Africa. *PLoS ONE*,7(6):e38443.

Lawl, B. (2003). *Categorical data analysis with SAS and SPSS applications.* Lawrence Erlbaum Associates, Inc.

Liu, X. (2012). *Survival Analysis: Models and Applications.* John Wiley and Sons Ltd.

Machin, D., Cheung, Y. B. P. M. (2006). *Survival analysis: a practical approach.* John Wiley and Sons Ltd, 2nd edition.

Ministry of Health (2011). Clinical management of HIV in children and adults.

Ministry of Health (2014). Integrated HIV program report. Technical report, Government of Malawi Ministry of Health.

Ministry of Health and ICF International (2015). Malawi service provision assessment (mspa) 2013-14. Technical report, Ministry of Health and ICF International, Lilongwe, Malawi, and Rockville, Maryland, USA: MoH and ICF International.

Odafe, S., Torpey, K., Khamofu, H., Ogbanufe, O., Oladele, E. A., Kuti, O., Adedokun, O., Badru, T., Okechukwu, E., and Chabikuli, O. (2012). The pattern of attrition from an antiretroviral treatment program in Nigeria. *PLoS ONE*, 7(12):e51254.

Rollins, N. C., Becquet, R., Orne-Gliemann, J., Phiri, S., Hayashi, C., Baller, A., and Shaffer, N. (2014). Defining and analyzing retention-in-care among pregnant and breastfeeding HIV-infected women: Unpacking the data to interpret and improve pmtct outcomes. *JAIDS Journal of Acquired Immune Deficiency Syndromes*, 67.

Schouten, E. J., Jahn, A., Chimbwandira, F., Harries, A. D., and Van Damme, W. (2013). Is Option B+ the best choice? *Lancet*, 381(9874):1272-3.

Schouten, E. J., Jahn, A., Midiani, D., Makombe, S. D., Mnthambala, A., Chirwa, Z., Harries, A. D., van Oosterhout, J. J., Meguid, T., Ben-Smith, A., Zachariah, R., Lynen, L., Zolfo, M., Van Damme, W., Gilks, C. F., Atun, R., Shawa, M., and Chimbwandira, F. (2011). Prevention of mother-to-child transmission of HIV and the health-related millennium development goals: time for a public health approach. *Lancet*, 378(9787):282{4.

Tenthani, L., Haas, A. D., Tweya, H., Jahn, A., van Oosterhout, J. J., Chimbwandira, F., Chirwa, Z., Ng'ambi, W., et al., (2014). Retention in care under universal antiretroviral therapy for HIV-infected pregnant and breastfeeding women ('Option B+') in Malawi. *AIDS*, 28(4):589-98.

Tweya, H., Feldacker, C., Estill, J., Jahn, A., Ng'ambi, W., Ben-Smith, A., Keiser, O., Bokosi, M., Egger, M., Speight, C., Gumulira, J., and Phiri, S. (2013). Are they really lost? "true" status and reasons for treatment discontinuation among HIV infected patients on antiretroviral therapy considered lost to follow up in urban Malawi. *PLoS One,* 8(9):e75761.

Vittinghoff, E., Gliden, D.V., Shiboski, S. C, McCulloch, C.E. (2005). Regression methods in biostatistics: linear, logistic, survival, and repeated measures models. Springer Science+Business Media, Inc., 233 Spring Street, New York, NY 10013, USA.

World Health Organization (2010). Antiretroviral drugs for treating pregnant women and preventing HIV infection in infants: recommendations for a public health approach. *World Health Organization*, 2010 version edition.

World Health Organization (2013). The use of antiretroviral drugs for treating and preventing HIV infection: recommendations for public health approach. *World Health Organization*, 20 Avenue Appia, 1211 Geneva 27, Switzerland, 2013 edition.

Yehia, B. R., Stephens-Shields, A. J., Fleishman, J. A., Berry, S. A., Agwu, A. L., Metlay, J. P., Moore, R. D., Christopher Mathews, W., Nijhawan, A., Rutstein, R., Gaur, A. H., Gebo, K. A., and Network, H. R. (2015). The HIV care continuum: Changes over time in retention in care and viral suppression. *PLoS ONE*, 10(6):e0129376.

EPILOGUE

Ngianga-Bakwin Kandala*
Department of Mathematics, Engineering and Physical Sciences,
Northumbria University, Newcastle, UK

Since Demographic and Health Survey (DHS) and secondary administrative such as Health Information System (HIS) and other electronic hospital registers administrative data has been made available and effectively used in developing countries, a number of developments have taken place in the field of statistical modelling based on hierarchical and spatial temporal analysis. These developments have sought new ways to understand the role of biosocial processes in shaping health inequalities. They have also served to draw lessons for population health research as well as health care planning and delivery. The aim of this book was to bring together the various methodological advances and make use of case studies in order to analyze such complex clustered and correlated household secondary data. The recent developments reflect our own teaching and research interests and showcase the outputs achieved by our students in the course of postgraduate training in biostatistics in a number of African countries.

In the absence of census data in many Sub-Saharan Africa countries, the analysis of good quality data such as the DHS and HIS is vital when making decisions about health priorities. However, for example using DHS and HIS data presents some challenges. Some of these challenges are: the complex sampling design of the surveys, the hierarchical nature of the data, the dependence of data at the sampled cluster, missing data among many more challenges. Firstly, for any given health outcome, DHS data are collected hierarchically at the family, household and community or state levels, which are inter-related. Various advanced spatial models have been implemented that control for

* Corresponding Author's Email: Ngianga-bakwin.kandala@northumbria.ac.uk.

the correlation structure and dependence in the data while exploring effects of other variables. Secondly, in order to analyse complex secondary data emanating from DHS and other health care data, we draw on our extensive experience in teachingbiostatistics in several African countries and developing appropriate methods for analysis of such data.

The authors have endeavoured to illustrate concepts with examples through use of real health outcomes from the DHS and other sources. Some of the data sets are openly available on http://www.measuredhs.com. While the book did not attempt to cover all recent developments in the area of complex clustered and corelated data, it is fully referenced in order to provide and encourage further reading.

The book presents both theoretical contributions and empirical applications of recent developments in complex clustered and correlated household data. The book covers a range of new developments emerging from both classical and Bayesian approaches. In the Bayesian framework, Monte Carlo techniques, in particular MCMC and the application to spatial and spatial-temporal data are covered. Within the conventional classical approach, the book has examined multilevel models that address the issue of clustering within family and household. It has also considered models that account for interdependencies over life-course events and attrition using Cox proportional hazards. Due consideration was also given to competing risks and multistate survival models in the context of initiation of women to antiretroviral therapy.

In the past, there were few statisticians who could apply, evaluate, and use the right advanced statistical methods. This book indicates that our efforts in capacity building in biostatistics in Africa have been successfully in showing clear application of what would otherwise be regarded as a difficult area of scholarship.

At a time when the proportion of articles written by Africa-based academics has declined from 30 to 15 per cent (citing Africa see http://www.lse.ac.uk/africa/citing-africa/citing-Africa), this book should provide practical guidance to young scholars who are keen to publish their own work. It should also encourage a critical thinking at the wider context of knowledge production about the African continent as highlighted by Citing Africa. This kind of initiative is worthwhile and should be accelerated in order to encourage young scholars from Africa to publish their work internationally.

The book is coherently organized and clearly written so that readers can follow its public health policy messages without having to master the technical parts.

About the Editors

Ngianga-Bakwin Kandala is Professor of Biostatistics and Head of Applied Statistics Research Group, Department of Mathematics, Physics and Electrical Engineering (MPEE) at Northumbria University, UK. Prior to this, he was the Director of Business and Enterprise at MPEE and worked as Head of the Health Economics and Evidence Synthesis Research Unit at the Luxembourg Institute of Health, Luxembourg and was an Associate Professor in Health Technology Assessment, a joint appointment with the University of Oxford and the University of Warwick. He has previously worked in many academic institutions and in various positions as a Medical Statistician at the University of Southampton and King's College London; a Mellon Foundation Fellow at the University of Montreal, Canada; Associate Professor at the University of Botswana and Associate Lecturer at the University of Lagos, Nigeria. He is also affiliated with the University of Witwatersrand, South Africa as a Distinguished Professor of Biostatistics. He obtained a PhD in Economics/Statistics from the University of Munich (LMU) in Germany in February 2002.

For the past 18 years, his main research interests are in Bayesian statistical methods and their application to epidemiology and health, in particular, addressing maternal and child health and a variety of health-related health inequalities both in the developing countries and command economies, using large scale household data. Kandala has published widely in high impact peer review journals in both the field of Statistics and health in diverse populations. His current research interests include, Bayesian modelling and diseases mapping, statistical methods applied to epidemiology, survival analysis, longitudinal data analysis, meta-analysis, Bayesian spatial Analysis, Health Economics and Health Technology Assessment.

In addition to his book published with Springer Science on 'Advanced Techniques in Modelling Maternal and Child health in Africa' (2014), his latest published book by the same publisher is titled "Female Genital Mutilation Around The World: Analysis of Medical Aspects, laws and Practice" (2018). His current research interests include,

Bayesian modelling and diseases mapping, statistical methods applied to epidemiology, survival analysis, longitudinal data analysis, meta-analysis, Bayesian spatial Analysis, Health Economics and Health Technology Assessment.

Lawrence N. Kazembe is a Professor of applied statistics at the University of Namibia. He joined the University of Namibia in 2012, and on leave from the University of Malawi, Chancellor College. He has previously served as senior biostatistician at Malawi Liverpool Wellcome Trust Clinical Research Programme in Blantyre, Malawi, in 2010-2012; as a research fellow at Medical Research Council of South Africa, Durban in the Malaria Lead Research Programme, 2005-2007. He has worked as a consultant statistician with Malaria Alert Centre, College of Medicine, Blantyre-Malawi since 2005.

He has published extensively, with over 80 peer-reviewed publications in population health including food and nutrition security, malaria, HIV and TB research. In 2008, he was recognized as a young scientist by the Malawi National Commission of Science and Technology. In 2009, in Durban-South Africa, he was accorded a young research affiliate award by the Academy of Sciences in Developing Countries (TWAS), and later in the year was acknowledged with an identical medal of young biometrician at the Sub-Saharan Network of International Biometrics Society (SUSAN-IBS) meeting in Nakuru, Kenya. He obtained his PhD in 2007 from the University of KwaZulu-Natal. His main research interests are in Bayesian statistical modeling and spatial analysis with applications in population health.

About the Contributors

Dr. Mavuto Mukaka is the Head of Statistics at the University of Oxford's Clinical Trials Support Group, MORU, Thailand. He holds a PhD in Biostatistics from University of Liverpool. His role includes: coordinating statisticians within his team, advising researchers in the design of the different types of studies that includes determining the appropriate sample sizes; writing statistical sections of the study protocols, reviewing methodological aspects of study protocols, data analysis, co-supervision of PhD/MSc students, and provision of statistical training to researchers at MORU. He teaches Biostatistics at University of Oxford (UK), University of Malawi and Mahidol University in Thailand. Dr. Mukaka is Honorary Professor of Biostatistics at Malawi University of Science and Technology. He is an active researcher in statistical methodology as well as medical research. He has co-authored more than 70 peer-reviewed publications on a diversity of health research topics. Dr. Mukaka and his colleagues have recently co-authored a book chapter on investigating risk factors for stunting in under-five children to be published in NOVA Science Publications. Dr. Mukaka is a Statistical Editor and reviewer in a number of biomedical journals.

Dr. Don Mathanga is an infectious diseases epidemiologist who is currently an Associate Professor in the Faculty of Public Health and Director of the Malaria Alert Centre, a research unit within the College of Medicine. His current research interests are focused on understanding the risks associated with malaria and how various interventions affect child health in Malawi. Dr. Mathanga is also interested in knowledge translation (he is an award winner of a Cochrane review) and currently supports the efforts of the South African Cochrane Centre to build knowledge translation capacity in sub Saharan Africa.

Dr. Lillian Pazvakawambwa is currently serving as a Senior Lecturer in the Department of Statistics and Population Studies. She holds a Ph.D. in Applied Statistics from the University of Namibia. Her research interests are in Biostatistics and Epidemiology.

Dr. Ndeyapo Nickanor is a senior lecturer in the Department of Statistics and Population Studies at the University of Namibia, Namibia. She holds a PhD from the University of Cape Town. She is an associate of the African Food Security Urban Network (AFSUN), Hungry Cities Partnership (HCP) and the country lead of the IDRC-funded Urban Food Systems Governance and Disease Prevention in Africa. Her work focuses on urban food security, food systems, urban poverty and livelihoods.

Misheck Luhanga is a Public Health Specialist working with US centers for Disease Control and Prevention (CDC) in Malawi. Prior to that, he worked as Technical Advisor in Monitoring and Evaluation with the Ministry of Health's National Malaria Control Programme. In this role Misheck supported and developed systems for monitoring and evaluation of malaria control interventions. These included determining and understanding the prevalence and incidence of malaria. He holds a Master of Public Health in Epidemiology and Biostatistics from the University of Pretoria's School of Health Systems and Public Health, a Master's degree in Biostatistics and a Bachelor's degree in Statistics, both from University of Malawi.

Tobias Shinyemba is a Demographer working as a Junior Lecturer in Statistics and Population Studies with the University of Namibia. He holds an MSc in Applied Statistics and Demography from the University of Namibia. His main research interest focuses on modelling of demographic data using indirect estimation methods, population health, monitoring and evaluation, and food security and child nutrition. Currently, Tobias is working as a research assistant for Nourishing Spaces study, a research project with the goal to examine the relationship between food consumptions, non-communicable diseases and local food systems in Windhoek and Oshakati.

Bamidele Oguntoyinbo is Nigeria citizen based in Lesotho. He is a biostatistician by profession and holds the Master of Science in Biostatistics from University of Malawi and both his BA in Statistics & Demography, and Certificate in Statistics from National University of Lesotho. He currently works as Evaluation Advisor for Elizabeth Glaser Pediatric AIDS Foundation (Lesotho), since January 2016. His main duties leading and coordinating all project evaluations that includes the design and development of evaluation questions, design and protocols, conducting and managing the evaluations, and conducting the data analysis, interpretation and reporting of all evaluation data. He is currently actively involved in more than 4 ongoing evaluations/research in the Foundation.

Andreas Itashipu Shipanga completed his honors degree in Statistics and Mathematics in 2011 from University of Namibia. Thereafter, Andreas worked for Ministry of Health and Social Services as a Regional Monitoring and Evaluation Officer for Erongo Region, Namibia. He also held a similar position at The Society of Family Health (SFH). Working in the area of Public Health with over 6 years of experience, this influenced his decision to persuade his MSc in Biostatistics the course he completed in 2018 at University of Namibia. His research topic focused on Spatial temporal analysis of survival among TB and HIV co-infected patients in Erongo Region, Namibia. Andreas research area of interest includes spatial statistics, survival data analysis, epidemiology and Bayesian analysis just, just to highlight some. He would like to continue his education in the field of infectious diseases modelling.

Jason Nakaluudhe completed his undergraduate in Statistics and Mathematics in 2017 and his Master of Science in Biostatistics in 2019 at University of Namibia (UNAM). His Master's project focused on "A Copula Approach to Sample Selection of Treatment Adherence and Viral Suppression". He has presented different papers at different local and international conferences. He has submitted various manuscripts for publication. His research interest in copula models, survival analysis, and longitudinal studies. He was one of the first Namibian to benefit from SSACAB DELTA Africa. Jason has worked as Statistician at National Council for Higher Education (NCHE). He is currently unemployed. Jason would like to continue his career in the field of infectious diseases.

Emmanuel Banda has just finalized his study for MSc in Biostatistics at Chancellor College of University of Malawi now for waiting graduation. He also possesses a BSc degree in Statistics and Demography from the same Chancellor College. He is currently working for Save the Children Somalia Country Office as Head of Research, Analysis, Learning and Monitoring. He has also previously worked for Save the Children Malawi as Senior Monitoring and Evaluation Manager and Karonga Prevention Study (KPS) as a Research Social Scientist where he coordinated and co-authored publications in a health and demographic research program affiliated to London School of Hygiene and Tropical Medicine.

Anastasia Johannes is currently working as a Chief Development Planner in population and statistics at Hardap Regional Council. She holds a Master of Science in Applied Statistics and Demography from the University of Namibia. She has 6 years' experience in the Central Government of Namibia in dealing with data analysis for socio-economic development and decision-making. In addition, she has experience in research and monitoring of project such as Human Trafficking and Gender Based Violence in collaboration with UNDP, UN Women, UNFPA and UNODC. Her main research focuses on WASH indicators using Copula approach.

Tommy R. Harris holds a Masters' degree in Applied Statistics and Demography. He is a senior statistician in the Department of Census and Demographic Surveys, at the Namibian Statistics Agency. Prior to that, for over 10 years, he has served in various area of research and data analysis to support evidence-based planning. These include: Development Planner at the Ministry of Gender Equality and Child Welfare; Regional Statistician: Survey and Field Operations at the Namibia Statistics Agency; Statistician: Vital and Migration Statistics at the Namibia Statistics Agency; and later as a Senior Statistician: Census and Demographic Surveys at the Namibia Statistics Agency.

George Traoli Mwenye-Phiri is a Senior Lecturer in Biostatistics and Epidemiology, Malawi College of Health Sciences. He holds a MSc in Biostatistics and BEd (Science) earned in 2016 and 1999 respectively, from the University of Malawi. His MSc thesis involved developing a mover-stayer model on future contraceptive use among married women as applied to 2010 MDHS data. Currently, together with two oncologists, he is in the process of conducting a systematic review on the risk of esophageal cancer in women on oral contraception.

INDEX

B

C

N

R

S